ACI 314R-11

Guide to Simplified Design for Reinforced Concrete Buildings

(For Buildings of Limited Size and Height, based on ACI 318-11 and ACI IPS-1, "Essential Requirements for Reinforced Concrete Buildings")

Reported by ACI Committee 314

JoAnn Browning,* Chair

Michael C. Mota, Secretary

Iyad M. Alsamsam
Kenneth B. Bondy
James R. Cagley
Omar D. Cardona
W. Gene Corley
Om P. Dixit
David A. Fanella
Yosef Farbiarz
Luis E. García*
Jose M. Izquierdo-Encarnación†
H. Rolfe Jennings
Mahmoud E. Kamara
Jason J. Krohn
James S. Lai

Lionel A. Lemay
Andres Lepage
Robert F. Mast
Adolfo B. Matamoros
Lila Gabriela Mendez Florez
Javeed Munshi
T. George Muste
Viral B. Patel
Santiago Pujol
William E. Rushing Jr.
Guillermo Santana
Jorge I. Segura
Larbi M. Sennour
Jairo Uribe

*Committee members responsible for the additions and revisions to ACI IPS-1.
†Chair of Task Group.

This guide presents simplified methods and design techniques that facilitate and speed the engineering of low-rise buildings within certain limitations. Material is presented in an order that follows typical design process with procedures introduced as the designer will need them in the course of a building design.

Keywords: analysis, frame; analysis, section; concrete quality, mixing, and placing; construction, low-rise building; design, foundation; design, seismic; design, simplified; design, structure; inspection; layout, structure; specifications; structures, low-rise.

ACKNOWLEDGMENTS

The original 2002 draft of the guide, published by ACI as document ACI IPS-1, was produced by a Joint Committee of *ICONTEC — Instituto Colombiano de Normas Técnicas y Certificación* (Colombian Institute for Technical Standards and Certification) and *AIS — Asociación Colombiana de Ingeniería Sísmica* (Colombian Association for Earthquake Engineering).

ICONTEC-AIS JOINT COMMITTEE MEMBERS

Luis Enrique García, Chair
Nélson Sánchez, Secretary

Guillermo Alonzo
Gilberto Areiza
Omar Darío Cardona
Augusto Espinosa
Diego Jaramillo
José Miguel Paz
Marco Puccini
Roberto Rochel
Carlos Alberto Rodríguez

ACI 314R-11 was adopted and published February 2012.

Daniel Rojas
Mauricio Sánchez
Jorge Segura
Pedro Theran
Luis Yamin

The following individual members of ACI offered useful comments and suggestions during the drafting of ACI IPS-1 (2002):

Sergio M. Alcocer
Shuaib H. Ahmad
John E. Breen
Juan P. Covarrubias
Werner Fuchs
John C. Glumb
David P. Gustafson
Bilal Hamad
Neil M. Hawkins
Kenneth C. Hover
James G. MacGregor
James O. Jirsa
Dov Kaminetzky
Richard E. Klingner
Leslie D. Martin
Jack P. Moehle
Vilas S. Mujumdar
James S. Pierce
Basile G. Rabbat
Julio A. Ramírez
Mete A. Sozen
Richard D. Stehly*
James K. Wight
Min-Hong Zhang

*Deceased

FOREWORD

Guide information presented is derived from:

1. "Building Code Requirements for Structural Concrete (ACI 318-11) and Commentary," by ACI;
2. "Minimum Design Loads for Buildings and Other Structures (ASCE 7-10)," by the American Society of Civil Engineers; and
3. "International Building Code (IBC 2009)," by the International Code Council.

The quality and testing of materials used in construction are covered by references to the appropriate ASTM standard specifications.

Whereas many of the tables, charts, and values included in this guide originated from the reference documents above, they have been modified or reorganized to be more conservative, to match design process flow, or better support the holistic and simplified design approach presented.

Although ACI 314R-11 is not written in mandatory language, the information is presented in such a manner that a structure designed following this guide will, in principle, comply with the codes and standards on which it was based. Although this guide is written in non-mandatory language, it is meant to be applied as a whole, because the simplified provisions are interdependent, and it would be unsafe to employ only a portion of this guide and disregard the remainder. This guide is not a code and is not deemed to satisfy ACI 318-11, ASCE 7-10, and IBC 2009. As this guide will be used as a design aid, it is the licensed design professional's responsibility to ensure that the requirements of ACI 318-11, ASCE 7-10, and IBC 2009 are satisfied.

PREFACE

Publication of ACI IPS-1 (2002) was the result of an agreement between ACI and the two Colombian Institutions: ICONTEC and AIS.

The initial drafting of ACI IPS-1 (2002) was motivated by frequent worldwide discussions that reinforced concrete codes might be unnecessarily sophisticated for some applications, such as small, low-rise buildings. Current knowledge on reinforced concrete behavior obtained through experimentation and experience, and its status and dissemination as a structural material used worldwide, made developing a simplified design and construction guide feasible. This guide used ACI IPS-1 (2002) as a basis, with information derived from the following: ACI 318-11, ASCE 7-10, and IBC 2009.

This guide presents simplified approaches to assist engineers in designing low-rise buildings within certain limitations. This guide is a design aid and educational tool, but not a code. It does not fully satisfy the requirements of ACI 318-11, ASCE 7-10, and IBC 2009. There are many options within these standards that are not considered in this guide, such as the use of supplementary cementitious materials in concrete mixtures. It is the licensed design professional's responsibility to ensure a structure's design satisfies the legal requirements of the local jurisdiction.

This guide provides simplified design information that is interdependent, and a user should not employ only a portion of this guide and disregard the remainder. This guide provides adequate design information when used as a whole and for structures within its scope. It is unsafe to use only selected sections of this guide or for structures outside its scope.

Many of the tables, charts, and values included in this guide are drawn from information in the reference standards, but are modified or reorganized to be more conservative, to match design process flow, or better support the holistic and simplified design approach. This guide is expected to be especially useful in the education and training of young engineers in reinforced concrete design of low-rise structures of small to medium floor areas.

This guide presents:

(a) Information in the order needed in the course of a design;

(b) Explanatory material at appropriate places;

(c) Computations only requiring a hand calculator;

(d) Graphs and graphical explanations;

(e) Design information based on simplified strength models;

(f) Other limit states accounted for by minimum dimensions;

(g) Conservative loads and simplified analysis guidelines;

(h) Simplified geotechnical information to help define soil-bearing capacity;

(i) Shear walls as the seismic-force-resisting system; and

(j) Material and construction guidelines based on commonly available steel grades and medium-strength concrete that can be site mixed.

CONTENTS

CHAPTER 1—GENERAL

1.1—Scope

This guide is intended for the planning, design, and construction of reinforced concrete structures in new low-rise buildings of restricted occupancy, number of stories, and area. Although the information presented was developed to produce, when properly used, a reinforced concrete structure with an appropriate margin of safety, this guide is not a replacement for a licensed design professional's experience and working knowledge. For the structure designed by the guide to attain the intended margin of safety, the guide should be used as a whole, and alternative procedures should be used only when explicitly permitted in the guide. The minimum dimensioning prescribed in the guide replaces, in most cases, more detailed procedures prescribed in the supporting codes and standards listed in 1.4.

1.2—Purpose

The guide provides a licensed design professional with sufficient information to design structural reinforced concrete members that comprise the structural framing of a low-rise building with the limits set in 1.3. Design rules set forth in this guide are simplifications that, when used together, comply with the more detailed requirements of the reference standards listed in 1.4.

1.3—Limitations

This guide is only meant for buildings meeting all the limitations set forth in 1.3.1 to 1.3.10.

These limits maintain the guide's scope in close adherence to the collective experience of the original drafting committee (ICONTEC-AIS). Buildings within this scope are expected to have a normal rectangular footprint with simple standard geometries and member dimensions in both plan and vertical directions. Such buildings also depend primarily on reinforced concrete structural walls for lateral load resistance. Observing these limits justifies the simplified analysis and design methods here without the need for special analyses including slenderness and second-order effects. Buildings with offsets, reentrant corners, and vertical or horizontal irregularities are outside the scope of this guide.

1.3.1 *Use and occupancy*

1.3.1.1 *Permitted uses and occupancies*—Table 1.3.10 lists building occupancy groups and subgroups, indicating for each whether the use of this guide is permitted.

1.3.1.2 *Mixed occupancy*—Recommendations described in this guide apply to cases involving combinations of the uses identified in 1.3.1.1.

1.3.2 *Maximum number of stories*—Recommendations described in this guide apply to buildings with five or fewer stories above ground and no more than one basement level.

1.3.3 *Maximum area per floor*—The area per floor should not exceed 10,000 ft^2 (1000 m^2).

1.3.4 *Maximum story height*—Story height, measured from floor finish to floor finish, should not exceed 13 ft (4 m).

Table 1.3.10—Permitted uses and occupancies

Occupancy group	Occupancy subgroup		Permitted
Group A—Assembly	A-1	Fixed-seating theaters, television, and radio studios	NO
	A-2 A-3	Building having an assembly room with capacity less than 100 persons and not having a stage	YES
	A-4	Arenas, skating rinks, swimming pools, and tennis courts	NO
	A-5	Amusement parks, bleachers, grandstands, and stadiums	NO
Group B—Business	B	Building for use as offices, or professional services containing eating and drinking establishments with less than 50 occupants	YES
Group E—Educational	E	Educational purposes	YES
Group F—Factory	F-1	Light industries not using heavy machinery	YES
	F-2	Heavy industries using heavy machinery	NO
Group H—Hazardous	H	Manufacturing, processing, generation, or storage of materials that constitute a physical or health hazard	NO
Group I—Institutional	I-1	Residential board and care facilities	YES
	I-2	Hospitals	NO
	I-3	Prisons, jails, reformatories, and detention centers	YES
	I-4	Daycare facilities	YES
Group M—Mercantile	M	Display and sale of merchandise	YES
Group R—Residential	R-1	Hotels having an assembly room with capacity less than 100 persons and not having a stage	YES
	R-2	Apartment buildings and dormitories	YES
	R-3	Houses	YES
	R-4	Residential care and assisted-living facilities	YES
Group S—Storage	S-1	Storage of heavy or hazardous materials	NO
	S-2	Storage of light materials	YES
Group U—Utility and miscellaneous	U	Utilities, water supply systems, and power-generating plants	NO
	U	Garages for vehicles with carrying capacity up to 4000 lb (1800 kg)	YES
	U	Garages for trucks of more than 4000 lb (1800 kg) carrying capacity	NO

1.3.5 *Maximum span length*—The span length for girders, beams, and slab-column systems, measured center-to-center of the supports, should not exceed 30 ft (10 m).

1.3.6 *Maximum difference in span length*—Spans should be approximately equal, and the shorter of two adjacent spans should be at least 80 percent of the larger span, except in elevator and stair cores. Refer to Section 7.9.1 for cores.

1.3.7 *Minimum number of spans*—There should be at least two spans in each of the two principal directions of the building in plan. Single spans may be permitted in one- and two-story buildings if the span length does not exceed 15 ft (5 m).

1.3.8 *Maximum overhang*—For girders, beams, and slabs with overhangs, length of the overhang should not exceed 1/3 of the length of the first interior span of the member.

1.3.9 *Maximum slope for slabs, girders, beams, and joists*—When sloping slabs, girders, beams, or joists are used, slope of the member should not exceed 15 degrees.

1.3.10 *Maximum slope of the terrain*—The slope of the terrain surrounding the building should not exceed:

(a) 30 degrees (Fig. 3.3.3); or

(b) The ratio of the height of the first story to the smaller dimension of the building in plan.

1.4—Supporting codes and standards

For cases within the limits described in 1.3, this guide is intended to be a simplification complying with the following supporting codes and standards:

1. Building Code Requirements for Structural Concrete (ACI 318-11) and Commentary;
2. Minimum Design Loads for Buildings and Other Structures (ASCE 7-10); and
3. International Building Code (IBC 2009).

Other cases are not covered by this guide. Please refer to Table A.1 in Appendix A for a guide by section to corresponding topics in the supporting codes and standards.

1.5—Notation and definitions

Terms for general use are presented in Chapter 2.

1.6—Design and construction procedure

1.6.1 *Procedure*—The design procedure comprises the steps listed in Table 1.6.1. Refer also to Fig. 1.6.1a and 1.6.1b. Note that by conforming to the dimensional limits and cover of this guide, a 1-hour fire rating is achieved. This rating is usually sufficient for the permitted occupancies in this guide. Other fire ratings are beyond the scope of this guide, and such designs should be performed using the codes and standards of 1.4.

1.6.2 *Design documentation*—The design steps should be recorded in a calculation record (1.6.2.1), a geotechnical report (1.6.2.2), the structural drawings (1.6.2.3), and the project specifications (1.6.2.4) as follows.

1.6.2.1 *Calculation record*—The licensed design professional should document all design steps in a calculation record. This record should contain, as a minimum, the:

Table 1.6.1—Design and construction procedure steps

Step	Description	Related chapter
A	Verification that the limitations for using the guide are met. Definition of the layout in plan and height of the structure.	1 and 3
B	Calculation of all gravity loads that act on the structure, excluding the self-weight of the structural members.	4
C	Definition of an appropriate floor system, depending on the span lengths and the magnitude of the gravity loads.	6
D	Selection of trial dimensions for the slab of the floor system. Calculation of the self-weight of the system and design of the members that comprise it, correcting the dimension if needed by the strength and serviceability limit states, complying with the limits for slab systems with beams, or slab-column systems.	6, 7, and 9
E	Trial dimensions for the beams and girders (if needed). Calculation of the self-weight of girders, beams, and joists. Flexural and shear design of the beams and girders, correcting the dimensions if needed by the strength and serviceability limit states.	6 and 8
F	Trial dimensions for the columns. Verification of column slenderness through the use of minimum dimensions. Calculation of the self-weight of the columns. Design of the columns for combination of axial load, moment, and shear. Correcting the dimensions if needed by the strength and serviceability limit states.	10
G	If lateral load, such as earthquake, wind, or lateral earth pressure, is beyond nominal, magnitude and application point are to be established; otherwise the designer may proceed to Step I.	4
H	Preliminary location and trial dimensions for reinforced concrete walls capable of resisting lateral loads. For earthquake loads, the influence of wall self-weight is evaluated. Flexure and shear design of the reinforced concrete walls.	11 and 12
I	Design of the stairways, ramps, small potable water tanks, and retaining walls.	13
J	Loads at the foundation level are determined. Definition of the foundation system is performed. Design of the structural members of the foundation.	14
K	Production of the structural drawings and specifications.	15
L	The structure is built complying with the construction and inspection requirements.	16

(a) General structural program, as defined in Chapter 3;
(b) Description of the structural system;
(c) Loads;
(d) Characteristics, strength, and fabrication standards for all structural materials;
(e) Justification of all design calculations; and
(f) Sketches of the reinforcement layout for all structural members.

1.6.2.2 *Geotechnical report*—The geotechnical report should record, as a minimum, the soil investigation performed, selected allowable bearing capacity of the soil, soil profile type, lateral soil pressures anticipated for design of any soil-retaining structures, and all other information indicated in Chapters 4 and 14.

1.6.2.3 *Structural drawings*—Structural drawings should include, as a minimum, all the plans indicated by Chapter 15 for construction of the building.

1.6.2.4 *Project specifications*—Project specifications should include, as a minimum, all the construction specifications described in Chapter 15.

1.6.3 *Precast concrete components*—Precast concrete components may be used, including prestressed concrete manufactured in off-site facilities. Such components should be designed by a licensed design professional in accordance with the codes and standards listed in 1.4. Calculations should be reviewed by the licensed design professional of record (1.2), and included in the calculation record (1.6.2.1). Detailing and placing drawings conforming to 15.2.2 should be furnished and included as part of the structural drawings (1.6.2.3). Manufacture of precast components should be done in a facility with demonstrated capability of producing quality products.

1.7—Limit states

The design approach of this guide is based on limit states, where a limit state is a condition beyond which a structure or member becomes unfit for service and is judged to be unsafe or no longer useful for its intended function. The designer should verify that the following limit states are accounted for in the resulting structure:

(a) Strength limit state; and
(b) Serviceability limit state.

The following are considered implicitly in the design procedure:

(a) Structural integrity;
(b) Lateral load story drift;
(c) Durability; and
(d) Fire resistance.

1.8—Strength design

1.8.1 *General*—In strength design, the structure and the structural members are dimensioned to have design strengths at all sections at least equal to the demands calculated for the combinations of factored loads described in Chapter 4.

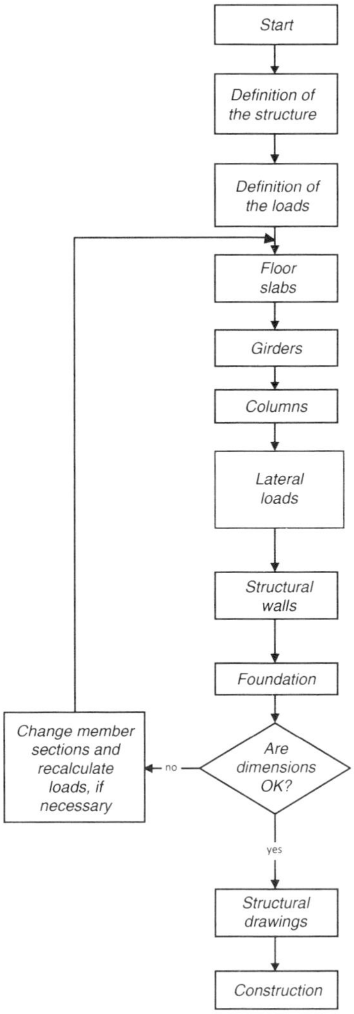

Fig. 1.6.1a—Design and construction procedure.

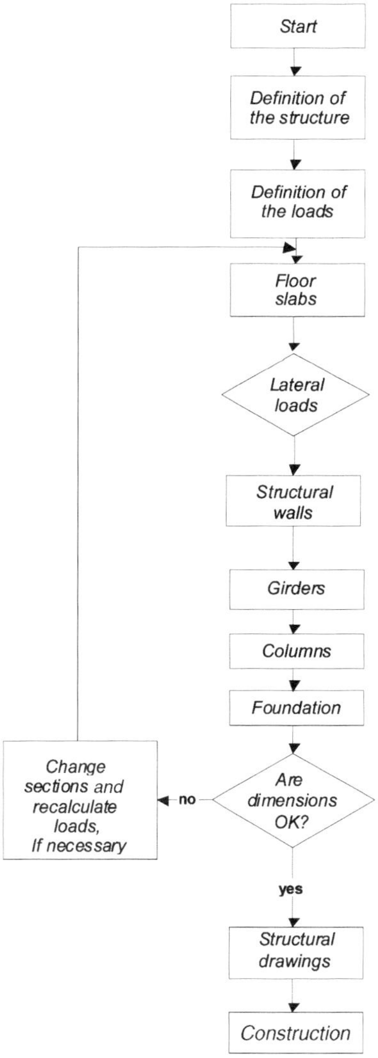

Fig. 1.6.1b—Design and construction procedure for earthquake regions.

The basic expression for the strength limit state is:

$$\text{resistances} \geq \text{load effects} \tag{1.8.1a}$$

Because resistances may be less than computed and the load effects could be larger than computed, strength reduction factors ϕ less than 1, and load factors γ generally greater than 1, are used:

$$\phi R_n \geq \gamma_1 S_1 + \gamma_2 S_2 + \cdots \tag{1.8.1b}$$

where R_n stands for nominal strength, and S stands for load effects based on the loads described in Chapter 4. Therefore, the strength design requires that:

$$\text{design strength} \geq \text{required strength} \tag{1.8.1c}$$

$$\phi\ (\text{nominal strength}) \geq U$$

where the required strength is $U = \gamma_1 S_1 + \gamma_2 S_2 + \ldots$

1.8.2 *Required strength*—The required strength U should be computed for the combinations of factored loads listed in 4.2.

1.8.3 *Design strength*—The design strength provided by a member, its connections to other members, and its cross sections in terms of flexure, axial load, and shear, is the nominal strength multiplied by a strength reduction factor ϕ. Nominal strength should be calculated for each particular force effect in each of the member types at the defined critical sections. The following strength reduction factors ϕ should be used:

Flexure, without axial load .... $\phi = 0.90$
Axial tension and axial tension with flexure..... $\phi = 0.90$
Axial compression and axial compression with flexure:
Columns with ties and reinforced concrete walls
.... $\phi = 0.65$
Columns with spiral reinforcement .... $\phi = 0.75$
Shear and torsion .... $\phi = 0.75$
Bearing of concrete....................................... $\phi = 0.65$

1.9—Serviceability design

To ensure adequate response during service, follow the recommendations for limiting dimensions, cover, detailing, and construction. These serviceability conditions include effects such as:

(a) Long-term environmental effects, including exposure to aggressive environment or corrosion of the reinforcement;
(b) Dimensional changes due to variations in temperature, relative humidity, and other effects;
(c) Excessive cracking of the concrete;
(d) Excessive vertical deflections; and
(e) Excessive vibration.

CHAPTER 2—NOTATION AND DEFINITIONS

2.1—Notation

a = depth of equivalent rectangular compressive stress block, in. (mm)
a_w = distance from edge of wall footing to the resultant of soil reaction in wall footing, in. (mm)
A_a = effective seismic peak ground horizontal acceleration in rock for short periods of vibration, expressed as a fraction of gravity g
A_b = area of an individual reinforcing bar or wire, in.^{2} (mm^2)
A_{cb} = bearing area of concrete, in.^{2} (mm^2)
A_{cs} = area of core of spirally reinforced compression member measured to outside diameter of spiral, in.^{2} (mm^2)
A_f = contact area of footing with soil, ft^2 (m^2)
A_g = gross area of section, or area of concrete only excluding area of voids, in.^{2} (mm^2)
A_i = area of additional hanger reinforcement where beams are supported by girders or other beams, in.^{2} (mm^2)
A_j = effective cross-sectional shear area within a joint, in.^{2} (mm^2)
A_p = component, or cladding, wind exposed surface area, ft^2 (m^2)
A_s = area of longitudinal tension reinforcement, in.^{2} (mm^2)
A_s' = area of longitudinal compression reinforcement, in.^{2} (mm^2)
A_{se} = steel area at the extreme face of column or reinforced concrete wall, in.^{2} (mm^2)
$A_{s,min}$ = minimum area of longitudinal tension reinforcement, in.^{2} (mm^2)
A_{ss} = steel area at the side face of column or reinforced concrete wall, in.^{2} (mm^2)
A_{st} = total area of longitudinal reinforcement, in.^{2} (mm^2)
A_{su} = wind-exposed surface area, ft^2 (m^2)
A_v = area of shear reinforcement, in.^{2} (mm^2)
b = width of compression flange of member, or width of member, in. (mm)
b_c = width of column section, and for punching shear evaluation, the smallest plan dimension of pedestal, column capital or drop panel, or thickness change in stepped footings, in. (mm)
b_f = width of compression face of member, in. (mm)
b_o = perimeter of critical section for two-way shear (punching shear) in slabs, in. (mm)
b_w = web width of section, or wall width, in. (mm)
B_f = short horizontal dimension of footing, in. (mm)
c_c = least distance from surface of reinforcement to the side face, in. (mm)
C_p = component, or cladding, wind surface pressure coefficient
C_{su} = wind surface pressure coefficient
C_{vx} = coefficient defined in 4.11.4 for design of seismic loads
d = effective depth of section, taken as distance from extreme compression fiber to centroid of tension reinforcement, in. (mm)
d' = distance from extreme compression fiber to centroid of compression reinforcement, in. (mm)
d_b = nominal diameter of reinforcing bar or wire, in. (mm)
d_c = distance from extreme tension fiber to centroid of tension reinforcement, in. (mm)
d_s = outside diameter of spiral reinforcement, in. (mm)
D = dead loads or related internal moments and loads
e_x = eccentricity, measured in x-direction, between story center of lateral stiffness and application point of story lateral loads acting in y-direction, in. (mm)
e_y = eccentricity, measured in y-direction, between story center of lateral stiffness and application point of story lateral loads acting in x-direction, in. (mm)
e_B = eccentricity of resultant applied to footing in direction parallel to B_f, in. (mm)
e_H = eccentricity of resultant applied to footing in direction parallel to H_f, in. (mm)
E = seismic loads or related internal moments and loads
E_c = modulus of elasticity of concrete, psi (MPa)
f_c' = specified compressive strength of concrete, psi (MPa)
$\sqrt{f_c'}$ = square root of specified compressive strength of concrete, the result has units of psi (MPa)
f_{cr}' = required average compressive strength of concrete used as the basis for selection of concrete proportions, psi (MPa)
f_{cu} = extreme fiber compressive stress due to factored loads at edges of structural walls, psi (MPa)
f_y = specified yield strength of reinforcement, psi (MPa)
f_{ypr} = probable specified maximum strength of reinforcement, psi (MPa) ($f_{ypr} = 1.25f_y$)
f_{yt} = specified yield strength of transverse or spiral reinforcement, psi (MPa)
F = loads due to weight and pressure of fluids with well-defined densities and controllable maximum heights or related internal moments and loads
F_a = seismic site coefficient for short periods of vibration
F_{ac} = total lateral active soil force, lb (kN)
F_i, F_x = wind or seismic force applied at level i or x, respectively, lb (kN)
F_o = total lateral at-rest soil force, lb (kN)

F_{pw} = equivalent static wind force for components and cladding acting normal to wind-exposed surface, lb (kN)
F_{su} = equivalent static wind force acting normal to wind-exposed surface, lb (kN)
F_{ui}, F_{ux} = factored lateral force applied to wall at level i or x, respectively, lb (kN)
g = acceleration due to gravity, 386 in./s^2 (9.8 m/s^2 ≈ 10 m/s^2)
h = overall depth or thickness of member, or height of section of member, or outside diameter of circular section, in. or ft (mm or m)
h_b = vertical distance measured from bottom of supporting girder to bottom of supported beam, ft (m)
h_c = depth of column, or dimension of column in direction parallel to girder span, and for punching shear evaluation the largest plan dimension of capital, drop panel, pedestal, or thickness change in stepped footings, in. (mm)
h_f = flange thickness, in. (mm)
h_g = total depth of supporting girder, in. (mm)
h_i, h_x = height above base to level i or x, respectively, ft (m)
h_n = clear vertical distance between lateral supports of columns and walls, in. (mm)
h_{pi} = story height of floor i measured from floor finish of story to floor finish of story immediately below, ft (m)
h_r = mean roof height for wind design, measured over terrain, ft (m)
h_s = depth of soil against retaining wall, in. (mm)
h_w = height of wall from base to top, in. (mm)
H = loads due to weight and pressure of soil, water in soil, or other materials, or related internal moments and loads
H_f = long horizontal dimension of footing, in. (mm)
I_c = moment of inertia of column section, in.^{4} (mm^4)
k_r = story total rotational stiffness
k_x, k_y = wall lateral stiffness in direction x or y, respectively, lb/in. (N/mm) (Eq. (4.14.5a(a))
K_a = active soil pressure coefficient
K_o = at-rest soil pressure coefficient
K_p = passive soil pressure coefficient
ℓ_a = length of clear span in short direction of two-way slabs or in direction of moments, measured face-to-face of beams or other supports, in. (mm)
ℓ_c = length of clear span in long direction of two-way slabs or systems, measured face-to-face of beams or other supports, in. (mm)
ℓ_d = development length, in. (mm)
ℓ_n = length of clear span, in long direction for two-way systems, measured face-to-face of beams or other supports, in. (mm)
ℓ_s = center-to-center span length; shortest distance between adjacent parallel column centerlines, in. (mm)
ℓ_w = horizontal length of wall, in. (mm)
ℓ_0 = column confinement length, in. (mm)
ℓ_1 = length of span in direction of moments, measured center-to-center of supports, in. (mm)
ℓ_2 = length of span transverse to ℓ_1, measured center-to-center of supports, in. (mm)
L = live loads or related internal moments and loads
L_r = roof live load or related internal moments and loads
M_a = factored moment in short direction in two-way slabs, lb·in. (N·m) per unit slab width
M_b = factored moment in long direction in two-way slabs, lb·in. (N·m) per unit slab width
M_{bn} = nominal moment strength at section at balanced conditions, lb·in. (N·m)
M_i, M_x = unfactored overturning moment due to lateral loads for story i or x, respectively, lb·in. (kN·m)
M_{iu}, M_{xu} = factored story moment due to lateral loads at story i or x, respectively, lb·in. (kN·m)
M_n = nominal moment strength at section, lb·in. (N·m)
M_o = total factored moment at section, lb·in. (kN·m)
M_{ot} = unfactored overturning moment due to lateral loads at base of structure, lb·in. (kN·m)
M_{otu} = factored overturning moment due to lateral loads at base of structure, lb·in. (kN·m)
M_{pr} = probable flexural strength of member at joint face, computed using f_{ypr} and ϕ of 1.0, lb·in. (N·m)
M_u = factored moment at section, lb·in. (N·m)
$M^+_{a \text{ or } b}$ = factored positive moment at section, lb·in. (N·m) per unit slab width
$M^-_{a \text{ or } b}$ = factored negative moment at section, lb·in. (N·m) per unit slab width
M_u^+ = factored positive moment at section, lb·in. (N·m)
M_u^- = factored negative moment at section, lb·in. (N·m)
M_{ux} = factored moment at section in x direction, lb·in. (N·m)
M_{uy} = factored moment at section in y direction, lb·in. (N·m)
M_{xu} = factored overturning moment due to lateral loads for story x, lb·in. (kN·m)
n_c = number of interior columns in story in direction under consideration, for the entire structure
n_e = number of edge columns in story in direction under consideration, for the entire structure
N = number of blows in standard penetration test (SPT)
$\bar{N}$ = average soil SPT resistance, measured in number of blows per ft (0.305 m) of penetration, averaged over upper 100 ft (30 m) of soil profile
p = unfactored pressure for braced retaining walls, psi or lb/ft^2 (kPa or kN/m^2)
p_a = unfactored active pressure, psi or lb/ft^2 (kPa or kN/m^2)
p_d = unfactored concentrated dead load applied directly to member, lb (kN)
p_ℓ = unfactored concentrated live load applied directly to member, lb (kN)
p_o = unfactored pressure at rest, psi or lb/ft^2 (kPa or kN/m^2)
p_p = unfactored passive pressure, lb/ft^2 (kPa or kN/m^2)
p_t = unfactored vertical surcharge pressure on top of retaining wall backfill, lb/ft^2 (kPa or kN/m^2)
p_u = factored concentrated load applied directly to member, lb (kN)

p_{utw} = factored design horizontal pressure for retaining walls caused by surcharge pressure on top of retaining wall backfill, lb/ft^2 (kPa or kN/m^2)
p_{uw} = factored design horizontal pressure for retaining walls, lb/ft^2 (kPa or kN/m^2)
p_z = at-rest, or active lateral soil pressure at depth x, lb/ft^2 (kPa or kN/m^2)
P_{bn} = nominal compression axial compression strength at section at balanced conditions, lb (N)
P_{cu} = factored compression load on wall boundary element, including seismic effects, lb (N)
P_d = unfactored dead load axial force at section, lb (kN)
P_l = unfactored live load axial force at section, lb (kN)
P_n = nominal axial compression strength at given eccentricity, lb (N)
$P_{n(max)}$ = maximum compression nominal axial compression strength at section, lb (N)
P_{on} = nominal compression, without flexure, or axial compression strength at section, lb (N)
P_{ov} = maximum vertical load applied to footing including wind or seismic overturning effects, lb (N)
P_{tn} = nominal tension, without flexure, or axial tension strength at section, lb (N)
P_{tu} = factored tension force on wall boundary element, including seismic effects, lb (N)
P_u = factored axial or concentrated load, or factored axial load at given eccentricity, lb (N or kN)
P_{ub} = factored axial load at base of column, lb (N)
P_v = maximum vertical load applied to footing not including wind or seismic, lb (N)
PI = soil plasticity index, equal to difference in percentage of moisture content at liquid limit and at plastic limit
q_a = unfactored allowable bearing capacity of soil, lb/ft^2 (kPa or kN/m^2)
q_c = standard cone penetration resistance in cone penetration test (CPT), lb/ft^2 (kPa or kN/m^2)
q_d = unfactored dead load per unit area, lb/ft^2 (kN/m^2)
q_h = wind velocity pressure at height h over terrain, lb/ft^2 (kN/m^2)
q_ℓ = unfactored live load per unit area, lb/ft^2 (kN/m^2)
q_o = overburden pressure, or unfactored gravity loads applied directly to slab in mat foundations, lb/ft^2 (kN/m^2)
q_u = factored load per unit area, lb/ft^2 (kN/m^2)
q_{uc} = unconfined compression strength of soil, lb/ft^2 (kPa or kN/m^2)
q_{un} = factored net soil reaction pressure on footing, lb/ft^2 (kN/m^2)
r_u = factored uniformly distributed reaction from slab on supporting girder, beam, or reinforced concrete wall, lb/ft (kN/m)
R = rain load or related internal moments and loads, lb or ft-lb and lb (N) or (N.m), (N)
R_n = nominal strength in terms of flexure, axial load, shear or bearing strength, lb (N)
R_{pi} = reaction from lateral soil pressure at story i, lb (kN)
R_s = response modification factor related to energy dissipation capacity in inelastic range of seismic-resistant structural system
R_u = factored reaction from supported structural member, lb (kN)
s = spacing of transverse reinforcement, stirrups, or wall reinforcement measured along axis of member, in. (mm)
s_j = center-to-center spacing between parallel joists, in. (mm)
s_s = sample standard deviation, psi (MPa)
s_{sk} = spacing of skin reinforcement, in. (mm)
s_u = shear strength of undrained cohesive soil, lb/ft^2 (kPa or kN/m^2)
S = snow load or related internal moments and loads
S_a = value of elastic acceleration design response spectrum, for damping ratio of 5 percent of critical, expressed as fraction of acceleration of gravity
S_i = nominal load effect based on load i
S_{DS} = value of design earthquake spectral acceleration parameter at short period
t_x, t_y = structural vertical member cross-sectional dimension in direction x or y, respectively, in. (mm)
T = cumulative effect of temperature, creep, shrinkage, differential settlement, and shrinkage-compensating concrete
T_i, T_x = unfactored story torsional moment due to lateral loads at story i or x, respectively, lb·in. (kN·m)
T_o = unfactored story torsional moment due to lateral loads at base of structure, lb·in. (kN·m)
T_{ou} = factored story torsional moment due to lateral loads at base of structure, lb·in. (kN·m)
T_u = factored torsional moment at section, lb·in. (N·mm)
T_{xu}, T_{iu} = factored story torsional moment due to lateral loads at story i or x, respectively, lb·in. (kN·m)
U = required strength to resist factored loads or related internal moments and loads
V = basic wind speed, mph (m/s), corresponding to 3-second gust speed at 33 ft (10 m) above ground
V_{bs} = seismic design base shear, lb (kN)
V_c = nominal shear strength provided by concrete, lb (N)
V_i, V_x = unfactored story shear due to lateral loads at story i or x, respectively, lb (kN)
V_{iu}, V_{xu} = factored story shear due to lateral loads at story i or x, respectively, lb (kN)
V_n = nominal shear strength at section, lb (N)
V_o = unfactored story shear due to lateral loads at base of structure, lb (kN)
V_{ou} = factored story shear due to lateral loads at base of structure, lb (kN)
V_s = nominal shear strength at section provided by transverse reinforcement, lb (kN)
V_{sw} = contribution of the horizontal wall reinforcement to the nominal shear strength at section, lb (kN)
V_u = factored shear, lb (N)
V_w = wind design base shear, lb (kN)
w = moisture content of the soil, percentage

w_d = unfactored dead load per unit member length applied directly to the member, lb/in. (kN/m)
w_{df} = unfactored dead load per unit member length applied directly to the member, lb/ft (kN/m)
w_i, w_x = part of W_s corresponding to story i or x, respectively, lb (kN)
w_ℓ = unfactored live load per unit member length applied directly to the member, lb/in. (kN/m)
w_u = factored load per unit member length applied directly to the member, lb/ft (kN/m)
W = wind loads or related internal moments and loads
W_s = total building weight for seismic design, lb (kN) (4.11.3.3)
W_u = total factored uniformly distributed design load per unit member length, lb/in. (kN/mm)
W_{uf} = total factored uniformly distributed design load per unit member length, lb/ft (kN/m)
$\bar{x}$, $\bar{y}$ = story lateral stiffness center coordinates in directions x and y, respectively, in. (mm)
z = depth of soil, ft (m)
α_a = fraction of load acting in short direction in two-way slabs-on-girders
α_b = fraction of load acting in long direction in two-way slabs-on-girders
α_f = parameter of Eq. (5.11.4.2)
α_{sh} = factor affecting equivalent shear force due to unbalanced moment at column-slab connection in Eq. (9.5.4.4)
α_w = horizontal angle between normal to wind-exposed surface and wind direction, degrees
β = ratio of clear spans in long to short direction of two-way slabs
β_f = ratio of long side to short side of footing
β_w = vertical angle between normal to wind-exposed surface and horizontal line, degrees
γ = unit weight of material or soil, lb/ft^3 (kN/m^3)
γ_i = load factor for load effect i
ΔM_u = factored unbalanced moment at column-girder joint or wall-girder joint, lb·in. (N·mm)
$\Delta M_{u\text{-}ad}$ = additional factored unbalanced moment at column-slab connection, lb·in. (N·mm)
ΔV_e = factored design shear force from development of probable flexural capacity of member at faces of joints, lb (N)
ΔV_u = factored shear force due to unbalanced moment at column-slab connection, lb (N)
ΔV_{ut} = increase in wall factored shear due to torsional effects, lb (kN)
λ_{ps} = factor to calculate punching-shear strength (9.5.4.3)
ρ = ratio of longitudinal tension reinforcement $A_s/(bd)$
ρ' = ratio of longitudinal compression reinforcement $A_s'/(bd)$
ρ_ℓ = ratio of total longitudinal reinforcement area to gross concrete section area $A_{st}/(bh)$
ρ_{max} = maximum ratio of longitudinal flexural tension reinforcement
ρ_{min} = minimum ratio of longitudinal flexural tension reinforcement
ρ_s = ratio of volume of spiral reinforcement to core volume confined by spiral reinforcement (measured out-to-out)
ρ_t = ratio of horizontal shear reinforcement area to gross concrete area of vertical section
ρ_{vw} = ratio of vertical reinforcement in reinforced concrete walls
ϕ = strength reduction factor (1.8.3)
ϕ_s = angle of internal friction of soil
ΣM_c = sum of nominal flexural strengths (M_n) of columns framing into joint, lb·in. (N·m)
ΣM_g = sum of nominal flexural strengths (M_n) of girders framing into joint, lb·in. (N·m)
ΣP_u = sum of factored concentrated loads within span, lb (kN)
ΣR_u = sum of factored reactions from supported structural members at same story, lb (kN)

2.2—Definitions

ACI provides a comprehensive list of definitions through an online resource, "ACI Concrete Terminology" (http://terminology.concrete.org.) Definitions provided herein complement that resource.

admixture—material other than water, aggregate, or hydraulic cement, used as an ingredient of concrete and added to concrete before or during its mixing to modify its properties.

aggregate—granular material, such as sand, gravel, crushed stone, and iron blast-furnace slag, used with a cementing medium to form a hydraulic cement concrete or mortar.

aggregate, lightweight—aggregate meeting the requirements of ASTM C330/C330M-09 and having a loose bulk density of 70 lb/ft^3 (1120 kg/m^3) or less, determined in accordance with ASTM C29/C29M-09. Concrete manufactured using lightweight aggregate is not covered by this guide.

allowable bearing capacity—the maximum pressure to which a soil or other material should be subjected to guard against shear failure or excessive settlement.

anchorage—a device embedded in concrete for the purpose of providing a connection to another member or structure.

bar diameter, nominal—value computed using the nominal bar area.

base of structure—level at which the horizontal earthquake ground motions are assumed to be imparted to a building. This level does not necessarily coincide with the ground level.

beam—a structural member subjected to axial load and flexure but primarily to flexure. See also **girder**.

bending moment—the bending effect at any section of a structural element; it is equal to the algebraic sum of the moments of the internal compression and tension forces acting at the section, with respect to the centroidal axis of a member, on a free body of the member.

boulders—coarse materials more than 8 in. (200 mm) in diameter.

boundary element—portion along wall edges strengthened by longitudinal and transverse reinforcement. Boundary elements do not necessarily require an increase in thickness of the wall.

cement—any of a number of materials that are capable of binding aggregate particles together.

clay—natural mineral material having plastic properties and composed of very fine particles; the clay mineral fraction of a soil is usually considered to be the portion consisting of particles finer than 8×10^{-5} in. (2 μm); clay minerals are essentially hydrous aluminum silicates or occasionally hydrous magnesium silicates.

column—member with a ratio of height-to-least lateral dimension exceeding 3 used primarily to support axial compressive load.

collector element—element that acts in axial tension or compression to transmit earthquake-induced loads between a structural diaphragm and a vertical element of the seismic-force-resisting system.

concrete—mixture of portland cement and any other hydraulic cement, fine aggregate, coarse aggregate, and water, with or without admixtures.

concrete mixture proportioning—the proportions of ingredients that make the most economical use of available materials to produce mortar or concrete of the required properties.

concrete, specified compressive strength of, f_c'—compressive cylinder strength of concrete at 28 days used in design and evaluated in accordance with the appropriate ASTM standards, expressed in psi (MPa). Whenever the quantity f_c' is under a radical sign ($\sqrt{f_c'}$), square root of numerical value only is intended, and the result maintains the units as psi (MPa).

concrete, lightweight—concrete containing lightweight aggregate and an equilibrium density, as determined by ASTM C567-05, between 90 and 115 lb/ft^3 (1440 to 1840 kg/m^3). This type of concrete is not covered by this guide.

confinement stirrup or tie—see also **hoop**.

construction documents—the written, graphic, electronic, and pictorial documents describing the design, locations, and physical characteristics of the project required to verify compliance with the standard.

contraction joint—formed, sawed, or tooled groove in a concrete structure to create a weakened plane and regulate the location of cracking resulting from the dimensional change of different parts of the structure.

corrosion—destruction of metal by a chemical, electrochemical, or electrolytic reaction within its environment.

cover, concrete—the least distance between the surface of embedded reinforcement and the closest outer surface of the concrete.

crosstie—a continuous reinforcing bar having a seismic hook at one end and a hook not less than 90 degrees with at least a six-diameter extension at the other end. The hooks shall engage peripheral longitudinal bars. The 90-degree hooks of two successive crossties engaging the same longitudinal bars shall be alternated end for end.

curing—action taken to maintain moisture and temperature conditions in a freshly placed cementitious mixture to allow hydraulic cement hydration and (if applicable) pozzolanic reactions to occur so that the potential properties of the mixture may develop.

curtain wall—walls that are part of the façade or enclosure of the building that do not form part of the gravity or lateral-load-resisting system. See also **partitions**.

depth of member, *h*—distance in a flexural member, measured from extreme compression fiber to the extreme tension fiber.

design load combinations—combinations of factored loads and loads.

design method, strength—a method of member proportioning based on ensuring that the design strength obtained by reducing the nominal strength is larger than the required strength obtained by applying load factors to service loads.

development length—length of embedded reinforcement required to develop the design strength of reinforcement at a critical section.

development length for a bar with a standard hook—shortest distance between the critical section (where the strength of the bar is to be developed) and a tangent to the outer edge of the 90- or 180-degree hook.

differential settlement—the lowering in elevation of various parts of a foundation by different amounts.

documents, contract—a set of documents supplied by the owner to the contractor as the basis for construction. These documents contain contract forms, contract conditions, specifications, drawings, addenda, and contract changes.

effective depth of section, *d*—distance measured from extreme compression fiber to centroid of tension reinforcement.

embedment length—length of embedded reinforcement provided beyond a critical section.

essential facilities—buildings and other structures intended to remain operational during and after an extreme environmental event from wind, snow, or earthquake.

flange, compression—the widened portion of an I, T, or similar cross section that is shortened or compressed by bending under normal loads, such as the horizontal portion of the cross section of a simple span T-beam.

flexural reinforcement—reinforcement provided to resist the tension and compression stresses induced by flexural moments acting on the member section.

floor system—structural members that comprise the floor of a story in a building, including girders, beams, joists, and the slab that spans between them, or the slab only where it is directly supported on columns, as in slab-column systems.

footing—a structural element of a foundation that transmits loads directly to the soil.

footing, combined—a structural unit or assembly of units supporting more than one column.

formwork—total system of support for freshly placed concrete including the mold or sheathing in contact with the concrete as well as supporting members, hardware, and necessary bracing.

foundation—a system of structural elements that transmit loads from the structure above to the earth.

foundation beam (grade beam)—a reinforced concrete beam, usually at ground level, that strengthens or stiffens the foundation or supports overlying construction.

foundation, mat—a continuous footing supporting an array of columns in several rows in each direction, having a slab-like shape with or without depressions or openings.

girder—a large beam, usually horizontal, that serves as a main structural member often supporting reactions from other beams or girders. See also **beam**.

grade beam—see **foundation beam**.

grade slab (slab-on-ground)—a shallow foundation consisting of a continuous concrete slab, placed over native soil or engineered subgrade, where loads are locally distributed and transmitted to the ground.

gravel—

1. Granular material predominantly retained on the No. 4 (4.75 mm) sieve and resulting either from natural disintegration and abrasion of rock or processing of weakly bound conglomerate; and
2. That portion of an aggregate retained on the No. 4 (4.75 mm) sieve resulting either from natural disintegration and abrasion of rock or processing of weakly bound conglomerate.

gravity loads—loads that act downward and are caused by the acceleration of gravity acting on the mass of the elements and content that cause the dead and live loads.

hook—a bend in the end of a reinforcing bar.

hoop—a closed tie or continuously wound tie. A closed tie can be made up of several reinforcement elements each having seismic hooks at both ends. A continuously wound tie shall have a seismic hook at both ends.

isolation joint—a separation between adjoining parts of a concrete structure, usually a vertical plane, at a designed location such as to interfere least with performance of the structure, yet such as to allow relative movement in three directions and avoid formation of cracks elsewhere in the concrete and through which all or part of the bonded reinforcement is interrupted.

integrity, structural—the design concept that after an overload event or after damage occurs to a major supporting member, the structure has sufficient toughness to confine the damage to a local area and sufficient overall stability to prevent immediate collapse.

joist—a comparatively narrow beam used in closely spaced parallel arrangements to support floor or roof slabs.

lap splice—a connection of reinforcing steel made by lapping the ends of bars.

lateral-force-resisting system—portion of the structure composed of members proportioned to resist loads related to lateral loads.

licensed design professional—an individual who is licensed to practice structural design as defined by the statutory requirements of the professional licensing laws of the state or jurisdiction in which the project is to be constructed and who is in responsible charge of the structural design; in other documents, also called licensed design professional.

limit state—a condition beyond which a structure or member becomes unfit for service and is judged either to be no longer useful for its intended function (serviceability limit state) or to be unsafe (strength limit state).

loads—loads or other actions that result from the weight of all building materials, occupants and other variable or permanent contents, environmental effects, differential movement, and restrained dimensional changes. Permanent loads are those loads in which variations over time are rare or of small magnitude. All other loads are variable loads.

load effects—loads and deformations produced in structural members by the applied loads.

load, factored—load, multiplied by appropriate load factors, used to proportion members by the strength design method of this guide.

load factor—a factor by which a service load is multiplied to determine a factored load used in the strength design method.

load, live—live load specified by general building code (without load factor).

load, permanent—load in which variations over time are rare or of small magnitude. See also **load, service**.

load, service—all loads, static or transitory, imposed on a structure, or element thereof, during operation of a facility; specified in Chapter 4 (dead, live, soil, wind, snow, rain, flood, and seismic).

longitudinal reinforcement—reinforcement parallel to the length of a concrete member.

modulus of elasticity—ratio of normal stress to corresponding strain for tensile or compressive stresses below proportional limit of material.

moment, negative—a condition of flexure in which top fibers of a horizontally placed member, or external fibers of a vertically placed exterior member, are subjected to tensile stresses.

moment, positive—a condition of flexure in which, for a horizontal simply supported member, the deflected shape is normally considered to be concave downward and the top fibers subjected to compression stresses; for other members and other conditions, consider positive and negative as relative terms.

nonstructural elements—the architectural, mechanical, and electrical components and systems permanently attached to the building.

occupancy—the purpose for which a building or other structure, or part thereof, is used or intended to be used.

partitions—a nonstructural interior wall that spans horizontally or vertically from support to support. The supports may be the basic building frame, subsidiary structural members, or other portions of the partition system.

pedestal—upright compression member with a ratio of unsupported height to average least lateral dimension not exceeding 3.

pile—a structural element that is driven, jetted, or otherwise embedded on end in the ground to resist loads or compact the soil.

pile cap—a concrete element that transfers load from a column or pedestal to the top of one or more supporting piles.

plain concrete—structural concrete with no reinforcement or with less reinforcement than the minimum amount specified for reinforced concrete. Plain concrete is not covered by the guide.

plain reinforcement—reinforcement without surface deformations, or having deformations that do not conform to the applicable requirements.

positive reinforcement—steel reinforcement to resist positive bending moment.

precast concrete—structural concrete element cast elsewhere than its final position in the structure. Precast concrete is not covered by the guide, except as described in 1.6.3.

prestressed concrete—structural concrete in which internal stresses have been introduced to reduce potential tensile stresses in concrete resulting from loads. Prestressed concrete is not covered by the guide, except as described in 1.6.3.

project drawings—drawings that, along with the project specifications, complete the descriptive information for constructing the work required in the contract documents.

project specifications—written documents that specify requirements for a project in accordance with the service parameters and other specific criteria established by the owner. See also **specifications**.

reinforced concrete—structural concrete reinforced with no less than the minimum amounts of reinforcement.

reinforced concrete wall—structural concrete wall reinforced with no less than the minimum amount of prestressing steel or nonprestressed reinforcement as specified in the applicable building code. A shear wall is a reinforced concrete wall.

reinforcement—deformed steel bars, wire, or wire mesh, embedded in concrete in such a manner that it and the concrete act together in resisting loads.

reinforcement, compression—reinforcement designed to carry compressive stresses.

reinforcement, deformed—metal bars, wire, or fabric with a manufactured pattern of surface ridges that provide a locking anchorage with surrounding concrete. The following steel reinforcement is considered deformed reinforcement: deformed reinforcing bars, welded plain wire fabric, and welded deformed wire fabric conforming to the appropriate ASTM standards.

reinforcement, negative—steel reinforcement for negative moment.

reinforcement, spiral—continuously wound reinforcement in the form of a cylindrical helix.

reinforcement, tension—reinforcement designed to carry tensile stresses such as those in the bottom of a simple beam.

reshores—shores placed snugly under a concrete slab or other structural member after the original forms and shores have been removed to allow the new slab or structural member to deflect and support its own weight.

resistance, fire—the property of a material or assembly to withstand fire or give protection from it; as applied to elements of buildings, it is characterized by the ability to confine a fire or, when exposed to fire, to continue to perform a given structural function, or both.

required strength—strength of a member or cross section required to resist factored loads or related internal moments and loads in specified load combinations.

retaining wall—wall built to hold back earth.

sand—

1. Granular material passing the 3/8 in. (9.5 mm) sieve and almost entirely passing the No. 4 (4.75 mm) sieve and predominantly retained on the No. 200 (75 μm) sieve, and resulting either from natural disintegration and abrasion of rock or processing of completely friable sandstone; and

2. That portion of an aggregate passing the No. 4 (4.75 mm) sieve and predominantly retained on the No. 200 (75 μm) sieve, and resulting either from natural disintegration and abrasion of rock or processing of completely friable sandstone.

seismic hook—a hook on a stirrup, or crosstie having a bend not less than 135 degrees, except that circular hoops shall have a bend not less than 90 degrees. Hooks shall have a $6d_b$ (but not less than 3 in. [75 mm]) extension that engages the longitudinal reinforcement and projects into the interior of the stirrup or hoop.

self-weight—weight of the structural member, caused by the material that composes the member.

service load—all loads, static or transitory, imposed on a structure, or element thereof, during operation of a facility (without load factors).

settlement—downward movement of the supporting soil.

shear—an internal force tangential to the plane on which it acts.

shear reinforcement—reinforcement designed to resist shear or diagonal tension stresses.

shell concrete—concrete outside the transverse reinforcement confining the concrete core.

shores—vertical or inclined support members designed to carry the weight of the formwork, concrete, and construction loads above.

shrinkage and temperature reinforcement—reinforcement provided to resist shrinkage and temperature stresses in concrete.

silt—a granular material resulting from the disintegration of rock, with grains largely passing a No. 200 (75 μm) sieve; alternatively, such particles in the range from 8×10^{-5} to 0.002 in. (2 to 50 μm) in diameter.

slab—a molded layer of reinforced concrete, flat, horizontal (or nearly so), usually of uniform but sometimes of variable thickness, either on the ground or supported by beams, columns, walls, or other framework.

soil—a generic term for unconsolidated natural surface material above bedrock.

soil, coarse-grained—soil in which the larger grain sizes, such as sand and gravel, predominate.

soil, fine-grained—soil in which the smaller grain sizes predominate, such as fine sand, silt, and clay.

soil-bearing capacity—the maximum stress under a foundation that provides adequate safety against soil failure and against excessive soil settlement.

solid slab—slab of uniform thickness.

span length—horizontal distance between supports of a horizontal structural member such as a slab, joist, beam, or girder, measured center-to-center of support.

specifications—written document describing in detail the scope of work, materials to be used, method of installation, and quality of workmanship. See also **project specifications**.

specified lateral seismic loads—lateral loads corresponding to the appropriate distribution of the base shear force, for seismic-resistant design.

specified wind loads—nominal pressure of wind to be used in design.

spread footing—a generally rectangular prism of concrete, larger in lateral dimensions than the column or wall it supports, to distribute the load of a column or wall to the subgrade. See also **footing**.

stirrup—reinforcement used to resist shear and torsion stresses in a structural member; typically bars, wires, or welded wire fabric (plain or deformed) either single leg or bent into L, U, or rectangular shapes and located perpendicular to or at an angle to longitudinal reinforcement. The term "stirrups" is usually applied to lateral reinforcement in beams and girders and the term "ties" to those in columns. See also **tie**.

story height—vertical distance from the floor finish of a story to the floor finish of the story below.

strength, design—nominal strength multiplied by a strength reduction factor ϕ.

strength, nominal—strength of a member or cross section calculated in accordance with provisions and assumptions of the strength design method of this guide before application of any strength reduction factors.

strength reduction factor ϕ—capacity reduction factor in structural design; a number less than 1.0 by which the nominal strength of a structural member or element in terms of load, moment, shear, or stress is required to be multiplied to determine design strength or capacity.

stress—intensity of force per unit area.

stress, allowable—maximum permissible stress used in design of members of a structure and based on a factor of safety against failure.

structural concrete—all concrete used for structural purposes including prestressed and reinforced concrete and, under special circumstances, plain concrete.

structural diaphragm—structural member, such as a floor or roof slab, that transmits loads acting in the plane of the member to the vertical elements of the seismic-force-resisting system.

tank—container for the storage of water or other fluids. The guide only covers tanks used for storage of potable water in locations where the water supply system is unreliable.

temporary facilities—buildings or other structures to be in service for a limited time and have a limited exposure period for environmental loadings.

tie—loop of reinforcing bar or wire enclosing longitudinal reinforcement. A continuously wound bar or wire in the form of a circle, rectangle, or other polygon shape without re-entrant corners is acceptable. See also **stirrup**.

tie elements—elements that serve to transmit inertia loads and prevent separation of building components such as footings and walls.

transverse reinforcement—reinforcement located perpendicular to the longitudinal axis of the member, comprising stirrups, ties, and spiral reinforcement.

wall—member, usually vertical, used to enclose or separate spaces. See also **reinforced concrete wall**.

web—thin vertical portion of an I-shaped section that connects the flanges. See also **flange, compression**.

wire—reinforcing bar of small diameter.

wire mesh—welded wire reinforcement.

work—the entire construction or separately identifiable parts thereof that are required to be furnished under the contract documents.

working stress—maximum permissible design stress using working-stress design methods.

yield strength f_y—specified minimum yield strength or yield point of reinforcement. Yield strength or yield point should be determined in tension according to applicable ASTM standards.

CHAPTER 3—STRUCTURAL SYSTEM LAYOUT

3.1—Description of structural components

The building structure should be divided into components as described by 3.1.1 through 3.1.5.

3.1.1 *Floor system*—Consists of structural members that comprise the floor of a story in a building. Chapter 6 describes different types of floor systems. The floor system can include girders, beams, joists, and the slab that spans between them or the slab only, where it is directly supported by columns, as in slab-column systems.

3.1.2 *Vertical supporting members*—Support the floor system at each story and carry accumulated gravity loads down to the foundation of the structure. Vertical supporting members should be either columns or reinforced concrete walls.

3.1.3 *Foundation*—Comprises structural elements through which the load of a structure is transmitted to the earth. It includes members such as spread footings, combined footings, foundation mats, basement and retaining walls, and grade beams. Foundation members are described in Chapter 14. Deep foundations, such as piles, caissons, pile footings, and caps, are beyond the scope of the guide.

3.1.4 *Lateral-force-resisting system*—Comprises the structural members that, acting jointly, resist and transmit to the ground the lateral loads arising from seismic motions, wind, and lateral earth pressure. The floor system acts as a diaphragm that carries in its plane the lateral force from the application point to the vertical members of the lateral-force-resisting system. Vertical members of the lateral-force-resisting system, in turn, collect the loads arising from all floors and transmit them down to the foundation and through the foundation to the underlying soil. For areas with moderate or high seismic risk, the main vertical members of the lateral-force-resisting system should be reinforced concrete walls.

3.1.5 *Other structural members*—Other parts of the building structure that are covered in this guide are stairways, ramps, small potable water tanks, and slabs-on-ground.

3.2—General

3.2.1 *Architectural program*—A general architectural program of the building should be coordinated with the licensed design professional before the structural design begins. The general architectural program should include the following items as a minimum:

(1) Plan geometry and dimensions of all building floors;

(2) Building elevation and the terrain, including the basement, if any;

(3) Type of roof, its shape and slopes, the type of waterproofing, the means to facilitate the runoff of water from rain and melting snow or hail, and the location of drainage gutters;

(4) Use of internal spaces of the building, its subdivision, and means of separation, in all stories;

(5) Minimum architectural clear height in all floors;

(6) Location and dimensions of stairways, ramps, and elevators;

(7) Type of building enclosure, internal partitions, architectural and nonstructural elements;

(8) Locations of ducts and shafts for utilities such as power supply, lighting, thermal control, ventilation, water supply, and wastewater, including enough information to detect interference with the structural members; and

(9) Architectural features that may reduce effective concrete cover of reinforcement.

3.2.2 *Structural program*—Based on the general architectural program information, the licensed design professional should define the general structural program for the building being designed. The general structural program includes the following items as a minimum:

(1) Intended use of the building;

(2) Nominal loads related to the use of the building;

(3) Special loads defined by the owner;

(4) Design seismic loads, if the building is located in a seismic zone;

(5) Wind loads appropriate for the site;

(6) Loads from snow, hail, or rain;

(7) Fire rating;

(8) Type of roof and associated loads when not built of reinforced concrete;

(9) Site information related to slopes and site drainage;

(10) Allowable soil-bearing capacity and recommended foundation system derived from the geotechnical investigation and additional restrictions related to expected settlement;

(11) Environmental conditions derived from local seasonal and daily temperature variations, humidity, presence of deleterious chemicals, and salts;

(12) Availability, type, and quality of materials such as reinforcing bars, cement, and aggregates;

(13) Availability of formwork materials;

(14) Availability of a testing lab for concrete mixture proportioning and quality control during construction;

(15) Availability of qualified work force; and

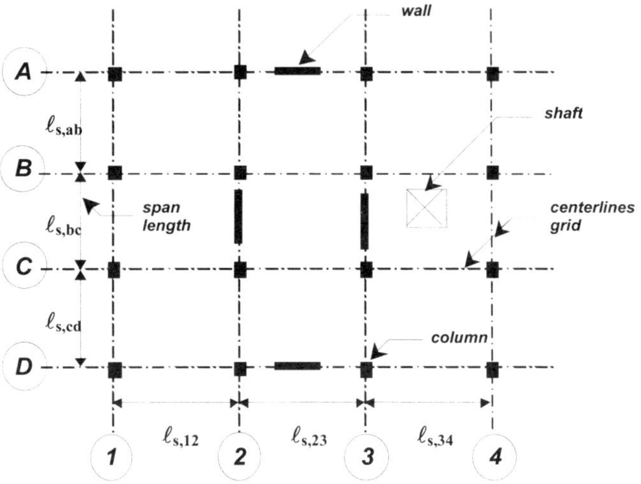

Fig. 3.3.1—General structural layout in plan.

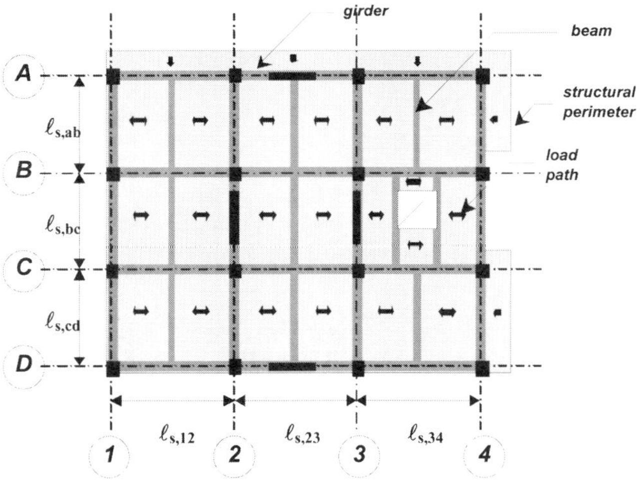

Fig. 3.3.2—Typical floor structural layout.

(16) General and local sustainable construction practices.

3.3—Structural layout

3.3.1 *General structural layout*—The licensed design professional should define a general structural layout in plan, including all information common to all structure levels of the structure (Fig. 3.3.1). The general structural layout in plan should include:

(1) A dimensioned axis grid, or centerlines, in both principal directions in plan, located at the intersection of the vertical supporting members (columns and reinforced concrete walls);

(2) Location in plan of all vertical supporting members, columns, and reinforced concrete walls. These vertical supporting members should be aligned vertically and be continuous to the foundation. Reinforced concrete walls are permitted if they are continuous to the foundation and have no openings for windows or doors. Future revisions of ACI 314R-11 may provide parameters that would allow limited openings;

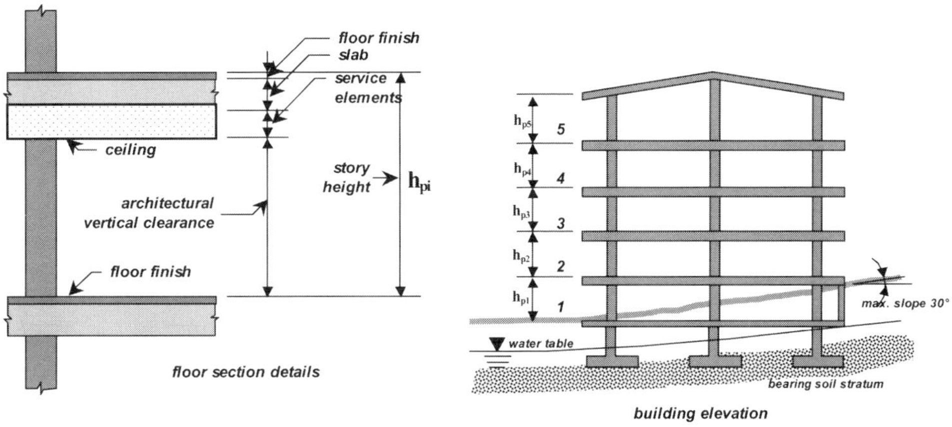

Fig. 3.3.3—General structural layout in elevation

(3) Location of all ducts, shafts, elevators, and stairways that are continuous from floor to floor;

(4) Horizontal distance between centerlines, ℓ_s, which correspond to the center-to-center span lengths of the floor system; and

(5) Location and distribution of all reinforced concrete walls.

3.3.2 *Floor layout*—For each floor, the licensed design professional should develop a structural floor layout (Fig. 3.3.2). The structural floor layout includes:

(1) Location of the floor perimeter on the general axis grid;

(2) Girder and beam locations, or column and middle strips for slab-column systems;

(3) All substantial architectural openings in the floor; and

(4) An approximate load path from all floor areas to the supporting beams and girders.

3.3.3 *Vertical layout*—The licensed design professional should define a general structural layout in elevation (Fig. 3.3.3). The general structural layout in elevation includes:

(1) Number of stories;

(2) Story height for each floor, defined as the vertical distance from floor finish to floor finish;

(3) Slope and shape of the roof;

(4) Architectural clearance from floor finish to ceiling for each floor;

(5) Space necessary to accommodate power distribution, water supply and drainage, and heating, ventilation, and air conditioning;

(6) Slope of the terrain and its relationship to the ground floor or basement; and

(7) Supporting soil stratum depth and water-table depth.

3.4—Feasibility of guide usage

If any of the conditions stated in Chapter 3 are not met, the structural design should be performed using the codes and standards given in 1.4.

CHAPTER 4—LOADS

4.1—General

The load information in this chapter is based on the requirements of the model code (IBC 2009) and the reference standard (ASCE 7-10).

In jurisdictions where the local governing authority has adopted other standards, they should be used rather than the load information in this chapter.

4.2—Load factors and load combinations

The largest required factored strength U, as defined in 1.8.1, for the structural member should be determined by using load factors and load combinations of this section. The following apply:

(a) Each relevant strength limit state should be investigated, including effects of one or more loads not acting simultaneously;

(b) In load combinations where the symbol ± is used in factoring alternating loads that act in one direction or the opposite, it should be interpreted as the sign that leads to the maximum (positive) or minimum (negative) value of U; and

(c) The most unfavorable effects from both wind and seismic loads should be investigated, but they need not be considered to act simultaneously.

4.2.1 *Dead and live load*—Required factored strength U to resist dead load D and live load L should be the greater of:

$$U = 1.4D \quad (4.2.1a)$$

$$U = 1.2D + 1.6L \quad (4.2.1b)$$

4.2.2 *Rain load, snow load, and sloping roof live load*—Required factored strength U to resist rain load R, snow load S, or sloping roof live load L_r should be evaluated based on the following load combinations:

$$U = 1.2D + 1.6L + 0.5(R \text{ or } S \text{ or } L_r) \quad (4.2.2a)$$

$$U = 1.2D + 1.0L + 1.6(R \text{ or } S \text{ or } L_r) \quad (4.2.2b)$$

4.2.3 *Wind loads*—Required factored strength U to resist wind loads W should be evaluated based on the following load combinations:

$$U = 1.2D + 1.0L \pm 1.0W + 0.5(R \text{ or } S \text{ or } L_r) \quad (4.2.3a)$$

$$U = 1.2D \pm 0.5W + 1.6(R \text{ or } S \text{ or } L_r) \quad (4.2.3b)$$

$$U = 0.9D \pm 1.0W \quad (4.2.3c)$$

Note: In ASCE 7-10, wind forces are defined at strength level. In previous versions of ASCE 7, wind forces were defined at working stress level. In jurisdictions where the local governing authority has adopted a standard that is different from ASCE 7-10, where wind forces are defined at working stress level, the factor for W in Eq. (4.2.3a) and (4.2.3c) should be 1.6 instead of 1.0, and in Eq. (4.2.3b) should be 0.8 instead of 0.5.

4.2.4 *Seismic loads*—Required factored strength U to resist seismic loads E should be evaluated based on the following load combinations:

$$U = 1.2D + 1.0L + 0.2S \pm 1.0E \quad (4.2.4a)$$

$$U = 0.9D \pm 1.0E \quad (4.2.4b)$$

4.2.5 *Earth pressure*—For members resisting lateral earth pressure, H, the required factored strength U should be at least equal to:

$$U = 1.2D + 1.6L + 1.6H \quad (4.2.5)$$

When the building structure as a whole resists permanent uncompensated horizontal loads due to lateral soil pressure, $1.6H$ should be added to the right side of Eq. (4.2.3c) and Eq. (4.2.4b).

4.2.6 *Weight and pressure of fluids*—For members subjected to weight and pressure of fluids with well-defined densities and controllable maximum heights, $1.4F$ in Eq. (4.2.1a) and $1.2F$ in Eq. (4.2.1b) should be added to the right side.

4.2.7 *Other effects*—Where structural effects T of differential settlement, shrinkage, or temperature change are significant in design, the design should not be performed using the guide, and the codes and standards listed in 1.4 should be used.

4.3—Mass and weight

A clear distinction between mass and weight should be made by the licensed design professional in all calculations.

4.4—Weight of materials

In determining dead loads, the actual weight of materials and constructions should be used.

In the absence of definite information, the approximated values listed in Table 4.4 may be used.

4.5—Dead loads

4.5.1 *Definition of dead loads*—Dead loads consist of the weight of construction material incorporated into the building, including, but not limited to: structure, walls, floors, roofs, ceilings, stairways, ramps, finishes, cladding, other incorporated architectural and structural systems, and fixed service equipment. When determining dead loads for design purposes, the actual weights of materials and constructions should be used. For design purposes, the weight of fixed service equipment, such as plumbing stacks and risers, electrical feeders, and heating, ventilating, and air conditioning systems, should be included with the dead loads.

4.5.2 *Self-weight of concrete*—The unit weight of reinforced concrete should be taken as 150 lb/ft^3 (2400 kg/m^3).

4.5.3 *Dead loads from nonstructural elements*—Dead loads from nonstructural elements should be divided into vertical and horizontal nonstructural elements.

4.5.3.1 *Flat nonstructural elements*—They should include construction material with a vertical dimension that is substantially less than the horizontal dimensions and are either applied, supported, attached, fixed, or suspended from the slabs or roof of the building. These elements should include, but are not limited to: permanent slab or joist forms, floor topping, floor fills, floor finishes, roof slope fills, roof coverings, tiles, waterproofing membranes, thermal insulation, skylights, ceilings, furring, and ducts for services.

For purposes of design, the dead loads due to flat nonstructural elements may be considered uniform vertical loads per unit area of horizontal surface or horizontal projection, applied to the corresponding zones or areas where the flat nonstructural elements are located. In determining the dead loads of flat nonstructural elements, the actual weight of materials and constructions and realistic thickness should be used. As a guide, Table 4.4 gives minimum unit weight. (Multiply the values provided by the corresponding thickness in ft [m] to obtain dead loads in lb/ft^2 [N/m^2].) In Tables 4.5.3.1a to 4.5.3.1d, values are given for dead loads of typical flat nonstructural element construction materials. The values provided in Tables 4.5.3.1a to 4.5.3.1d correspond to average minimum values. The licensed design professional should consider the possibility of variation of these values from different local materials and local construction practices.

4.5.3.2 *Standing nonstructural elements*—Standing nonstructural elements should include construction material whose vertical dimension is substantially greater than the least horizontal dimension and are either free standing, supported by vertical structural members or attached to them, or span vertically from floor to floor. These elements include, but are not limited to: façades, nonstructural walls, partitions, wall coverings, veneer, architectural ornaments, windows, doors, and vertical ducts for services. In office buildings or other buildings where partitions can be erected or rearranged, provisions for partition weight should be made, whether or not partitions are shown on the architectural drawings. Dead load should be included as uniform vertical loads per unit area of horizontal surface, in lb/ft^2 (kN/m^2).

Dead loads from standing nonstructural elements may be considered as concentrated loads or as uniform line loads per unit of length of nonstructural element. As a guide, in Table 4.4 minimum unit weights are given. Multiply values by the corresponding thickness in ft (m) and by the element height in ft (m) to obtain uniform line dead loads in lb/ft (N/m). In Tables 4.5.3.2a to 4.5.3.2e, values are given for dead loads of typical standing nonstructural elements construction materials, in lb (kN) per unit vertical area in ft^2 (m^2). To obtain uniform line dead loads in lb/ft (kN/m), multiply values suggested in Tables 4.5.3.2a to 4.5.3.2e by the height of the standing nonstructural element in ft (m).

Table 4.4—Minimum unit weight γ for evaluation of dead and live loads from materials

Material	Unit weight γ, lb/ft³	Mass density μ, kg/m³	Material	Unit weight γ, lb/ft³	Mass density μm, kg/m³
Aluminum	170	2700	Iron		
Bituminous products			Cast	450	7200
Asphalt and tar	81	1300	Wrought	480	7700
Gasoline	42	700	Lead	710	11400
Graphite	135	2160	Lime		
Paraffin	56	900	Hydrated, loose	32	500
Petroleum	53	850	Hydrated, compacted	45	800
Brass	526	8430	Masonry, brick (solid portion)	115	1850
Bronze	552	8850	Masonry, concrete (solid portion)	125	2150
Cement, portland, loose	90	1440	Masonry grout	140	2250
Ceramic tile	150	2400	Masonry, stone	162	2600
Charcoal	12	200	Mortar cement or lime	130	2100
Cinder fill	57	920	Particleboard	45	750
Coal, piled	50	800	Plywood	36	600
Concrete, plain	144	2300	Sand		
Concrete, reinforced	150	2400	Clean and dry	90	1440
Copper	556	9000	River, dry	106	1700
Cork, compressed	14	250	Steel	488	7800
Earth			Stone		
Clay, dry	63	1100	Basalt, granite, gneiss	169	2700
Clay, damp	110	1750	Limestone, marble, quartz	179	2850
Clay and gravel, dry	100	1600	Sandstone	169	2700
Silt, moist, packed	96	1550	Shale	163	2600
Silt, moist, loose	78	1250	Terra cotta		
Sand and gravel, dry, loose	100	1600	Voids filled	120	1950
Sand and gravel, dry, packed	110	1750	Voids unfilled	72	1150
Sand and gravel, wet	120	1900	Tin	459	7360
Glass	160	2600	Water		
Gravel, dry	104	1660	Fresh	62	1000
Gypsum, loose	70	1150	Sea	64	1030
Gypsum, wallboard	50	800	Wood, seasoned	28 to 47	450 to 750
Ice	57	920	Zinc, rolled sheet	449	7200

The values provided in Tables 4.4 and 4.5.3.2a to 4.5.3.2e correspond to typical average minimum values; the licensed design professional should consider the variation of these values due to local materials and local construction practices.

Dead load of internal standing nonstructural elements, such as internal walls and partitions, may be considered uniform vertical dead loads per unit area when secondary structural members of the floor system are capable of supporting the resulting concentrated or uniform line loads, without impairment to strength and serviceability of the floor system or the nonstructural element. If standing nonstructural elements are more than one story tall, their dead loads should be considered concentrated loads or line loads. The dead load of façades and elements of the building enclosure should be considered line loads on the slab edge.

4.5.3.3 *Alternative minimum values for dead loads of nonstructural elements*—For buildings with story heights not exceeding 10 ft (3 m), the minimum values in lb/ft^2 (kN/m^2) of horizontal floor area given in Table 4.5.3.3 for the occupancies listed, may be used instead of a detailed analysis of dead loads caused by nonstructural elements.

4.5.4 *Fixed equipment*—Fixed equipment dead load should be provided by the equipment manufacturer.

4.6—Live loads

Live loads are produced by the use and occupancy of the building and do not include construction or environmental loads such as wind load, snow load, rain load, seismic load, or dead load. Live loads should be the maximum loads expected for the intended use or occupancy but should not be less than the minimum uniformly distributed unit loads given in Table 4.6 for the occupancies listed.

4.7—Roof live loads

Roof live loads should not be less than the maximum live load used for the rest of the building, and when of mixed occupancy, the maximum of the different occupancies. While this roof live load is not required by the standards in 1.4, buildings in some countries commonly add an additional floor level during the life of the structure. This live load value addresses this possibility. When the roof is a non-concrete roof with a slope greater than 15 degrees, supported

Table 4.5.3.1a—Flat nonstructural elements minimum dead loads: ceilings

Component	Load (lb/ft²) per ft² of floor area	Load (kN/m²) per m² of floor area
Ceilings		
Acoustical fiberboard	1	0.05
Gypsum board	0.55 (per 1/8 in. thickness)	0.0080 (per mm thickness)
Mechanical duct allowance	4	0.20
Sprinkler system allowance	6	0.30
Luminaries and electrical system allowance	1	0.05
Plaster on tile or concrete	5	0.25
Plaster on wood lath	8	0.40
Suspended steel channel system	2	0.10
Suspended metal lath and cement plaster	15	0.70
Suspended metal lath and gypsum plaster	10	0.50
Wood furring suspension system	2.5	0.15

Table 4.5.3.1b—Flat nonstructural elements minimum dead loads: floor fill

Component	Load (lb/ft²) per ft² of floor area	Load (kN/m²) per m² of floor area
Floor fill	(per in. thickness)	(per mm thickness)
Cinder concrete	9	0.20
Lightweight concrete	8	0.015
Sand	8	0.015
Stone concrete	12	0.025

Table 4.5.3.1c—Flat nonstructural elements minimum dead loads: floors

Component	Load (lb/ft²) per ft² of floor area	Load (kN/m²) per m² of floor area
Floors and floor finishes		
Asphalt block (2 in. [50 mm]), 1/2 in. (12 mm) mortar	30	1.50
Cement finish (1 in. [25 mm]) on stone-concrete fill	32	1.50
Ceramic or quarry tile (3/4 in. [20 mm]) on 1/2 in. (12 mm) mortar bed	16	0.80
Ceramic or quarry tile (3/4 in. [20 mm]) on 1 in. (25 mm) mortar bed	23	1.10
Concrete fill finish	12 (per in. thickness)	0.020 (per mm thickness)
Hardwood flooring, 7/7 in. (25 mm)	4	0.20
Linoleum or asphalt tile, 1/4 in. (6 mm)	1	0.05
Marble and mortar on stone-concrete fill	33	1.60
Slate	15 (per in. thickness)	0.030 (per mm thickness)
Solid flat tile on 1 in. (25 mm) mortar base	23	1.10
Sub flooring, 3/4 in. (20 mm)	3	0.15
Terrazzo 1-1/2 in. (40 mm) directly on slab	19	0.90
Terrazzo 1 in. (25 mm) on stone-concrete fill	32	1.50
Terrazzo 1 in., (25 mm), 2 in. (50 mm) stone concrete	32	1.50
Wood block 3 in. (75 mm) on mastic, no fill	10	0.50
Wood block 3 in. (75 mm) on 1/2 in. (12 mm) mortar base	16	0.80

on a steel or wood structure, it is permitted to use a roof live load of 20 lb/ft² (0.80 kN/m²).

4.8—Rain load

Each portion of the roof should be designed to sustain the accumulated rainwater if the drainage system for that portion is blocked.

4.9—Snow load

Where snow is expected, the loads caused by its accumulation should be taken into account when designing the roof. The loads should be calculated using the local code provisions. In the absence of a governing local code, the minimum snow load at latitudes greater than 30 degrees in lb/ft² (kN/m²) is given in Fig. 4.9. Where records or experience near the site indicate that the value with a 2 percent annual probability of being exceeded (50-year mean recurrence interval) exceeds the value from Fig. 4.9 or the building is subject to significant ice loads, the design should not be performed using the guide, and the codes and standards listed in 1.4 should be used.

Table 4.5.3.1d—Flat nonstructural elements minimum dead loads: roof coverings

Component	Load (lb/ft^2) per ft^2 of floor area	Load (kN/m^2) per m^2 of floor area
Roof covering		
Asbestos-cement shingles	4	0.20
Asphalt shingles	2	0.10
Cement tile	16	0.80
Clay tile including mortar	25	1.20
Copper or tin	1	0.05
Corrugated asbestos-cement roofing	4	0.20
Deck, metal, 20 gauge (0.9 mm nominal thickness)	2.5	0.12
Deck, metal, 18 gauge (1.2 mm nominal thickness)	3	0.15
Decking, 2 in. (50 mm) wood	5	0.25
Decking, 3 in. (75 mm) wood	8	0.40
Fiberboard, 1/2 in. (12 mm)	0.75	0.05
Gypsum sheathing, 1/2 in. (12 mm)	2	0.10
Insulation, roof boards:		
Fibrous or cellular glass	1.0 (per in. thickness)	0.0020 (per mm thickness)
Fiberboard	1.5 (per in. thickness)	0.0030 (per mm thickness)
Perlite	0.8 (per in. thickness)	0.0015 (per mm thickness)
Polystyrene foam	0.2 (per in. thickness)	0.0005 (per mm thickness)
Urethane foam with skin	0.5 (per in. thickness)	0.0010 (per mm thickness)
Plywood (per 1/8 in. [3 mm] thickness)	0.4 (per 1/8 in. thickness)	0.0100 (per mm thickness)
Skylight, metal frame, 3/8 in. (10 mm) wire-glass	8	0.40
Waterproofing membranes:		
Bituminous, gravel-covered	5.5	0.25
Bituminous, smooth surface	1.5	0.10
Liquid applied	1	0.05
Single-ply, sheet	0.7	0.03
Wood sheathing	3.0 (per in. thickness)	0.0060 (per mm thickness)
Wood shingles	3	0.15

Table 4.5.3.2a—Standing nonstructural elements minimum dead loads: coverings for walls

Component	Load (lb/ft^2) per ft^2 of vertical surface (multiply by the height of element in ft to obtain line loads in lb/ft)	Load (kN/m^2) per m^2 of vertical surface (multiply by the height of element in m to obtain line loads in kN/m)
Covering for walls		
Cement tile	16	0.80
Fiberboard, 1/2 in. (12 mm)	0.75	0.05
Gypsum sheathing, 1/2 in. (12 mm)	2	0.10
Insulation, wall boards		
Fibrous or cellular glass	1.0 (per in. thickness)	0.0020 (per mm thickness)
Fiberboard	1.5 (per in. thickness)	0.0030 (per mm thickness)
Perlite	0.8 (per in. thickness)	0.0015 (per mm thickness)
Polystyrene foam	0.2 (per in. thickness)	0.0005 (per mm thickness)
Urethane foam with skin	0.5 (per in. thickness)	0.0010 (per mm thickness)
Plywood (per 1/8 in. (mm) thickness)	0.4 (per 1/8 in. thickness)	0.0100 (per mm thickness)
Wood sheathing	3.0 (per in. thickness)	0.0060 (per mm thickness)

4.10—Wind loads

4.10.1 *General*—Section 4.10.1 provides a method to calculate wind loads on a building and on individual structural members and cladding. The procedures and values provided by a local building code should be used to obtain wind loads. In the absence of a local building code, for buildings with regular geometric shapes, response characteristics, and site locations, the following should apply. Buildings that do not comply with these features should be considered outside the scope of the guide (1.4).

4.10.1.1 *Procedure*—Wind pressures and wind loads at strength level should be determined in accordance with Steps 1 through 6, keeping in mind that building wind loads are based on the gross exposed areas in the direction of the wind whereas component and cladding wind loads are based on the element area under study.

Table 4.5.3.2b—Standing nonstructural elements minimum dead loads: light partitions

Component	Load (lb/ft²) per ft² of vertical surface (multiply by the height of element in ft to obtain line loads in lb/ft)	Load (kN/m²) per m² of vertical surface (multiply by the height of element in m to obtain line loads in kN/m)
Light partitions		
Movable steel partitions (non-full height)	6	0.30
Movable steel partitions (full height)	4	0.20
Wood or steel studs, 1/2 in. (12 mm) gypsum board each side	8	0.80
Wood studs, 2 x 4 in. (50 x 100 mm), unplastered	4	0.30
Wood studs, 2 x 4 in. (50 x 100 mm), plastered one side	12	0.60
Wood studs, 2 x 4 in. (50 x 100 mm), plastered two sides	20	1.00

Table 4.5.3.2c—Standing nonstructural elements minimum dead loads: veneer

Component	Load (lb/ft²) per ft² of vertical surface (multiply by the height of element in ft to obtain line loads in lb/ft)	Load (kN/m²) per m² of vertical surface (multiply by the height of element in m to obtain line loads in kN/m)
Veneer		
Granite	9 (per in. thickness)	0.017 (per mm thickness)
Limestone	8 (per in. thickness)	0.015 (per mm thickness)
Sandstone	7 (per in. thickness)	0.013 (per mm thickness)
Ceramic	8 (per in. thickness)	0.015 (per mm thickness)

Table 4.5.3.2d—Standing nonstructural elements minimum dead loads: walls

Component	Load (lb/ft²) per ft² of vertical surface (multiply by the height of element in ft to obtain line loads in lb/ft)	Load (kN/m²) per m² of vertical surface (multiply by the height of element in m to obtain line loads in kN/m)
Walls		
Exterior stud walls (steel or wood studs):		
2 x 4 in. (50 x 100 mm) at 16 in. (400 mm), 5/8 in. (15 mm) gypsum, insulated, 3/8 in. (10 mm) siding	11	0.55
2 x 6 in. (50 x 150 mm) at 16 in. (400 mm), 5/8 in. (15 mm) gypsum, insulated, 3/8 in. (10 mm) siding	12	0.60
Exterior stud walls with brick veneer	48	2.30
Clay solid masonry unit walls:	Wall thickness	Wall thickness (mm)
	4 in. 8 in. 12 in. 16 in.	100 150 200 250 300
Unplastered	39 79 115 155	1.90 2.90 3.80 4.70 5.50
Clay tile masonry unit walls:	Wall thickness	Wall thickness (mm)
	4 in. 6 in. 8 in. 10 in. 12 in.	100 150 200 250 300
Plaster in both faces	37 52 64 79 91	1.80 2.50 3.10 3.80 4.40
Unplastered	27 41 54 68 81	1.30 2.00 2.60 3.30 3.90
Hollow concrete masonry unit walls:	Wall thickness	Wall thickness (mm)
	4 in. 6 in. 8 in. 10 in. 12 in.	100 150 200 250 300
No grout	29 30 39 47 54	1.40 1.45 1.90 2.25 2.60
48 in. o. c. (1.2 m) grout spacing	36 47 57 66	1.70 2.25 2.70 3.15
40 in. o. c. (1.0 m) grout spacing	37 48 59 69	1.80 2.30 2.80 3.30
32 in. o. c. (0.8 m) grout spacing	38 50 62 72	1.80 2.40 3.00 3.45
24 in. o. c. (0.6 m) grout spacing	41 54 67 78	2.00 2.60 3.20 3.75
16 in. o. c. (0.4 m) grout spacing	46 61 76 90	2.20 2.90 3.60 4.30
Full grout	62 83 105 127	3.00 4.00 5.00 6.10
Solid concrete masonry unit walls:	Wall thickness	Wall thickness (mm)
	4 in. 6 in. 8 in. 10 in. 12 in.	100 150 200 250 300
Unplastered	41 64 87 110 133	2.00 3.10 4.20 5.30 6.40

Table 4.5.3.2e—Standing nonstructural elements minimum dead loads: windows

Component	Load (lb/ft²) per ft² of vertical surface (multiply by the height of element in ft to obtain line loads in lb/ft)	Load (kN/m²) per m² of vertical surface (multiply by the height of element in m to obtain line loads in kN/m)
Windows		
Glass curtain walls, glass, frame and sash	10	0.50
Windows, glass, frame and sash	8	0.40

Table 4.5.3.3—Alternative minimum values for dead load of nonstructural elements when no detailed analysis is performed

Occupancy			Façade and partitions, lb/ft² (kN/m²) per ft² (m²) of floor area	Floor finish and ceiling, lb/ft² (kN/m²) per ft² (m²) of floor area
Group A—Assembly	A-2 A-3	Building having an assembly room with capacity of less than 100 persons and not having a stage	20 (1.0)	40 (1.8)
Group B—Business	B	Movable full height partitions	20 (1.0)	40 (1.8)
		Fixed masonry partitions	45 (2.0)	40 (1.8)
Group E—Educational	E	Classrooms	45 (2.0)	35 (1.5)
Group F—Factory	F-1	Light industries	18 (0.8)	35 (1.6)
Group I—Institutional	I-1	Residential board and care facilities	45 (2.0)	35 (1.6)
	I-3	Prisons, jails, reformatories, detention centers	55 (2.5)	40 (1.8)
	I-4	Daycare facilities	45 (2.0)	35 (1.6)
Group M—Mercantile	M	Display and sale of merchandise	35 (1.5)	30 (1.4)
Group R—Residential	R	Masonry façade and partitions	65 (3.0)	35 (1.6)
		Light façade and partitions	45 (2.0)	30 (1.4)
Group S—Storage	S-2	Storage of light materials	35 (1.5)	35 (1.5)
Group U—Utility and Miscellaneous	U	Garages for vehicles with carrying capacity up to 4000 lb (2000 kg)	5 (0.2)	20 (1.0)

Table 4.6—Minimum uniformly distributed live loads

Occupancy or use		Uniform load (lb/ft²) per ft² of floor area	Uniform load (kN/m²) per m² of floor area
Group A—Assembly	Balconies	125	6.0
	Corridors and stairways	125	6.0
	Gymnasiums	125	6.0
	Lobbies	125	6.0
	Movable seats	125	6.0
	Recreational areas	100	5.0
	Platforms	125	6.0
Group B—Business	Corridors and stairways	115	5.5
	Offices	50	2.5
	Restaurants	115	5.5
Group E—Educational	Classrooms	65	3.0
	Corridors and stairways	125	6.0
	Libraries		
	Reading rooms	75	3.5
	Stack rooms	170	8.0
Group F—Factory	Light industries	150	7.0
Group I—Institutional	Operating rooms, laboratories	85	4.0
	Wards and private rooms	50	2.5
	Corridors and stairways	115	5.5
Group M—Mercantile	Retail	115	5.5
	Wholesale	150	7.0
Group R—Residential	Balconies	115	5.5
	Private rooms and corridors serving them	50	2.2
	Public rooms and corridors serving them	115	5.5
	Stairways	125	6.0
Group S—Storage	Light	140	6.5
Group U—Utility and Miscellaneous	Garages for passenger cars only	60	2.8

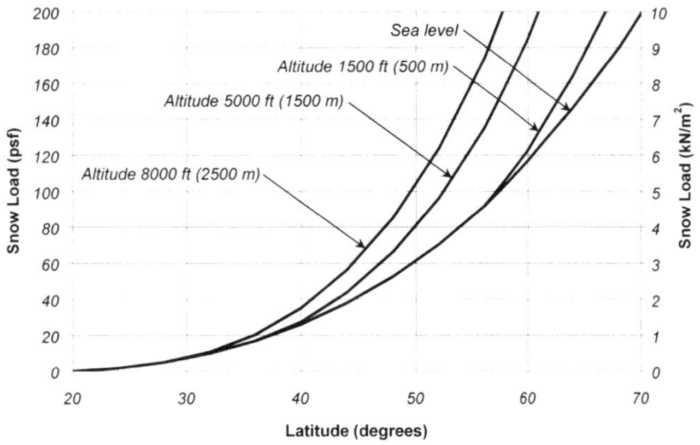

Fig. 4.9—Minimum snow load.

Step 1: Velocity pressure q_h should be determined in accordance with 4.10.2.1 and 4.10.2.2.

Step 2: Appropriate C_{su} pressure coefficients should be selected from 4.10.3 depending on exposed building surfaces, assuming that the wind acts in the positive and negative sense, in both principal directions in plan of the building.

Step 3: Velocity pressure q_h affected by the force coefficients and multiplied by the exposed area of the surface, A_{su}, should be used to determine the equivalent static wind design force F_{su} that acts in a direction normal to the exposed building surfaces.

Step 4: Components in the direction of the wind of the static wind loads that act on the surfaces F_{su} should be algebraically added to obtain the wind base shear V_w for the building in each principal direction in plan, in both senses.

Step 5: Static wind loads that act on the structural members and cladding should be used to design the individual elements and their attachments to the structure.

Step 6: Static wind loads that act on the building should be used to design the members of the lateral-load-resisting system and the overall stability of the structure.

4.10.1.2 *Application of pressures and loads*—Static wind design loads F_{su} should be assumed to act in a direction normal to the surface considered at the centroid of its area. F_{su} used in calculating building loads should be expressed as the resultant along the principal axis of the building (Eq. (4.10.2.3b)).

4.10.2 *Calculation of the wind loads*

4.10.2.1 *Velocity pressure calculation*—Velocity pressure q_h at the mean roof height h_r over the terrain should be calculated from Eq. (4.10.2.1a) and (4.10.2.1b):

$$q_h = \frac{1}{400}V^2\left(\frac{h_r}{33}\right)^{\frac{2}{7}} \quad (q_h \text{ in lb/ft}^2,\ h_r \text{ in m, and } V \text{ in mph}) \qquad (4.10.2.1a)$$

$$\left[q_h = \frac{1}{1630}V^2\left(\frac{h_r}{10}\right)^{\frac{2}{7}} \text{(SI)}\right] \quad (q_h \text{ in kN/m}^2,\ h_r \text{ in m, and } V \text{ in m/s}) \qquad (4.10.2.1b)$$

Table 4.10.2.2—Typical values of strength level basic wind speed *V*

Region	*V*, mph	*V*, m/s
Islands and coastal regions exposed to tropical storms	150 to 200	67 to 89
Islands and coastal regions exposed to subtropical storms	110 to 150	49 to 67
Islands and coastal regions not exposed to tropical or subtropical storms	90 to 110	40 to 49
Tornado-prone regions	90 to 160	40 to 70
Continental regions, more than 100 miles (160 km) from the coast	70 to 110	30 to 50

4.10.2.2 *Selection of basic wind speed*—Basic wind speed at strength level, V, corresponds to a 3-second gust speed at 33 ft (10 m) above ground measured at open terrain with scattered obstructions having heights less than 33 ft (10 m), such as open country, with a 300-year mean return period. The 15 percent exceedance probability in a 50-year period or the annual probability that the basic wind speed will be exceeded is 0.00333.

Table 4.10.2.2 gives typical values of basic wind speed at strength level given as a guide for regions outside of the United States and countries where wind speed velocity information is not available. The user should refer to historical records or local regulations for actual design speed. A distinction should be made between the type of basic wind speed (3-second gust or fastest mile), the mean return period, and the type of exposure where the wind velocity is defined. Given the lateral-load-building stiffness obtained from the guide, a gust effect factor is not included in the procedure.

4.10.2.3 *Wind loads on a building*—The equivalent static strength level wind force that acts in the direction normal to the wind-exposed surface should be calculated using Eq. (4.10.2.3a). Regardless of the openness of the exposed surface, the gross area should be used.

$$F_{su} = q_h A_{su} C_{su} \qquad (4.10.2.3a)$$

where A_{su} is the exposed surface area, and C_{su} is the value given in 4.10.3.

A positive value of F_{su} indicates that the force is acting toward the surface, and a negative value indicates that the force is acting away from the surface (suction). The wind design force at floor level x, F_x, due to wind in a principal direction in plan should be calculated as the sum of all wind-exposed surfaces of the story using Eq. (4.10.2.3b).

$$F_x = \sum_{w=1}^{n}\left(F_{su}^{(w)}\cos\alpha_w \cos\beta_w\right) \qquad (4.10.2.3b)$$

where α_w is the horizontal angle between the normal to the surface and the principal direction in plan under consideration, and β_w is the vertical angle between the normal to the surface and a horizontal line.

Table 4.10.3.1—Pressure coefficients for the building as a whole

Type of surface	C_{su} Total building	C_{su} Windward	C_{su} Leeward
Vertical surfaces			
Prismatic rectangular buildings	1.30	0.80	–0.50
Partially open surfaces (C_{su} applies over gross area)			
10 percent to 80 percent open	1.30	0.80	–0.50
>80 percent open	0.35	0.25	–0.10
Horizontal roof surfaces			
Enclosed buildings		–1.30	–0.70
Buildings with one or two open sides		–1.50	–1.25
(C_{su} applies over 1/3 of the roof area for windward)			
(C_{su} applies over 2/3 of the roof area for leeward)			
Sloping roof surfaces			
Roof angle measured with respect to a horizontal line			
10		–1.30	–0.70
15		–1.00	–0.60
20		–0.70	–0.60
25		–0.50	–0.60
30		–0.30	–0.60
35		–0.20	–0.60
45		0.30	–0.60
$Q > 60$		$0.01Q$	–0.60

In Eq. (4.10.2.3b), care should be used in defining the sign to be used for F_{su}, with respect to the direction in which the force is acting.

The wind base shear V_w in a principal direction in plan corresponds to the sum of all story wind loads F_x in the principal direction in plan under study.

4.10.2.4 *Wind loads on components and cladding*—Calculation of the strength level wind force on components and cladding should be performed using Eq. (4.10.2.4).

$$F_{pw} = q_h A_p C_p \qquad (4.10.2.4)$$

where A_p is the component exposed surface area, and C_p is the component surface pressure.

4.10.3 *Pressure coefficients*

4.10.3.1 *Pressure coefficients for a building*—Pressure coefficients in Table 4.10.3.1 should be used for the calculation of the wind loads on a building. Note that coefficients reflect the direction of the force relative to the direction of the members. In terms of the building total load effect for vertical surfaces, the absolute values should be added.

4.10.3.2 *Pressure coefficients for components and cladding*—Pressure coefficients in Table 4.10.3.2 should be used for the calculation of the strength level wind loads on the components and cladding, positive for pressure and negative for suction.

Table 4.10.3.2—Pressure coefficients for components and cladding

Type of surface	C_p
Vertical surfaces	
Exterior walls of enclosed buildings	1.80
Doors and windows	1.80
Exterior walls of buildings with one or more sides open	1.80
Horizontal surfaces	
Away from edges and corners	1.40
Edges of building and away from corners	2.30
Corners of building	3.20

Edge distance is 10 percent of length of surface in consideration

4.11—Seismic loads

4.11.1 *General*—Resistance to seismic loads should be provided by using a sufficient number of reinforced concrete walls continuous from the roof to the foundation in both principal directions in plan. Reinforced concrete walls produce stiff structures with a short fundamental period of vibration, and the seismic loads given in the guide reflect this type of structure. The calculation of seismic loads for more flexible structures is beyond the scope of the guide. Additional seismic information is presented in Chapter 11.

4.11.2 *Seismic ground motions*

4.11.2.1 *Seismic loads and seismic design*—It is intended that where the governing authority has adopted by local ordinance the model code (IBC 2009) and reference standard (ASCE 7-10), seismic design loads and seismic design prescribed under the model code (IBC 2009) and reference standard (ASCE 7-10) should be used meeting 4.11.2.2 and 4.11.2.3. Where these standards have not been adopted by the governing authority, 4.11.2.4 through 4.11.2.6 should be used.

4.11.2.2 *Design spectral acceleration parameters*—The design earthquake spectral acceleration parameter at short period, S_{DS}, should be determined from the procedure estab-

Table 4.11.2.4.2—Values of the site coefficient F_a

Soil profile	Site coefficient F_a for short periods of vibration				
	$A_a \le 0.1$	$A_a = 0.2$	$A_a = 0.3$	$A_a = 0.4$	$A_a \ge 0.5$
S_A	0.8	0.8	0.8	0.8	0.8
S_B	1.0	1.0	1.0	1.0	1.0
S_C	1.2	1.2	1.1	1.0	1.0
S_D	1.6	1.4	1.2	1.1	1.0
S_E	2.5	1.7	1.2	0.9	0.9

lished in reference standard ASCE 7-10. The design spectral ordinates for determining seismic loads, S_a, of the guide are equal to S_{DS}.

4.11.2.3 *Seismic Design Categories*—In the guide. the following distinction of Seismic Design Categories as associated with seismic risk should be used in those cases where reference standard ASCE 7-10 has not been adopted by local authorities:

(a) Low or no seismic risk zones are where the Seismic Design Category assigned to the structure in the procedure in the referenced standards for Short-Period Response Acceleration is either A or B.

(b) Moderate or high seismic risk zones are where the Seismic Design Category assigned to the structure in the procedure in the references standards for Short-Period Response Acceleration is C, D, E, or F.

4.11.2.4 *Seismic loads and seismic design when the referenced standards have not been adopted*—Where the governing authority has not adopted the model code (IBC) and reference standard (ASCE 7), the ground motion should be described through an effective peak ground horizontal acceleration in rock for short periods of vibration, A_a, expressed as a fraction of gravity g. The values given in the general building code having jurisdiction shall be used for obtaining the corresponding values (1.4).

Where the building code having jurisdiction defines a maximum considered seismic ground motion based on spectral response accelerations at 5 percent of critical damping, the short period (0.2 second) values S_s should be employed and the value of A_a may be obtained as $A_a = S_s/3.75$. Where the building code having jurisdiction defines a seismic zone factor Z, the value of A_a should be taken equal to Z.

4.11.2.4.1 *Soil profile types*—Based on the soil type present at the building site, the soil profile should be classified as one of the following:

(a) Soil Profile S_A: hard rock with a measured shear wave velocity $\bar{v}_s > 5000$ ft/s (1500 m/s);

(b) Soil Profile S_B: rock with moderate fracturing and weathering with a measured shear wave velocity in the range (5000 ft/s $\ge \bar{v}_s >$ 2500 ft/s [1500 m/s $\ge \bar{v}_s >$ 760 m/s]);

(c) Soil Profile S_C: soft weathered or fractured rock, or dense or stiff soil, where the measured shear wave velocity is in the range (2500 ft/s $\ge \bar{v}_s >$ 1200 ft/s [760 m/s $\ge \bar{v}_s >$ 370 m/s]), or, in the upper 100 ft (30 m), the SPT resistance has an average value of $\bar{N} > 50$ or a shear strength for clays $s_u \ge$ 2000 lb/ft^2 ($s_u \ge$ 100 kPa);

(d) Soil Profile S_D: predominately medium-dense to dense, or medium-stiff to stiff soil, where the measured shear wave velocity is in the range (1200 ft/s $\ge \bar{v}_s >$ 600 ft/s [370 m/s $\ge \bar{v}_s >$ 180 m/s]), or where, in the upper 100 ft (30 m), the SPT resistance has an average value in the range ($15 < \bar{N} \le 50$), or a shear strength for clays in the range (1000 lb/ft$^2 \le s_u <$ 2000 lb/ft^2 [50 kPa $\le s_u <$ 100 kPa]);

(e) Soil Profile S_E: a soil profile where the measured shear wave velocity $\bar{v}_s \le$ 600 ft/s (180 m/s), or the SPT resistance has an average value $\bar{N} \le 15$ in the upper 100 ft (30 m), or has more than 10 ft (3 m) of plastic (PI > 20), high moisture content ($w >$ 40 percent), and low shear strength ($s_u <$ 500 lb/ft^2 [$s_u <$ 25 kPa]) clays; and

(f) Seismically vulnerable soils: sites where the soil profile contains soil having one or more of the following characteristics are beyond the scope of the guide: (i) soils vulnerable to potential failure or collapse under seismic motions, such as liquefiable soils, quick and highly sensitive clays, collapsible weakly cemented soil; (ii) peats, highly organic clays, or both, with more than 10 ft (3 m) of thickness; (ii) very-high-plasticity clays ($PI > 75$) with more than 25 ft (8 m) of thickness; and (iv) soft to medium-stiff clays with more than 120 ft (40 m) of thickness.

When the soil profile and properties are not known well enough in detail to determine the soil profile type, a soil profile S_D should be used.

4.11.2.4.2 *Site effects*—Site effects should be described through the site coefficient for short periods of vibration, F_a. Values of the site coefficient for short periods of vibration, F_a, should be determined from Table 4.11.2.4.2 as a function of the effective peak ground horizontal acceleration in rock for short periods of vibration, A_a, and the soil profile type from 4.11.2.4. Linear interpolation can be used between values of A_a in Table 4.11.2.4.2.

Site effect of seismically vulnerable soils, as described in 4.11.2.4, is beyond the scope of the guide, and designs should not be made under the guide for buildings located on such soil.

4.11.2.5 *Seismic risk zones when referenced standards have not been adopted*—The following distinction of seismic risk should be used:

(a) *Low and no seismic risk zones:* zones where A_a is less than or equal to 0.10; and

(b) *Moderate or high seismic risk zones*: zones where A_a exceeds 0.10.

4.11.2.6 *Design response spectral ordinates when referenced standards have not been adopted*—Ordinates of the elastic design response spectrum S_a for a damping ratio of 5 percent of critical, expressed as a fraction of the acceleration of gravity, should be calculated as:

$$S_a = 2.5A_aF_a \quad (4.11.2.6)$$

4.11.3 *Seismic design base shear*

4.11.3.1 *Seismic-resistant structural system*—The seismic-resistant structural system used in this guide is classified as a dual building frame system, where an essentially complete moment-resistant space frame supports gravity loads, and resistance to lateral loads is provided by reinforced concrete walls, with the moment-resisting space frame providing a minimum collateral lateral load resistance. Reinforced concrete walls resisting lateral loads are not permitted to carry vertical axial loads greater that the balanced point axial strength (Eq. (5.12.4.1)).

4.11.3.2 *Energy dissipation capacity of the seismic-resistant structural system*—The energy dissipation capacity in the inelastic range of the seismic-resistant structural system, described by the response modification factor, should have a value of $R_s = 5.0$.

4.11.3.3 *Calculation of the seismic design base shear*—The seismic design base shear V_{bs}, equivalent to the total horizontal inertial effects caused by seismic ground motions, should be determined using Eq. (4.11.3.3).

$$V_{bs} = \frac{S_a W_s}{R_s} \quad (4.11.3.3)$$

where S_a should be determined from Eq. (4.11.2.6); R_s is the response modification factor determined from 4.11.3.2; and W_s corresponds to the total weight of the building. W_s should include the total weight of the structure plus the weight of all nonstructural elements, such as walls and partitions, permanent equipment, tanks and the contained liquid. In storage occupancies, W_s should also include 25 percent of the live load and the snow load when the snow load exceeds 30 lb/ft^2 (1.4 kPa).

4.11.4 *Vertical distribution of design seismic loads*—Total seismic design base shear should be distributed over the building height using Eq. (4.11.4a) and Eq. (4.11.4b). At each floor level designated as x, F_x should be applied over the building area in accordance with the mass distributions at each level.

$$F_x = C_{vx} V_{bs} \quad (4.11.4a)$$

$$C_{vx} = \frac{w_x h_x}{\sum_{i=1}^{n} (w_i h_i)} \quad (4.11.4b)$$

4.12—Soil weight and lateral pressure

The design of basement walls and similar vertical elements below grade should include the lateral pressure of adjacent soil following 4.13.2.3.

4.13—Lateral loads

4.13.1 *General*—All applicable lateral loads as described in Chapter 4 should be used in the design. The simultaneous occurrence of lateral loads with other loads should be evaluated using the load combination of 4.2. A continuous load path from the lateral force application point to the lateral-force-resisting structural elements should be identified and all elements along the load path should have adequate strength.

4.13.2 *Application of lateral loads*

4.13.2.1 *Wind loads*—Wind loads should be determined by 4.10. Horizontal strength level wind loads, F_x, at each level x and in each main direction in plan should be the sum of the windward and leeward wind components as determined by 4.10.2.3. A wind base shear V_o and its distribution over the building height should be determined for both main directions in plan, as determined by 4.13.3. The overturning moment M_{ot}, caused by the wind loads in both main directions in plan, should be determined as determined by 4.13.4. The application of the resultant wind loads on each floor and principal direction should be calculated, and the torsional moments on the structure, T_i, about a vertical axis should be evaluated as determined by 4.13.5 for each story.

4.13.2.2 *Seismic loads*—Seismic loads should be determined by 4.11. Horizontal seismic loads F_x at each level x and in each main direction in plan should be determined by 4.11.4. A lateral base shear V_o and its distribution over the building height should be determined by 4.11.3.3 and 4.13.3 for both main directions in plan. The overturning moment M_{ot} due to the seismic loads in both main directions in plan should be determined by 4.13.4. Application point of the seismic force at each floor should be the center of mass of the floor diaphragm. The effect of an eccentricity between the center of mass of the floor diaphragm and the center of stiffness corresponding to the lateral-force-resisting structural members should be calculated, and the torsional moments on the structure, T_i, about a vertical axis should be evaluated by 4.13.5 for each story.

4.13.2.3 *Soil lateral loads*—Lateral loads due to soil pressure should be determined by Chapter 14. Due to the restrictions of 1.3.2 and 1.3.10, retaining walls laterally supported by the building should be only one story in height. The reaction from the lateral soil pressure on the first story per unit horizontal length of retaining wall should be computed by Eq. (4.13.2.3).

$$R_{p1} = \frac{p_z h_s^3}{6 h_{p1}} \quad (4.13.2.3)$$

where p_z is the lateral pressure at depth of soil, h_s (14.11.2), and h_{p1} is the first-story height (Fig. 4.13.2.3).

Two different cases should be considered when designing the structure as a whole for the effects of lateral soil pressure:

1. Case 1—A building located in flat or nearly flat terrain does not need a lateral analysis of the overall structure due to soil pressure, because lateral reactions due to soil pressure on basement walls in opposite sides are approximately equal, compensate each other, and introduce a nearly uniform in-plane compression on the first-floor slab. In this case, lateral soil pressure should be considered only when designing the soil-retaining foundation members, as noted in Chapter 14, and the members framing into it.

2. Case 2—A building located in sloping terrain supports uncompensated lateral loads, where lateral reaction due to soil pressures on basement walls in opposite sides are different. In this case, a lateral base shear V_o should be determined by 4.13.3 for both main directions in plan from the first-floor reaction. The overturning moment M_{ot} due to soil lateral pressure in both main directions in plan should be determined by 4.13.4. The location in plan of the resultant lateral soil loads on the first floor should be calculated and

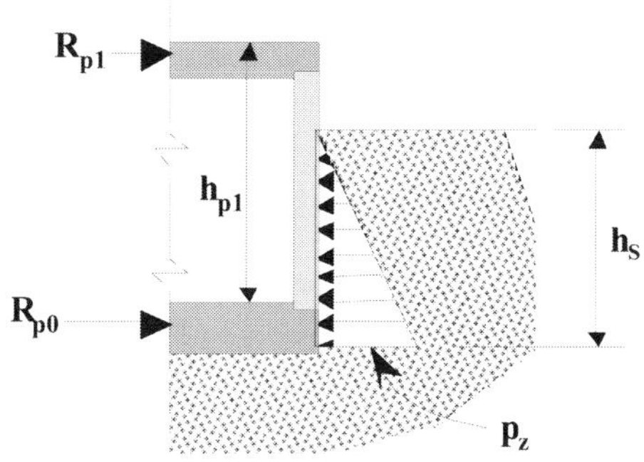

Fig. 4.13.2.3—*Lateral reaction due to soil pressure.*

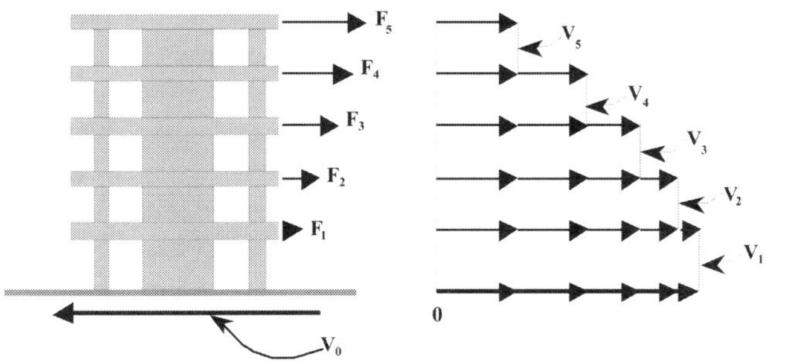

Fig. 4.13.3—*Story shear and base shear calculation.*

the torsional moments on the structure, T_i, about a vertical axis should be evaluated by 4.13.5. The occurrence of wind or seismic loads simultaneously with this soil pressure permanent lateral force should be evaluated using load combinations prescribed for this case in 4.2.5.

4.13.2.4 *Fluid pressure*—Auxiliary structures subjected to fluid pressure, such as water tanks, should be self-contained and fluid pressure should be compensated within the auxiliary structure. The main building structure should not be used to resist lateral loads derived from contained liquids. Weight from auxiliary tank and liquid content should be considered in the calculation of weight in the roof level.

4.13.3 *Factored story shear and base shear*—Base shear due to wind, seismic force, and lateral soil pressure should be determined independently for the two main directions in plan. The story shear at level x for each lateral force (wind, seismic, and soil) corresponds to the sum of the lateral loads applied to the structure in the main direction under consideration, from story x to the roof, as determined from Eq. (4.13.3) (Fig. 4.13.3).

$$V_x = \sum_{i=x}^{n} F_i \tag{4.13.3}$$

The base shear V_o for each lateral force (wind, seismic, and soil) corresponds to the sum of the lateral loads applied to the structure in the main direction under study (Fig. 4.13.3).

Factored story shear V_{xu} and factored base shear V_{ou} should be determined by multiplying V_x and V_o by the appropriate load factor from 4.2.

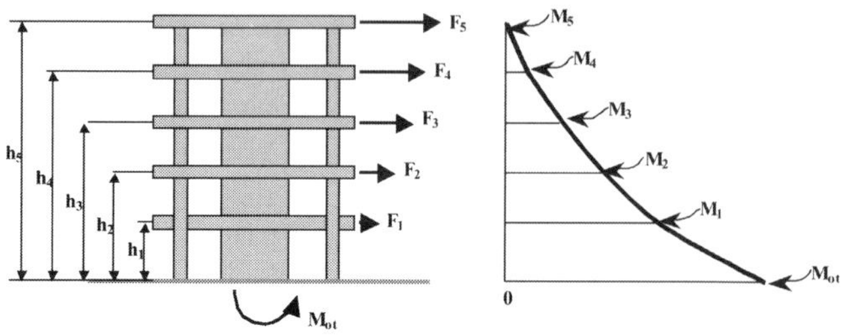

Fig. 4.13.4—*Overturning moment calculation.*

4.13.4 *Factored overturning moment*—The overturning moment due to wind, seismic force, and lateral soil pressure should be determined independently for the two main directions in plan. The story overturning moment at level x for each lateral force (wind, seismic, and soil) corresponds to the sum of the moments due to lateral loads at that story from level x to the roof, as determined from Eq. (4.13.4a) (Fig. 4.13.4).

$$M_x = \sum_{i=x}^{n}\left[F_i\left(h_i - h_x\right)\right] \tag{4.13.4a}$$

Base overturning moment M_{ot} for each lateral force (wind, seismic, and soil) corresponds to the sum of the product of the floor level lateral loads by the height to floor level measured from the base of the structure (Fig. 4.13.4), as defined by Eq. (4.13.4b).

$$M_{ot} = \sum_{i=1}^{n}\left[F_i h_i\right] \tag{4.13.4b}$$

Factored story overturning moment M_{xu} and factored base overturning moment M_{otu} should be determined by multiplying M_x and M_{ot} by the appropriate load factor from 4.2.

4.13.5 *Torsional moments*—For each story and in each principal direction, the unfactored torsion T_i should be computed as the product of the story lateral force F_i times an eccentricity equal to the transverse distance between the line of action of the lateral force and the story lateral stiffness center (Fig. 4.13.5). The story stiffness center should be determined by 4.14.5. For wind and seismic loads, T_i should be taken as the greater value determined for both principal directions. For lateral soil pressure, T_i should be taken as the algebraic sum of the values determined for both principal directions.

The story total unfactored torsion at level x for each lateral force (wind, seismic, and soil) corresponds to sum of the story unfactored torsion applied to the structure from story x to the roof, as determined by Eq. (4.13.5).

$$T_x = \sum_{i=x}^{n} T_i \tag{4.13.5}$$

Base total unfactored torsion T_o for each lateral force (wind, seismic, and soil) corresponds to sum of the story unfactored torsion applied to the structure.

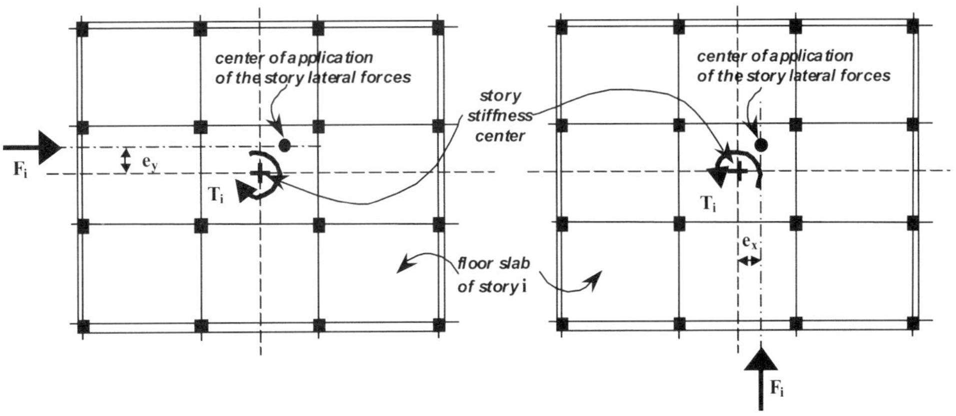

Fig. 4.13.5—Story torsion.

Factored story torsional moment T_{iu} and factored base torsional moment T_{ou} should be computed by multiplying T_i and T_o by the appropriate load factor from 4.2.

4.14—Lateral-force-resisting system

4.14.1 *General*—The lateral-force-resisting system is composed of structural members acting jointly that resist and transmit lateral loads resulting from seismic motions, wind, and lateral earth pressure to the ground (Fig. 4.14.1).

The floor system should act as a diaphragm that carries in its plane the lateral force from the application point to the vertical members of the lateral-force-resisting system. Vertical members of the lateral-force-resisting system, in turn, collect loads arising from all floors affected and transmit them down to the foundation and to the underlying soil. The main vertical members of the lateral-force-resisting system should be a number of reinforced concrete walls in both principal directions in plan. These reinforced concrete walls should have no openings for windows or doors.

4.14.2 *Restrictions on structural system for seismic resistance*

4.14.2.1 *General*—The following restrictions apply to structural systems used for the seismic risk zones as defined in 4.11.2.3.

4.14.2.2 *Zones of low and no seismic risk*—There are no restrictions on which structural systems can be used in zones of low or no seismic risk, as defined in 4.11.2.3. In these zones, structural walls for resistance to earthquake seismic loads may not be needed, but their need to resist wind and soil pressure lateral loads should be verified.

4.14.2.3 *Moderate or high seismic risk zone*—Lateral force resistance should be provided by reinforced concrete walls, continuous from the foundation to the roof, in both principal directions in plan. The special reinforcement details of Chapter 11 should be used.

4.14.3 *Diaphragm effect*

4.14.3.1 *General*—To obtain proper lateral force resistance, the building structure should act as a single unit through the diaphragm effect of the floor slabs. The lateral loads should be transferred from the application point to the lateral-force-resisting structural members (structural walls and moment-resisting frames) as in-plane loads in the diaphragm.

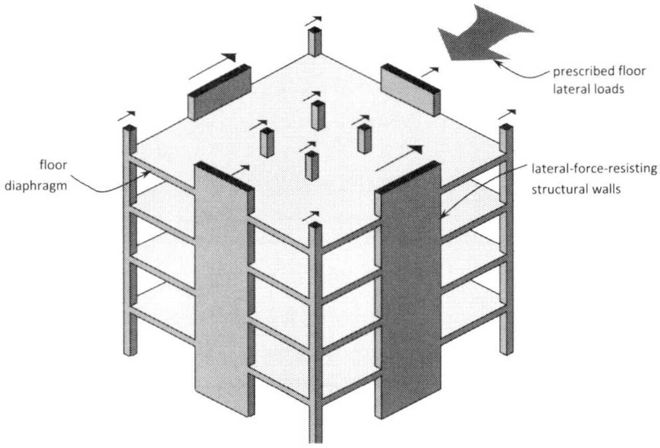

Fig. 4.14.1—Lateral-force-resisting structural system.

4.14.3.2 *Diaphragm*—The diaphragm should comply with (a) through (e):

(a) The shape in plan of the floor slab diaphragm should be as regular and symmetrical as practicable. Preferred forms should be square or rectangular. Rectangular diaphragms should have a maximum ratio of shorter to longer side of 4;

(b) Floor slab structural members should comply with the minimum thickness of Chapter 6;

(c) No opening in the slab should be greater than 25 percent of the floor area, and the sum of all openings in the slab should not exceed 50 percent of the floor area;

(d) Diaphragm collector elements should be provided, as indicated in 4.15.5, and be capable of transmitting lateral loads from the floor slab system to the lateral-force-resisting structural walls; and

(e) Diaphragm members can be at a slope but should comply with 1.3.9.

4.14.4 *Floor mass centroid*—The floor mass centroid should correspond to the centroid of the area of the floor slab. The area for computing each floor mass centroid should be bounded by slab borders, minus the area of all openings having a side greater than 6 ft (1.8 m). The lateral seismic

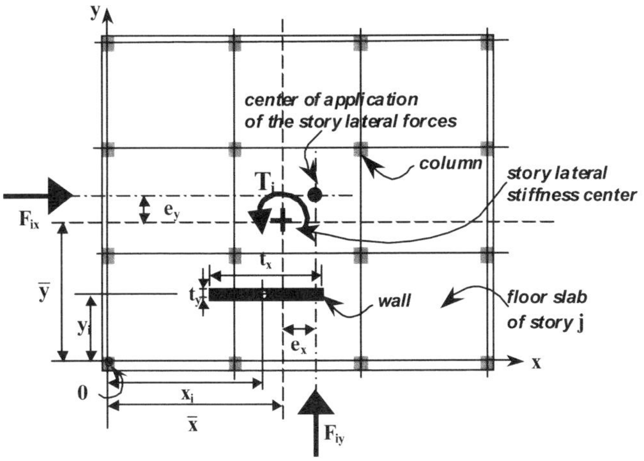

Fig. 4.14.5—Calculation of the story lateral stiffness center.

loads should be applied at the floor mass centroid for each story.

4.14.5 *Story lateral stiffness center*—Story lateral stiffness center corresponds to the point where the floor diaphragm will tend to rotate about a vertical axis when subjected to a story torsional moment. The location of the story lateral stiffness center in plan should be calculated using the lateral stiffness of the walls only, disregarding the stiffness contribution of the columns of the structure. If the location of the structural walls is symmetric with respect to the center of mass of every story in the structure, it should be assumed that the lateral stiffness center for all stories is located at the center of mass. Where the walls are not symmetric, procedures (a) through (d) should be used to calculate the location of the story lateral stiffness center (Fig. 4.14.5).

(a) An arbitrary origin location is assumed at point 0;

(b) Lateral stiffnesses, k_x and k_y, of structural walls should be computed using Eq. (4.14.5a(a)) and (4.14.5a(b)), for both principal directions x and y.

For structural walls with lengths parallel to the x axis:

$$k_x = \frac{b_w^3 \ell_w}{h_{pi}} \text{ and } k_y = \frac{\ell_w^3 b_w}{h_{pi}} \qquad (4.14.5a(a))$$

or, for structural walls with lengths parallel to the y axis:

$$k_x = \frac{\ell_w^3 b_w}{h_{pi}} \text{ and } k_y = \frac{b_w^3 \ell_w}{h_{pi}} \qquad (4.14.5a(b))$$

where b_w is the web width of section, or wall width; ℓ_w is the horizontal length of the wall; and h_{pi} is the story height;

(c) The story lateral stiffness center coordinates, $\bar{x}$ and $\bar{y}$, with respect to origin 0, should be:

$$\bar{x} = \frac{\Sigma(k_y x_i)}{\Sigma k_y} \text{ and } \bar{y} = \frac{\Sigma(k_x y_i)}{\Sigma k_x} \qquad (4.14.5b)$$

where x_i is the distance from origin 0 to the wall cross-sectional centroid measured in a direction parallel to x, and x_i is the distance from origin 0 to the wall cross-sectional centroid measured in a direction parallel to y; and

(d) Total story rotational stiffness about the story lateral stiffness center should be computed using Eq. (4.14.5c).

$$k_r = \left[\Sigma\left(k_y x_i^2 + k_x y_i^2\right)\right] - \left[\bar{x}^2 \Sigma k_y\right] - \left[\bar{y}^2 \Sigma k_x\right] \qquad (4.14.5c)$$

4.15—Minimum amount of reinforced concrete structural walls

4.15.1 *General*—A minimum amount of reinforced concrete structural walls should be provided for factored lateral force resistance. These structural walls should comply with (a) through (g):

(a) Structural walls should have rectangular cross sections. Other cross sections are beyond the scope of this guide, except core walls as prescribed in 12.8;

(b) Structural walls should be vertically continuous from foundation to roof, except where walls are needed only for resistance for unbalanced lateral soil pressure in a basement. In this case, the wall may be stopped at the first story level;

(c) Structural walls should be aligned vertically. Where section reductions occur, the wall cross section underneath should be greater than the one above and the centroid of the wall cross section above should be in the middle third, in both directions, of the one below;

(d) Structural walls should not have openings for windows or doors;

(e) In both principal directions in plan, there should be at least two parallel walls in different planes, and the planes should be as far apart as practicable. The walls should be placed as close to the periphery of the building as practicable;

(f) Walls should be located as symmetrically as possible with respect to the centers of mass and stiffness of each floor; and

(g) Structural wall dimensions should comply with 4.15.2 and 4.15.3.

4.15.2 *Minimum wall area for shear strength*—At any floor i, for the two principal directions, x and y, the minimum cross-sectional area ($A_g = \ell_w b_w$) for all reinforced concrete walls acting in the principal direction under consideration should be determined from Eq. (4.15.2).

$$\Sigma(\ell_w b_w) \geq \frac{3V_{iu}}{4\sqrt{f_c'}} \quad \left[\Sigma(\ell_w b_w) \geq \frac{9V_{iu}}{\sqrt{f_c'}} \text{ (SI)}\right] \qquad (4.15.2)$$

In Eq. (4.15.2), only walls where the horizontal length of wall ℓ_w is parallel to the direction under consideration should be included; b_w corresponds to the wall thickness, and V_{iu} should be determined from 4.13.3. As noted by 4.15.1(e), each principal direction should have at least two walls.

4.15.3 *Minimum wall dimensions for lateral stiffness*—Slenderness ratio h_w/ℓ_w for any individual wall should comply with Eq. (4.15.3), and the wall thickness b_w should comply with Chapter 12.

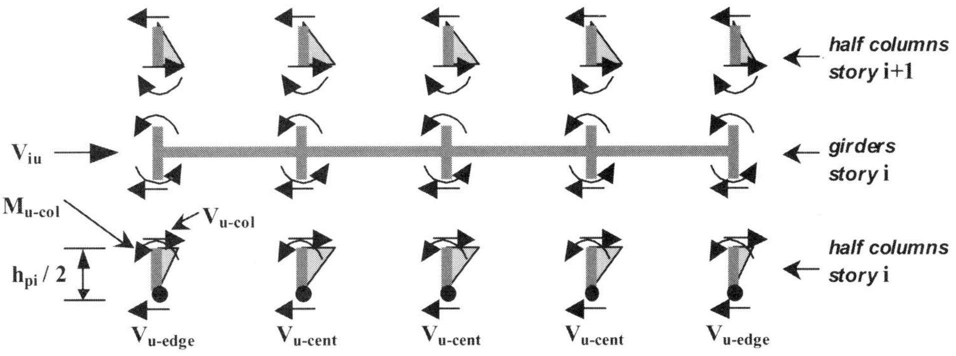

Fig. 4.15.4.2a—Column lateral shears and moments in frames.

$$\frac{h_w}{\ell_w} \leq 4 \qquad (4.15.3)$$

where h_w is the wall height from the foundation to the roof, and ℓ_w is the horizontal length of wall.

4.15.4 *Lateral strength in vertical structural members*

4.15.4.1 *Structural walls*—At any story i, the factored lateral shear V_u that a wall should sustain is determined from Eq. (4.15.4.1a). The summation in Eq. (4.15.4.1a) should be performed for all walls with ℓ_w parallel to V_{iu}.

$$V_u = V_{iu} \frac{b_w \ell_w^3}{\Sigma\left(b_w \ell_w^3\right)} + \Delta V_{ut} \qquad (4.15.4.1a)$$

where V_{iu} is the factored story shear (4.13.3) in the direction parallel to ℓ_w; b_w and ℓ_w are the wall cross-sectional dimensions; and ΔV_{ut} is the increase in shear due to torsion, determined from Eq. (4.15.4.1b). The value of ΔV_{ut} from Eq. (4.15.4.1b), parallel to ℓ_w, should be used in Eq. (4.15.4.1a).

$$\Delta V_{utx} = T_{iu} \frac{\bar{y} k_x}{k_r} \text{ and } \Delta V_{uty} = T_{iu} \frac{\bar{x} k_y}{k_r} \qquad (4.15.4.1b)$$

where T_{iu} is determined from 4.13.5; $\bar{x}$ and $\bar{y}$ are determined from Eq. (4.14.5b); k_x and k_y are determined from Eq. (4.14.5a); and k_y is determined from Eq. (4.14.5c).

4.15.4.2 *Frames*—The following procedure should be used to assign factored story shears to the structure's frame through the columns.

(a) In zones of low and no seismic risk, where reinforced concrete walls are not mandatory (4.14.2), total lateral story shear at any story should be distributed to the frames through the columns;

(b) In moderate or high seismic risk zones, 100 percent of factored lateral loads should be resisted by reinforced concrete walls. Frames should be proportioned to resist a minimum lateral force equal to 25 percent of the factored lateral force in each direction in plan to account for effects such as base rotation of the walls or a decrease in stiffness and strength due to inelastic response;

(c) The fraction of the story shear V_{iu} as described by (a) or (b), carried by an individual column in one plan direction for any story i, should be determined from Eq. (4.15.4.2a) for interior columns, and by Eq. (4.15.4.2b) for edge columns. An interior column is a column whose ends frame into two girders parallel to the direction under study, whereas edge columns frame into one girder on one side only.

Fig. 4.15.4.2b—Girder lateral force moments in frames.

$$V_u = \frac{2V_{iu}}{n_e + 2n_c} \qquad (4.15.4.2a)$$

$$V_u = \frac{V_{iu}}{n_e + 2n_c} \qquad (4.15.4.2b)$$

In Eq. (4.15.4.2a) and (4.15.4.2b), n_e and n_c are the total number of edge and central columns, respectively, in story i for the direction under consideration.

(d) The factored column moment due to lateral loads (Fig. 4.15.4.2a) should be determined from Eq. (4.15.4.2c).

$$M_u = V_u \frac{h_{pi}}{2} \qquad (4.15.4.2c)$$

where h_{pi} corresponds to the story height of floor i, and V_u is the corresponding value determined in (c).

(e) For column-slab systems, the girder or slab moment due to lateral loads should be taken as equal to the factored column moment for interior column joints and twice the factored column moment for edge column joints. This results in the moments at all girders at the face of the column-girder joints. This moment acts in both the positive and negative sense in the direction under study, and it should be taken as an increase or decrease of the gravity load factored moments (Fig. 4.15.4.2b).

4.15.5 *Collector elements*—Collector elements within the diaphragm, usually consisting of beams or girders, transmit lateral loads in the story from application point to the lateral-force-resisting structural members. Collector element longitudinal reinforcement should be confined by closed stirrups that comply with the limits for tied columns of 10.4.3.2. The force p_u to be transferred by the collector element connected to a wall should be the difference in shear between the wall above and below the collector element. The wall shear difference is obtained from Eq. (4.15.4.1a).

The longitudinal reinforcement area in the collector element A_{st} should be equal to or greater than the area indicated by Eq. (5.12.3.1) and Eq. (5.12.5). For application of Eq. (5.12.3.1), A_g should be the gross cross-sectional area of the collector element.

CHAPTER 5—GENERAL REINFORCED CONCRETE INFORMATION

5.1—Scope

Chapter 5 contains general information for reinforced concrete structural members, including allowable materials, concrete cover to reinforcement, and procedures for defining design strength of members subjected to flexure, axial loads with or without flexure, and shear. Additional information for each individual member type is presented in Chapters 6 to 16.

5.2—Materials for reinforced concrete

5.2.1 *Material standards*—All materials used in construction of a structure designed following the guide should conform to the ASTM standards listed in Chapter 17.

5.2.2 *Cement*—Cement should conform to one of the following: "Specification for Portland Cement" (ASTM C150/C150M-11) and "Specification for Blended Hydraulic Cements" (ASTM C595/C595M-11), excluding Types S and SA. ASTM C595/C595M-11, Types S and SA are not intended as principal cementing constituents of structural concrete.

5.2.3 *Aggregates*—Aggregates should conform to ASTM C33/C33M-10. Concrete with lightweight aggregates are not within the scope of this guide.

5.2.4 *Water*—Water should conform to ASTM C1602/C1602-06.

5.2.5 *Steel reinforcement*—Steel reinforcement should be deformed reinforcement, with exceptions noted in 5.2.5.3, conform to 5.2.5.1 through 5.2.5.3, and comply with the corresponding ASTM standards. Welded wire reinforcement should be considered deformed reinforcement.

5.2.5.1 *Deformed reinforcement*—Maximum specified yield strength for deformed reinforcement should be 60,000 psi (420 MPa). Deformed reinforcing bars should conform to ASTM A615/A615M-09 or ASTM A706/A706M-09.

ASTM A615/A615M-09 covers Grades 40, 60, and 75, with yield strengths of 40,000, 60,000, and 75,000 psi (280, 420, and 520 MPa), respectively, whereas ASTM A706/A706M-09 covers Grades 60 (420) and 75 (520) with yield strengths of 60,000 and 75,000 psi (420 and 520 MPa). Under this guide, only yield strength of 40,000 and 60,000 psi (280 and 420 MPa) are permitted.

Both standards cover the nominal diameters shown in Table 5.2.5.1. Under this guide, the maximum nominal diameter of deformed reinforcing bars is No. 8 (1 in.) (25 mm). Refer to 5.3. ASTM A615/A615M-09 and A706/A706M-09 standards do not include No. 2 (1/4 in.) (6 mm) diameter bars.

Table 5.2.5.1—Nominal characteristics for reinforcing bars

Bar denomination*	Nominal Diameter, in. (mm)	Area, in.^{2} (mm^2)	Weight, lb/ft (kg/m)
No. 2 (6)†	0.250 (6.4)	0.05 (32)	0.170 (0.250)
No. 3 (10)	0.375 (9.5)	0.11 (71)	0.376 (0.560)
No. 4 (13)	0.500 (12.7)	0.20 (129)	0.668 (0.994)
No. 5 (16)	0.625 (15.9)	0.31 (199)	1.043 (1.552)
No. 6 (19)	0.750 (19.1)	0.44 (284)	1.502 (2.235)
No. 7 (22)	0.875 (22.2)	0.60 (387)	2.044 (3.042)
No. 8 (25)	1.000 (25.4)	0.79 (510)	2.670 (3.973)
No. 9 (29)‡	1.125 (28.7)	1.00 (645)	3.400 (5.060)
No. 10 (32)‡	1.250 (32.3)	1.27 (819)	4.303 (6.404)
No. 11 (35)‡	1.375 (35.8)	1.56 (1006)	5.313 (7.907)
No. 14 (43)‡	1.750 (43.0)	2.25 (1452)	7.650 (11.380)
No. 18 (57)‡	2.250 (57.3)	4.00 (2581)	13.600 (20.240)

*Bar No. indicates the number of 1/8 in. (3 mm) corresponding to the bar diameter and the metric denominations of the rounded bar diameter in mm as designated by ASTM.

†No. 2 (1/4 in. [6 mm]) bars are not manufactured routinely in the United States.

‡The maximum nominal bar diameter permitted by this guide is No. 8 (1 in. [25 mm]).

5.2.5.2 *Welded wire reinforcement*—Maximum specified yield strength for wires in welded wire reinforcement should be 70,000 psi (500 MPa). Welded wire reinforcement should conform to ASTM A1064/A1064M-10. Under the guide, the maximum nominal diameter of wire for welded wire reinforcement is 3/8 in. (10 mm) (Section 5.3).

5.2.5.3 *Plain reinforcement*—Plain reinforcement may be used only for stirrups, ties, spirals, and when it is part of welded wire reinforcement, and should conform to the specifications listed in 5.2.5.1. Maximum specified yield strength for plain reinforcement should be 40,000 psi (280 MPa) and under this guide, the maximum nominal diameter of plain reinforcing bars is 5/8 in. (16 mm) (Section 5.3). Plain reinforcement is not allowed for use in the United States.

5.2.6 *Admixtures*—Admixtures should conform to the ASTM standards in Table 5.2.6.

Additional admixtures may be used if approved by a licensed design professional. Calcium chloride or admixtures containing chlorides should not be used in concrete having prestressed steel, or in concrete that is in contact with galvanized steel.

The use of admixtures in concrete containing expansive cement, as specified by ASTM C845-04, should be used with caution. The admixture should be proven compatible with the cement without producing harmful effects.

5.2.7 *Storage of materials*—It is important that cement and aggregates be stored to prevent contamination, deterioration, and intrusion of foreign matter. Any material that has deteriorated or has been contaminated should not be used for concrete.

5.3—Minimum and maximum reinforcing bar diameter

Reinforcement used in structures designed under this guide should have a nominal diameter d_b not less than the

Table 5.2.6—ASTM standard references for admixtures

Purpose	Standard
Water reduction and setting time modification	ASTM C494/494M-11
Produce flowing concrete	ASTM C1017/1017M-07
Air-entrainment	ASTM C260/C260M-10

Table 5.3—Minimum and maximum reinforcing bar diameters used in structures

Reinforcement type (reference section)	Minimum bar diameter d_b	Maximum bar diameter d_b
(a) Deformed reinforcing bars (5.2.5.1)	3/8 in. (10 mm)	1 in. (25 mm)
(b) Wire for welded wire reinforcement (5.2.5.2)	0.16 in. (4 mm)	3/8 in. (10 mm)
(c) For stirrups and ties (5.2.5.1)	3/8 in. (10 mm)	5/8 in. (16 mm)
(d) Plain reinforcing bars (5.2.5.3)	3/8 in. (10 mm)	5/8 in. (16 mm)

minimum diameter, nor greater than the maximum diameter given in Table 5.3.

5.4—Concrete cover for reinforcement

5.4.1 *Minimum concrete cover*—The minimum concrete cover to reinforcement that should be provided is shown in Table 5.4.1.

5.4.2 *Special fire protection of the structure*—When the designated fire rating, in hours, for the building is greater than 1 hour, the concrete cover of 5.4.1 should be increased by 1/4 in. (6 mm) per each additional hour of fire rating.

5.4.3 *Additional corrosion protection*—In very aggressive environments, such as direct contact with seawater, chemical facilities, or sewage water facilities, additional corrosion protection of the reinforcement should be specified, such as reduced water-cementitious material ratio (*w*/*cm*), epoxy-coated bars, air-entrained concrete, and other means. This type of protection is beyond the scope of this guide, and the standards of 1.4 should be used for design in that case.

Table 5.4.1—Minimum concrete cover to reinforcement

(a) Members cast against and permanently exposed to earth	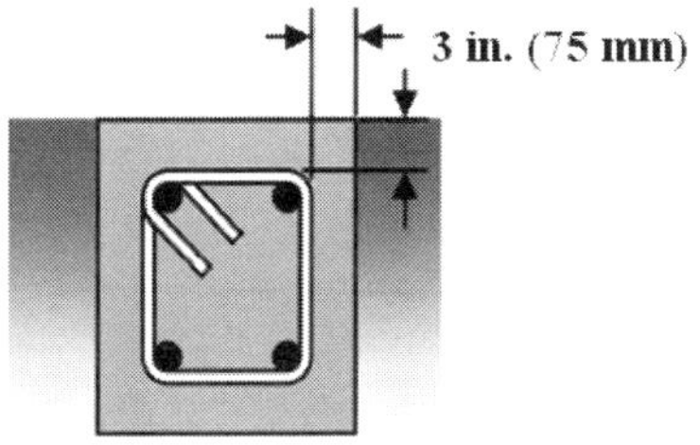 *Minimum concrete cover 3 in. (75 mm)*
(b) Members exposed to weather or earth	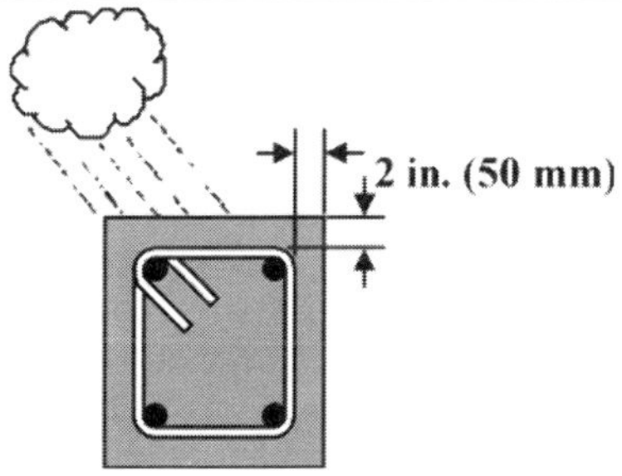 *Minimum concrete cover 2 in. (50 mm)*
(c) Girders, beams, or columns, when not exposed to weather or in contact with ground	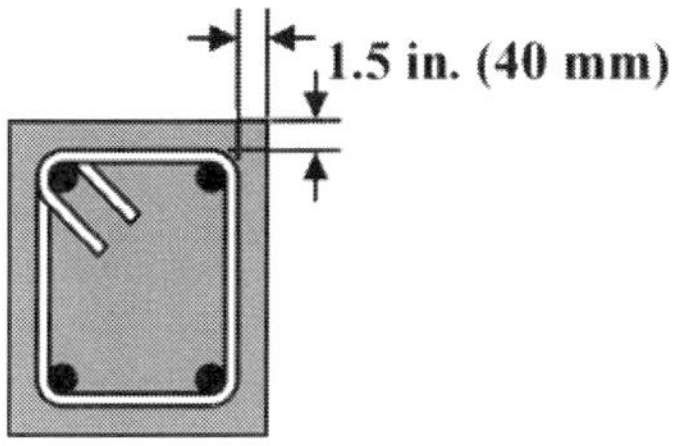 *Minimum concrete cover 1-1/2 in. (40 mm)*
(d) Solid slabs, reinforced concrete walls, or joists, when not exposed to weather or in contact with ground	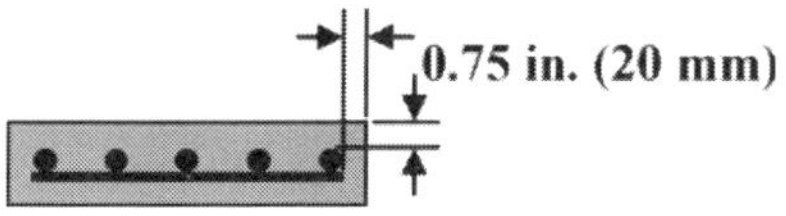 *Minimum concrete cover 3/4 in. (20 mm)*
(e) Solid slabs on grade	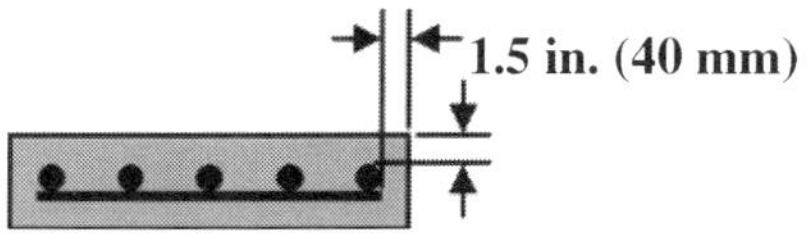 *Minimum concrete cover 1-1/2 in. (40 mm)*

Table 5.5—Bend diameter minimum values

Reinforcement type	Diameter of bend	
(a) Deformed reinforcing bars	$6d_b$	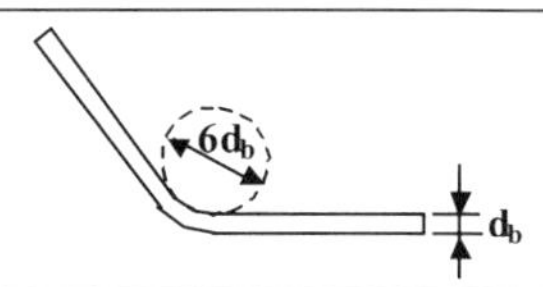
(b) Plain reinforcing bars	$6d_b$	
(c) For stirrups and ties	$4d_b$	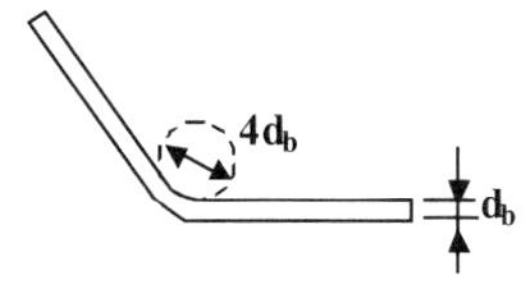
(d) Wire for welded wire reinforcement with d_b > 1/4 in. (6 mm)	$4d_b$	
(e) Wire for welded wire reinforcement with $d_b \leq$ 1/4 in. (6 mm)	$2d_b$	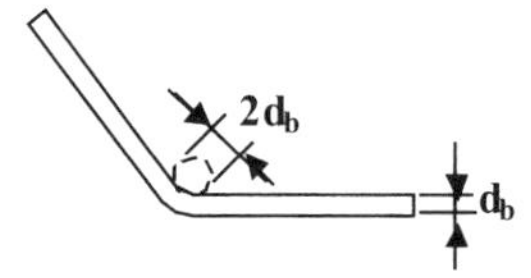

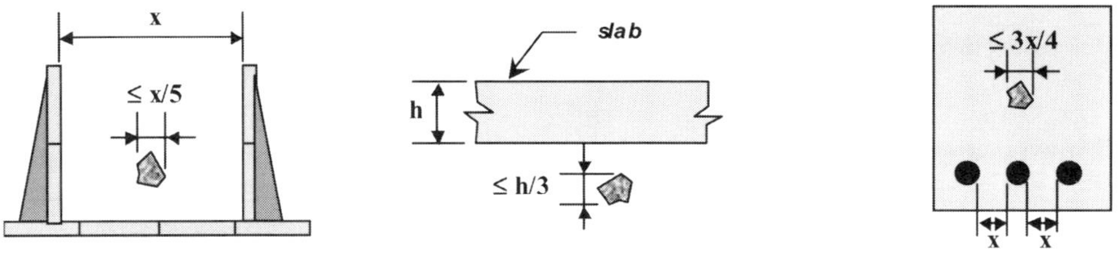

Fig. 5.7—Maximum nominal coarse aggregate size.

5.5—Minimum reinforcement bend diameter

Bend diameter, measured on the inside of the bar, should not be less than the values shown in Table 5.5.

5.6—Standard hook dimensions

The term "standard hook" means one of the following in Table 5.6.

5.7—Maximum aggregate size

Maximum nominal coarse aggregate size should not be larger than (1), (2), or (3) (Fig. 5.7).

(1) 1/5 the narrowest dimension between sides of forms;

(2) 1/3 the depth of slabs; and

(3) 3/4 the minimum clear spacing between parallel reinforcing bars.

A maximum nominal size of 3/4 in. (19 mm) is recommended for columns, girders, beams, and joists. Except for structural slabs, a larger size can be used if it meets the limits given in (1) through (3).

5.8—Development length, lap splicing, and anchorage of reinforcement

5.8.1 *Development length*

5.8.1.1 *Reinforcing bars*—The minimum length of embedment, ℓ_d, on each side of a critical section for a reinforcing bar to develop its full strength should be $50d_b$. Development length on one side of the critical section may be replaced by a standard hook complying with the minimum standard hook anchorage distance of 5.8.3 (Fig. 5.8.1.1).

5.8.1.2 *Welded wire reinforcement*—Development length ℓ_d of welded wire reinforcement measured on each side of the critical section to the end of wire should contain two crosswires, and should not be less than 8 in. (200 mm) (Fig. 5.8.1.2).

5.8.2 *Lap splice dimensions*

5.8.2.1 *Reinforcing bars*—Minimum lap length for splicing reinforcing bars should be $50d_b$. For buildings within the scope of this guide, this lap length calculation is a satisfactory simplification of the ACI 318-11 detailed calculation for splice length (Fig. 5.8.2.1).

The maximum nominal bar diameter permitted by the guide is 1 in. (25 mm).

5.8.2.2 *Welded wire reinforcement*—A welded wire reinforcement splice should contain two crosswires from each sheet and should not be less than 10 in. (250 mm) (Fig. 5.8.2.2).

5.8.3 *Minimum standard hook anchorage distance*—The bent portion of the hook should be located as close to the outer face of concrete as cover permits. Minimum distance between the outer face of concrete and the critical section where the hooked bar develops its full strength is $25d_b$ (Fig. 5.8.3).

5.9—Longitudinal reinforcement

Longitudinal reinforcement in reinforced concrete members should be provided to resist axial tension, axial compression, flexural-induced tension and compression, and stresses induced by temperature variation and drying shrinkage of concrete.

The area of longitudinal reinforcement placed in reinforced concrete members should be sufficient to resist the factored

Table 5.6—Standard hook description and dimensions

Hook designation	Description and dimensions	
(a) 90° hook	A 90° bend plus minimum $12d_b$ extension at free end of bar	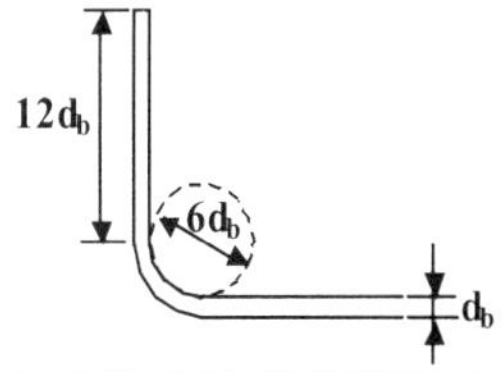
(b) 180° hook	A 180° bend plus minimum $4d_b$ extension at free end of bar	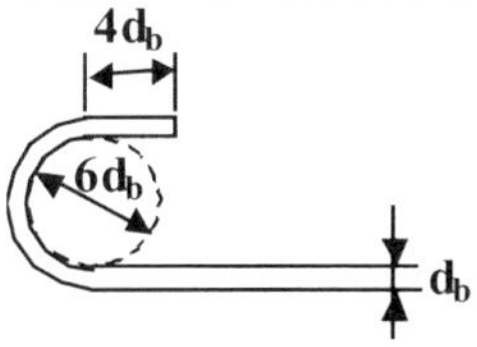
(c) For stirrup and tie hooks	A 90° bend plus minimum $6d_b$ extension at free end of bar, or	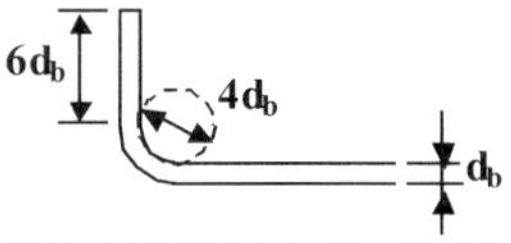
	A 135° bend plus minimum $6d_b$ extension at free end of bar	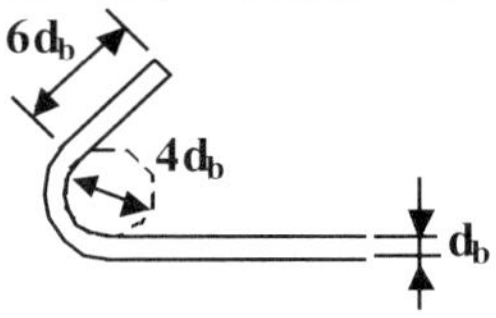
(d) For hoops (confinement stirrups and ties) in seismic zones	A 135° bend plus minimum $6d_b$ extension at free end of bar, but not less than 3 in. (75 mm)	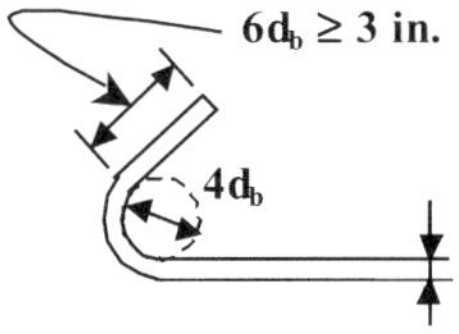
(e) For crossties in seismic zones	A 135° bend plus minimum $6d_b$ extension at one free end of bar, but not less than 3 in. (75 mm), and	A 90° bend plus minimum $6d_b$ extension at the other free end of bar
	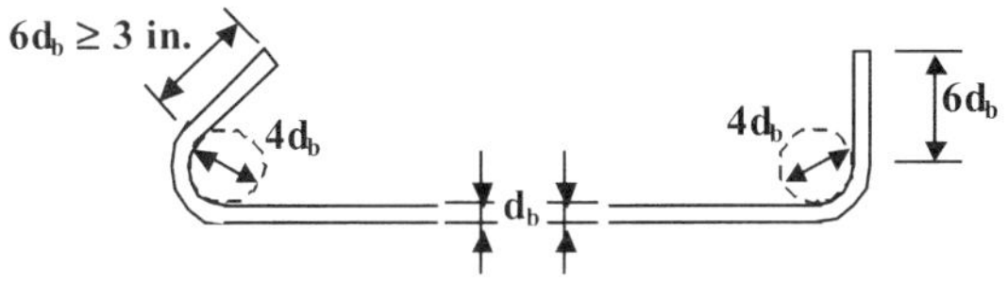	

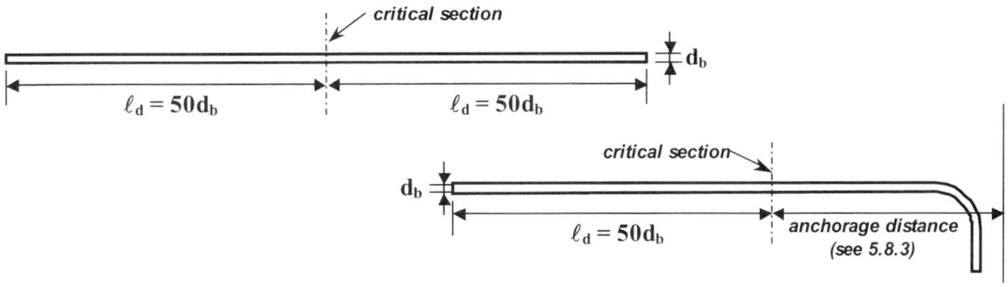

Fig. 5.8.1.1—Development length for reinforcing bars.

loads but not less than the minimum value nor greater than the maximum value.

When the calculated area of longitudinal reinforcement is less than the minimum value, no less than the minimum value should be provided. If the calculated area exceeds the maximum limit, reinforced concrete member dimensions should be appropriately modified.

5.10—Transverse reinforcement

Transverse reinforcement in reinforced concrete members resists shear and torsion loads, confines compression bars against buckling out of the concrete, and prevents displacement of longitudinal reinforcement during construction operations. In seismic zones, transverse reinforcement provides concrete confinement in special regions subjected to large strains. The area of transverse reinforcement placed in members should be sufficient to resist factored loads but not less than the minimum values.

When the calculated area of transverse reinforcement is less than the minimum value, no less than the minimum value should be placed. If the calculated area of transverse reinforcement exceeds the maximum limit, member dimensions should be appropriately modified.

5.11—Flexure

5.11.1 *General*—The design moment strength of sections should be computed by 5.11. Where the factored axial load on the member, P_u, exceeds $0.10f_c'A_g$ or produces axial tension, design strength should be computed by 5.12.

5.11.2 *Required moment strength*—The factored moment M_u (required moment strength) due to the applied factored loads should be determined for the particular member type from Chapters 7 to 14.

5.11.3 *Design moment strength*—The design moment strength of the section, ϕM_n, computed using the assumed dimensions, material strengths, and reinforcement, should be greater than or equal to the required moment strength M_u:

$$\phi M_n \geq M_u \tag{5.11.3}$$

Longitudinal reinforcement should be placed near the member face where flexural moment creates tension (Fig. 5.11.3).

5.11.4 *Design moment strength for rectangular sections with tension reinforcement only*

5.11.4.1 *Design moment strength*—Based on the assumption that the steel yields, for a section with tension reinforcement only, the design moment strength should be determined using Eq. (5.11.4.1), with ϕ = 0.90 (Fig. 5.11.4.1).

$$\phi M_n = \phi A_s f_y \left(d - \frac{a}{2}\right) \text{ and } a = \frac{A_s f_y}{0.85 f_c' b} \tag{5.11.4.1}$$

In slabs, girders, beams, and joists, the design moment strength is computed as permitted in Chapters 7 to 14.

5.11.4.2 *Flexural tension reinforcement area*—The ratio of flexural reinforcement, $\rho = A_s/(bd)$, should be determined using the factored moment M_u as:

$$\rho = \frac{A_s}{bd} = \alpha_f - \sqrt{\alpha_f^2 - \left(\frac{M_u}{\phi bd^2} \cdot \frac{2\alpha_f}{f_y}\right)} \text{ where } \alpha_f = \frac{f_c'}{1.18 f_y} \tag{5.11.4.2}$$

If the computed ρ is less than the minimum, ρ_{min}, as established in Chapters 7 to 14, A_s should not be less than the minimum. When the computed ρ is greater than the maximum, ρ_{max}, member dimensions and the self-weight moment should be appropriately modified. For girders, beams, and joists, when the computed ρ is greater than ρ_{max} (Table 7.3.4.3), use of compression reinforcement should be investigated.

A note on employment of U.S. customary units. The required factored moments generally are in units of lb·ft because they are determined from concentrated loads in lb, distributed loads in lb/ft, or loads per unit area in lb/ft^2, and spans in ft. They should be converted to lb·in. (12 lb·in. = 1 lb·ft) for use with f_y in psi, d and b in inches, and A_s in in.^{2}. For slabs in which the units of moment are lb·ft per ft of slab width (lb·ft/ft), the same conversion (12 lb·in./ft = 1 lb·ft/

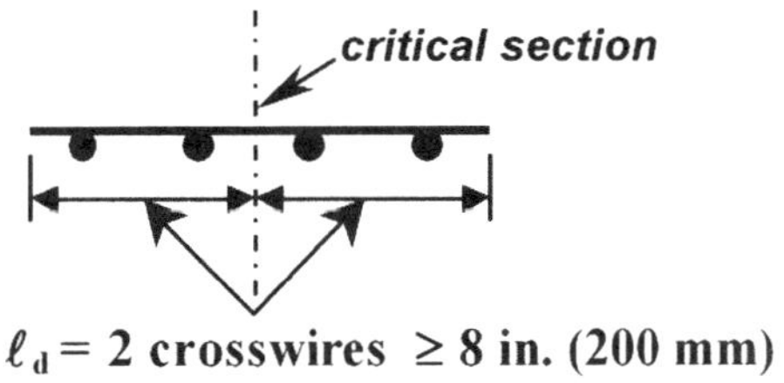

Fig. 5.8.1.2—Development length for welded wire reinforcement.

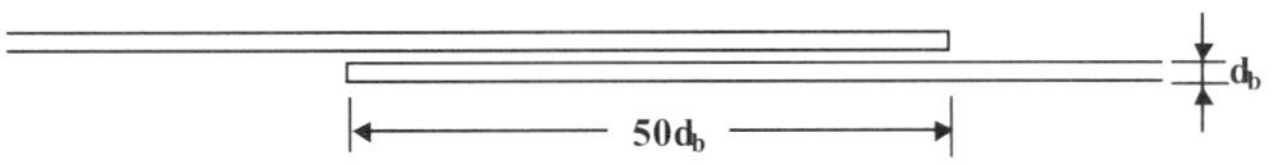

Fig. 5.8.2.1—Minimum lap splice length for reinforcing bars.

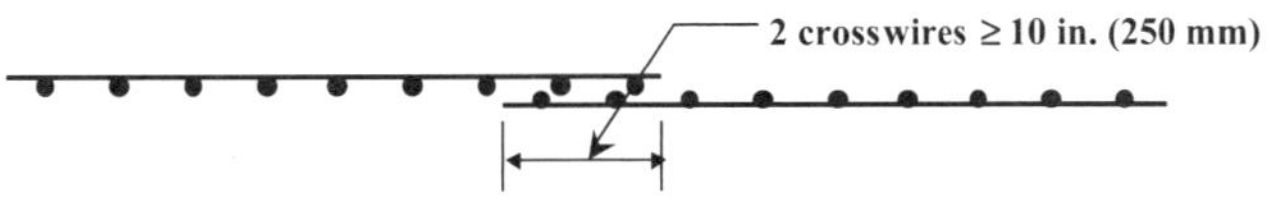

Fig. 5.8.2.2—Minimum lap splice length for welded wire reinforcement.

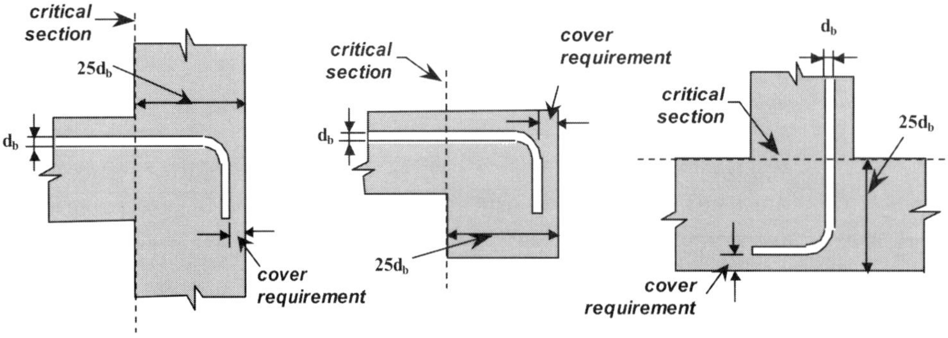

Fig, 5.8.3—Minimum standard hook anchorage distance.

ft) should be used and the A_s determined will be in in.^{2}/ft of slab width.

A note on employment of SI units. The required factored moments generally are in units of kN·m because they are determined from concentrated loads in kN, distributed loads in kN/m, or loads per unit area in kN/m^2, and spans in m. They should be converted to N·mm (1 kN·m = 10^6 N·mm) for use with f_y in MPa (1 MPa = 1 N/mm^2), d and b in mm, and A_s in mm^2. For slabs in which the units of moment are kN·m per meter of slab width (kN·m/m), the same conversion (1 kN·m/m = 10^6 N·mm/m) should be used and the A_s determined will be in mm^2/m of slab width.

5.12—Axial loads with or without flexure

5.12.1 *General*—Calculation of the design strength of columns and reinforced concrete walls subjected to axial loads alone or axial loads accompanied by flexure should be in accordance with 5.12.

5.12.2 *Combined axial load and moment*—The factored axial load P_u and the factored moment M_u, due to applied factored loads, should be determined for the particular member type from Chapters 7 to 14.

5.12.3 *Design axial compression strength*

5.12.3.1 *Design axial compression strength without flexure*—Equation (5.12.3.1) should be used to determine the design axial strength without flexure, ϕP_{on}.

$$\phi P_{on} = \phi\left[0.85 f_c'\left(A_g - A_{st}\right) + A_{st} f_y\right] \quad (5.12.3.1)$$

where $\phi = 0.65$ for columns with ties and reinforced concrete walls, and $\phi = 0.70$ for columns with spiral reinforcement.

5.12.3.2 *Maximum design axial compression strength*—Design axial strength ϕP_n in columns and reinforced concrete walls subjected to compression, with or without flexure, should not exceed the following:

(a) Columns with ties and reinforced concrete walls:

$$\phi P_{n(\max)} \le 0.80\phi P_{on} \text{ (with } \phi = 0.65\text{)} \quad (5.12.3.2a)$$

(b) Columns with spiral reinforcement:

$$\phi P_{n(\max)} \le 0.85\phi P_{on} \text{ (with } \phi = 0.70\text{)} \quad (5.12.3.2b)$$

5.12.4 *Balanced design strength for axial load with flexure*

5.12.4.1 *Square and rectangular tied-columns, and reinforced concrete walls*—Values for design axial compression strength ϕP_{bn} and design moment strength ϕM_{bn} at the balanced design strength point should be determined using Eq. (5.12.4.1a) and Eq. (5.12.4.1b), respectively, with ϕ = 0.65:

$$\phi P_{bn} = \phi 0.42 f_c' h b \quad (5.12.4.1a)$$

$$\phi M_{bn} = \phi P_{bn} 0.32h + \phi\left[0.6A_{se} + 0.15A_{ss}\right] f_y \left(\frac{h}{2} - d'\right)$$

(5.12.4.1b)

For Eq. (5.12.4.1b), the total longitudinal reinforcement area A_{st} should be divided into steel at the extreme face, A_{se}, and steel at the side face, A_{ss}, in such a manner that $A_{se} + A_{ss} = A_{st}$ (Fig. 5.12.4.1).

5.12.4.2 *Circular columns with spiral reinforcement*—Values for design axial compression strength ϕP_{bn} and design moment strength ϕM_{bn} at the balanced design strength point should be determined using Eq. (5.12.4.2a) and Eq. (5.12.4.2b), respectively, with h taken as the column diameter, and ϕ = 0.70 (Fig. 5.12.4.2).

$$\phi P_{bn} = \phi 0.5 f_c' A_{cs} \quad (5.12.4.2a)$$

$$\phi M_{bn} = \phi P_{bn} 0.2h + \phi 0.6 A_{st} f_y \left(\frac{h}{2} - d'\right) \quad (5.12.4.2b)$$

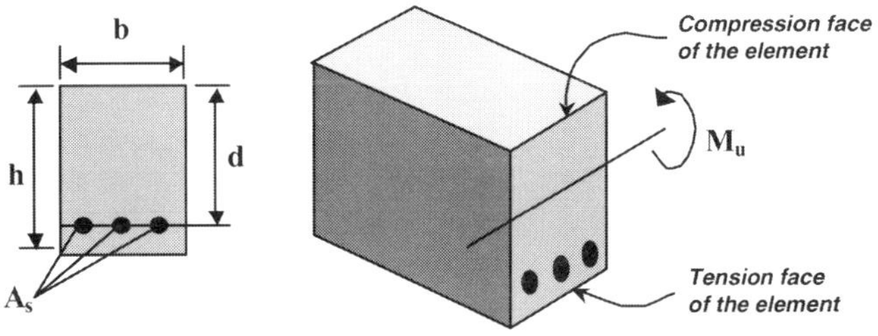

Fig. 5.11.3—Design dimensions for design moment strength.

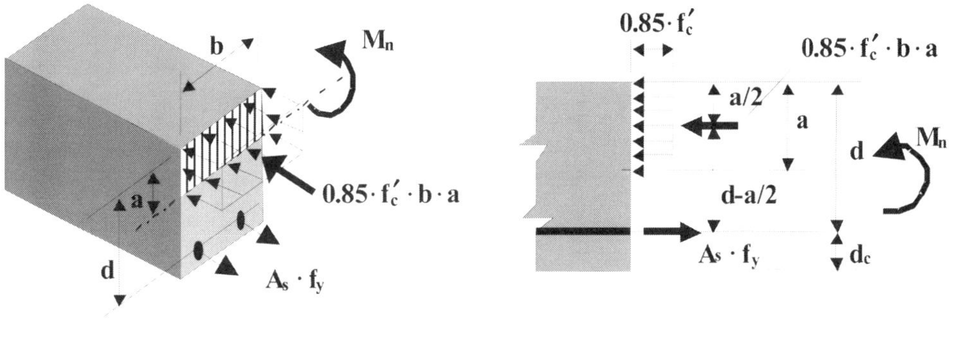

Fig. 5.11.4.1—Flexural nominal moment strength.

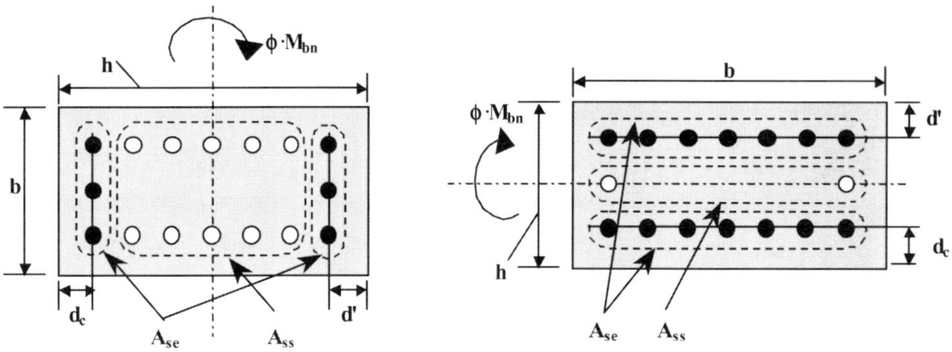

Fig. 5.12.4.1—Dimensions for calculation of balanced design strength for rectangular tied columns and reinforced concrete walls.

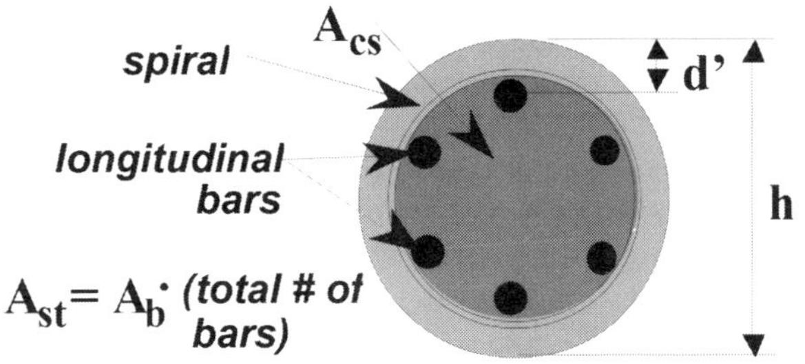

Fig. 5.12.4.2—Dimensions for calculation of balanced design strength for circular section columns with spiral reinforcement.

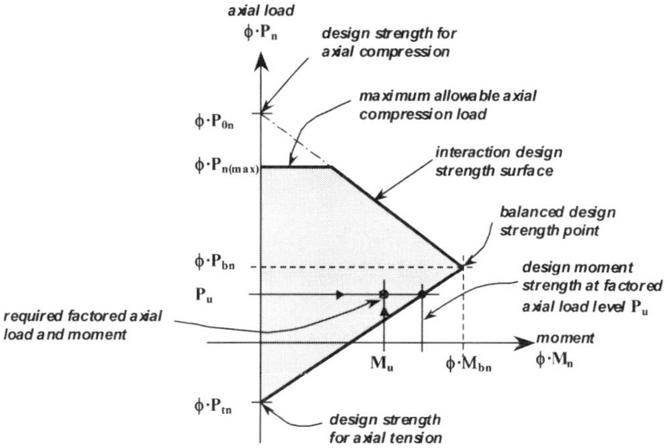

Fig. 5.12.6—Interaction diagram for (ϕM_n, ϕP_n).

5.12.5 *Design strength for axial tension without flexure*—The design axial tension strength without flexure, ϕP_{tn}, should be determined using Eq. (5.12.5) with $\phi = 0.90$.

$$\phi P_{tn} = \phi A_{st} f_y \quad (5.12.5)$$

5.12.6 *Combined axial load and moment strength*—The design moment strength at the section, ϕM_n, at the level of applied factored axial load, P_u, should be equal to or greater than the maximum factored moment M_u that can accompany the factored axial load P_u, as shown in Eq. (5.12.6a).

$$\phi M_n \geq M_u \quad (5.12.6a)$$

The compliance with Eq. (5.12.6a) should be accomplished by proving that the coordinates of M_u—P_u in a moment versus axial load interaction diagram relating ϕM_n and ϕP_{bn}—are inside the interaction design strength surface, the shaded portion in Fig. 5.12.6.

The following conditions should be met for all couples of M_u and P_u that act on the column section:

$$P_u \leq \phi P_{n(\max)} \quad (5.12.6b)$$

$$P_u \geq -(\phi P_{tn}) \quad (5.12.6c)$$

For values of $P_u \geq \phi P_{bn}$:

$$M_u \leq \phi M_n = \frac{\phi P_{on} - P_u}{\phi P_{on} - \phi P_{bn}} \phi M_{bn} \quad (5.12.6d)$$

For values of $P_u < \phi P_{bn}$:

$$M_u \leq \phi M_n = \frac{P_u + \phi P_{tn}}{\phi P_{bn} + \phi P_{tn}} \phi M_{bn} \quad (5.12.6e)$$

5.12.7 *Use of interaction diagrams*—Interaction diagrams for columns from authoritative sources can be used; however, use the strength reduction factors ϕ as set forth in the guide.

5.12.8 *Biaxial moment strength*—Corner columns and other columns subjected to moments about each principal section axis simultaneously should comply with Eq. (5.12.8).

$$\frac{(M_u)_x}{(\phi M_n)_x} + \frac{(M_u)_y}{(\phi M_n)_y} \leq 1.0 \quad (5.12.8)$$

where $(M_u)_x$ and $(M_u)_y$ are the factored moments that act about axis x and y (Fig. 5.12.8) simultaneously with the factored axial load P_u. $(\phi M_n)_x$ and $(\phi M_n)_y$ correspond to values of the design moment strength determined from Eq. (5.12.6d) or Eq. (5.12.6e) for the factored axial load P_u and for the appropriate direction *x* or *y*.

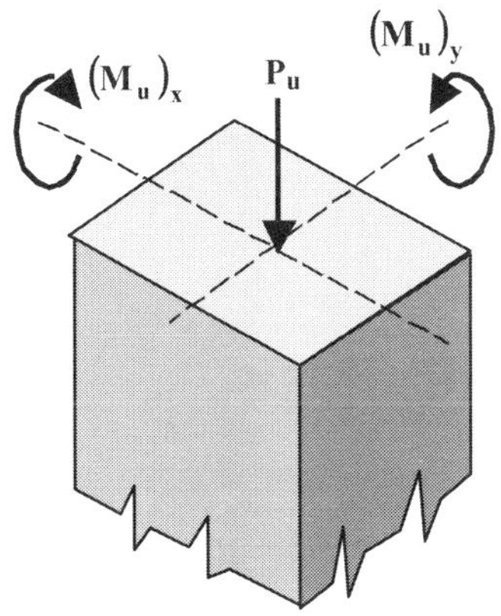

Fig. 5.12.8—Column subjected to biaxial moments.

5.13—Shear

5.13.1 *General*—Calculation of the design shear strength of sections subjected to shear loads should be performed using 5.13. There are two types of shear effects (Fig. 5.13.1):

(a) Beam-action shear that accompanies flexural moments and occurs in girders, beams, joists, solid slabs, and reinforced concrete walls, in the vicinity of supports and concentrated loads; and

(b) Punching shear or two-way action shear, which occurs in solid slabs and footings, near supports and concentrated loads.

Other types of shear effects, such as special effects in deep flexural members, shear-friction used when designing brackets and corbels, and strut-and-tie models, are beyond the scope of this guide and the standards of 1.4 should be used for design in those cases.

5.13.2 *Required shear strength*—The factored shear force V_u (required shear strength) due to the applied factored loads should be determined, for the particular member type, from Chapters 7 to 14.

5.13.3 *Design shear strength*—The design shear strength at a section, ϕV_n, should be equal to or greater than the required shear strength V_u, as shown in Eq. (5.13.3).

$$\phi V_n \geq V_u \quad (5.13.3)$$

where $\phi = 0.75$.

5.13.4 *Design beam-action shear strength*—Section 5.13.4 should be used when designing members for beam-action shear. Sections 5.13.4.1, 5.13.4.2, and 5.13.4.3 apply.

5.13.4.1 *Location of critical section*—Where the support reaction, in the direction of the applied shear, introduces compression into the end regions of the member, and no concentrated load occurs between the support face and a distance *d* from the support face for girders, beams, joists, columns, slabs, and footings, the sections in between may be designed for the factored shear force V_u, computed at distance *d* from the support.

5.13.4.2 *Where shear reinforcement is not permitted*—Where shear reinforcement is not permitted in Chapters 7 to 14, the design shear strength ϕV_n should be computed using Eq. (5.13.4.2).

$$\phi V_n = \phi V_c \quad (5.13.4.2)$$

In Eq. (5.13.4.2), $\phi = 0.75$, and ϕV_c is the concrete contribution to the design shear strength.

5.13.4.3 *Where shear reinforcement is permitted*—Where shear reinforcement is permitted in Chapters 7 to 14, the design shear strength ϕV_n should be computed using Eq. (5.13.4.3) with $\phi = 0.75$.

$$\phi V_n = \phi(V_c + V_s) \quad (5.13.4.3)$$

In Eq. (5.13.4.3), ϕV_c is the concrete contribution to the design shear strength, and ϕV_s is the shear reinforcement contribution to the design shear strength.

5.13.5 *Two-way action (punching) design shear strength in solid slabs and footings*

5.13.5.1 *Location of critical section*—The design shear strength for punching shear, or two-way action shear, should be investigated in slabs at column edges, concentrated loads, supports, thickness changes such as edges of capitals and drop panels, and where columns transfer load to a footing.

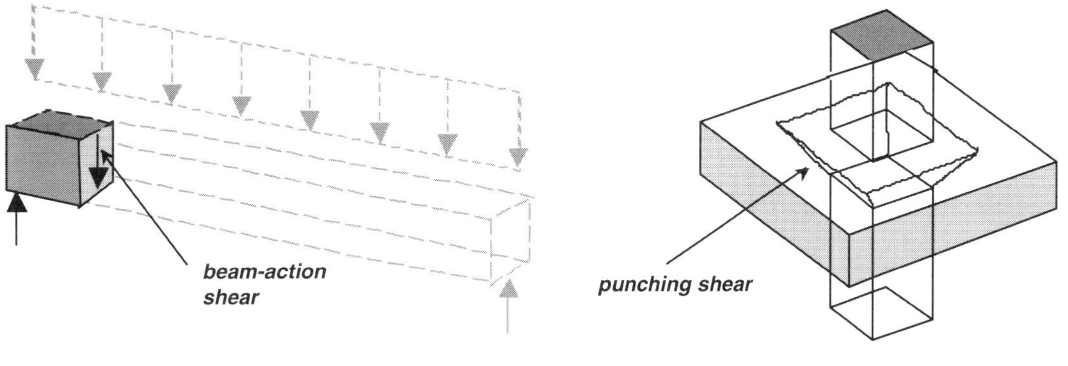

Fig. 5.13.1—Beam-action shear and punching shear.

5.13.5.2 *Two-way action design shear strength*—The design shear strength ϕV_n should be computed using Eq. (5.13.5.2).

$$\phi V_n = \phi V_c \tag{5.13.5.2}$$

where $\phi = 0.75$.

In Eq. (5.13.5.2), ϕV_c is the concrete contribution to the design shear strength. Procedures for design of transverse or shear reinforcement in slabs and footings are beyond the scope of the guide. The alternate standards of 1.4 should be used for design in that case.

5.13.6 *Torsion*—Design for torsion is beyond the scope of the guide, and torsion effects should be neglected when the calculated required torsion strength T_u is less than the value determined from Eq. (5.13.6):

$$T_u \leq \phi\left[\frac{\sqrt{f'_c}}{2}\right]\left[\frac{h^2b^2}{h+b}\right] \quad \left[T_u \leq \phi\left[\frac{\sqrt{f'_c}}{24}\right]\left[\frac{h^2b^2}{h+b}\right] \text{(SI)}\right] \tag{5.13.6}$$

where $\phi = 0.75$.

In members where torsion is smaller than the value given by Eq. (5.13.6), closed stirrups with a minimum bar diameter d_b of 3/8 in. (10 mm) should be provided near the supports. The spacing of these closed stirrups, measured along the member length, should not be greater than the smaller of $b/4$ or $d/4$, for a distance equal to 1/4 of the clear span of the member. At least one longitudinal bar with diameter d_b of 1/2 in. (13 mm) or greater should be located at each internal corner of these stirrups.

5.14—Bearing

The factored compression normal load P_u, applied concentrically on an area A_{cb}, should not exceed the design bearing strength on concrete, ϕP_n, determined using Eq. (5.14).

$$\phi P_n = \phi 0.85 f'_c A_{cb} \tag{5.14}$$

where $\phi = 0.65$.

CHAPTER 6—FLOOR SYSTEMS

6.1—Types of floor systems

6.1.1 *General*—The floor system used by a building designed under the guide should be one of the systems covered in 6.1 or their permitted variations.

The selection of an appropriate floor system should be made after studying several alternatives.

6.1.2 *Slab-on-girder system*

6.1.2.1 *Description*—This system consists of a grid of girders in both plan directions with a solid slab spanning between girders. These girders are located on the column lines or axis, spanning the distance between columns. The solid slab is shallower than the girders, and is supported by them (Fig. 6.1.2.1). The slab can cantilever out over the edge girder. This system should comply with the structural integrity of 6.3.

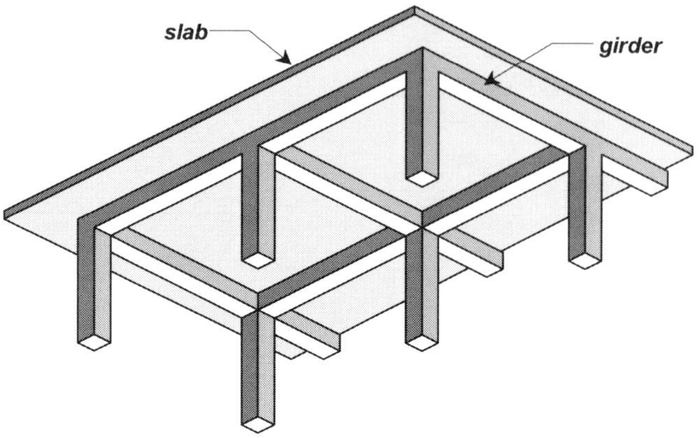

Fig. 6.1.2.1—Slab-on-girder floor system.

6.1.2.1.1 *Use of intermediate beams*—One of the main system variations is the use of intermediate beams supported on the girders. One or several beams may be used per girder span. The intermediate beams may be of the same depth of the girders or shallower. These intermediate beams may be used in one direction, as shown in Fig. 6.1.2.1.1a, or in two directions, as shown in Fig. 6.1.2.1.1b. The use of many intermediate beams will make the system gravitate to the joist system, described in 6.1.3.

6.1.2.1.2 *Advantages of slab-on-girder system*—For the slab-on-girder system, each component has the appropriate minimum depth and width to comply with design strength or serviceability; therefore, having a relatively low self-weight. The system can accommodate spans of any size, can easily be adapted to any plan shape, and large perforations, ducts, and shafts can be located without major problems.

6.1.2.1.3 *Disadvantages of slab-on-girder system*—For the slab-on-girder system, the depth of the girders, beams, and the slab can be large if the minimum number of members is used; however, trying to reduce it causes more members to be used and the formwork becomes more elaborate. A suspended ceiling may be needed for apartment and office occupancies.

6.1.3 *Joist systems*

6.1.3.1 *Description and restrictions*—The joist system consists of a series of parallel ribs or joists supported by girders. The girders are located on the column lines, and span between columns. Joists are usually the same depth as the girders but can be shallower. A solid slab spans between joists (Fig. 6.1.3.1a). This system should comply with the structural integrity of 6.3. The top slab should not cantilever out over the edge joist. The clear separation between parallel joists, measured at the underside of the slab, should not exceed 30 in. (750 mm). Joist web width should not be less than 5 in. (125 mm), measured at the underside of the slab, nor 4 in. (100 mm) measured at its thinnest section. Clear depth of the joist should not exceed 3.5 times its minimum width (Fig. 6.1.3.1b). The top slab should comply with 6.5.2.1.

6.1.3.1.1 *Type of formwork*—When joists have the same depth as the girders, a flat formwork decking supported on shores can be used. Joists shallower than the girders may result in more complex formwork. To create voids, permanent

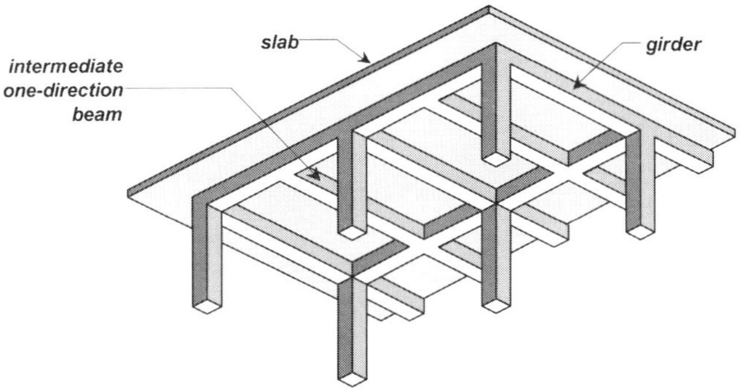

Fig. 6.1.2.1.1a—Use of one-direction intermediate beams in the slab-on-girder floor system.

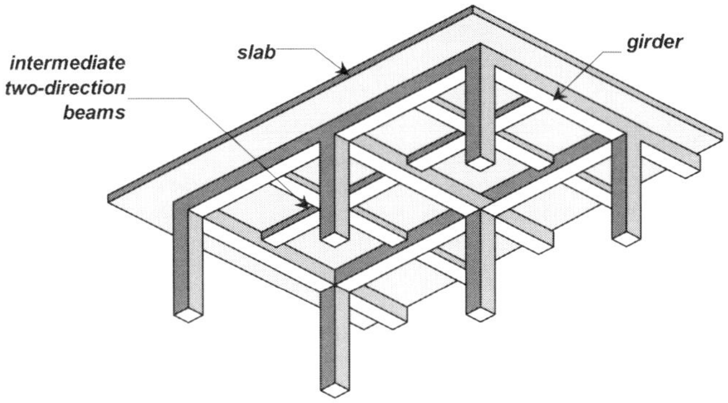

Fig. 6.1.2.1.1b—Use of two-direction intermediate beams in the slab-on-girder floor system.

and removable pans or domes of different shape and material are used. Among those commonly used are permanent and removable wood pans; removable metal, fiberglass, plastic, light polystyrene plastic pans; or permanent cement, cinder, or clay filler blocks.

6.1.3.2 *Distribution ribs*—To improve load distribution and avoid creating a concentrated load on a solitary joist in one-way joist systems, transverse rib spacing should not exceed the smaller of 10 times the total joist depth h and 12 ft (4 m) (Fig. 6.1.3.2).

6.1.3.3 *Two-way joist systems*—For spans approximately equal in both directions, it may be advantageous to use two-way joists. For the system to be classified as a joist system, joists should be supported on girders. This is called a waffle-slab-on-beams system (Fig. 6.1.3.3). When the beams are omitted, the system is called the waffle-slab system, described in 6.1.4.4.

6.1.3.3.1 *Advantages of joist systems*—The joist system can accommodate medium to large spans with relatively low self-weight. It is easy to locate small perforations, ducts, and shafts. For heavy live loads or large permanent loads, the serviceability deflection limits can be readily met because of the relatively large depth of the system. Clear spacing between joists is a tradeoff between a thinner top slab and a larger number of joists, thus allowing the designer great freedom in choosing appropriate dimensions.

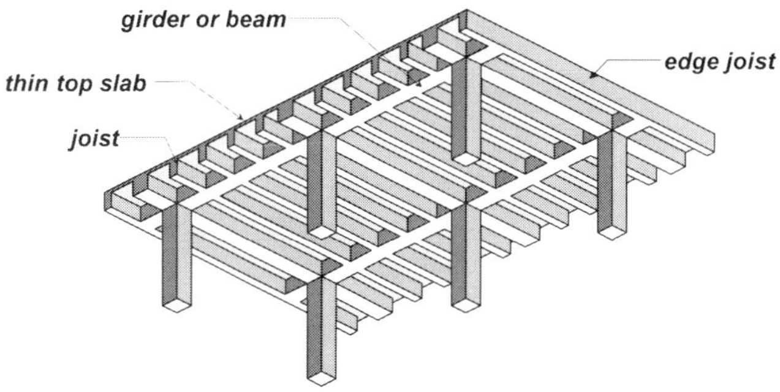

Fig. 6.1.3.1a—Joist floor system.

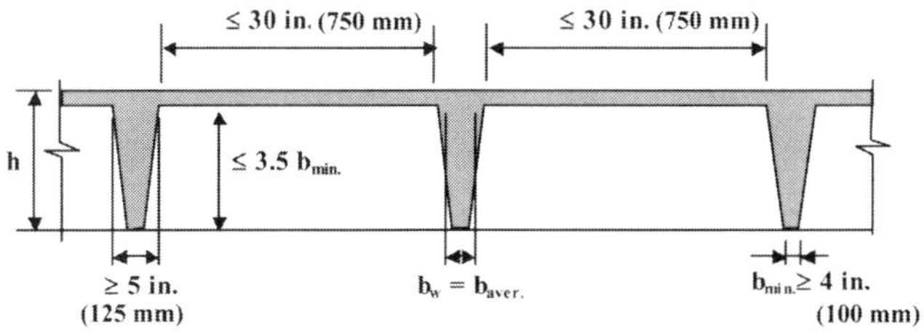

Fig. 6.1.3.1b—Joist dimensional limits.

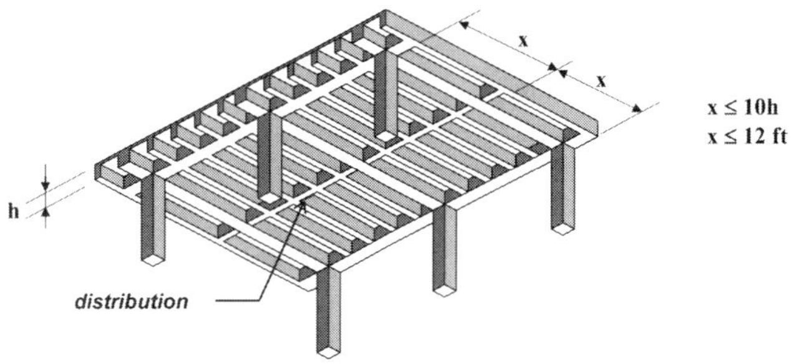

Fig. 6.1.3.2—Distribution ribs.

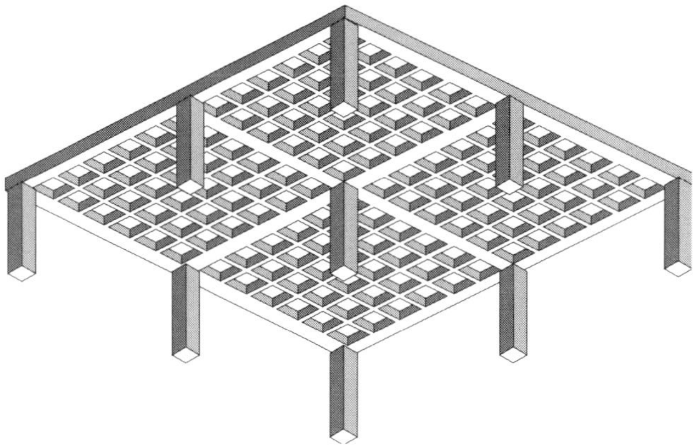

Fig. 6.1.3.3—Two-way joist system or waffle-slab-on-beams system.

6.1.3.3.2 *Disadvantages of joist systems*—The joist system demands more workmanship than other systems. For skew plan layouts, both design and construction procedures are more complex than for other systems. For apartment and office occupancies, the system may need a false ceiling or an

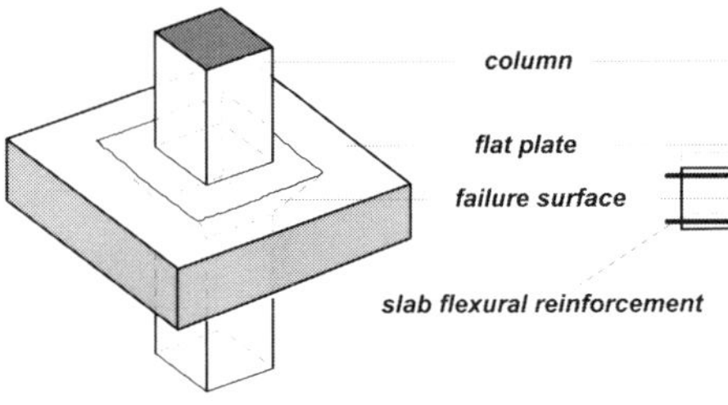

Fig. 6.1.4.1—Punching shear failure.

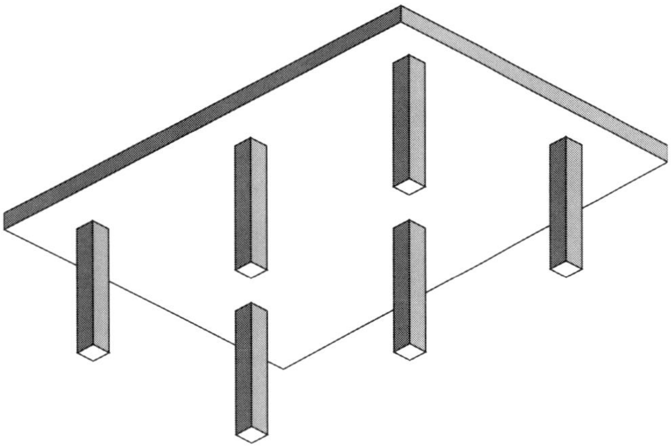

Fig. 6.1.4.2—Flat-plate system.

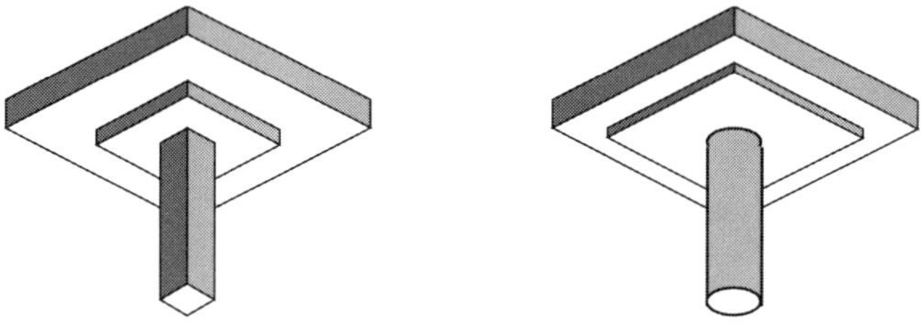

Fig. 6.1.4.3a—Drop panels.

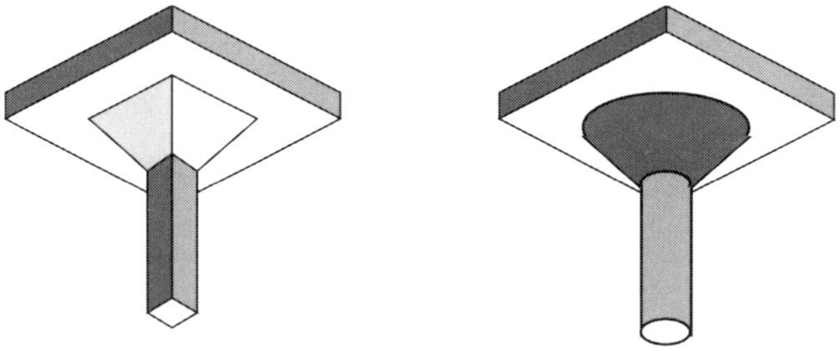

Fig. 6.1.4.3b—Column capital.

underneath concrete soffit, which in turn requires permanent pans or domes. Large perforations, ducts, or shafts interrupt several joists, whose tributary load should be transferred to other joists, thus making design and construction more intricate. If the joist depth is different than the girder depth, the advantage of a flat formwork decking is lost.

6.1.4 *Slab-column systems*

6.1.4.1 *Description*—In the slab-column system, the slab is supported directly by the columns, without beams or girders. The system has several variations listed in 6.1.4.2 to 6.1.4.4. A problem associated with this type of system, known since early development of reinforced concrete, is the punching shear failure of the slab, described in Fig. 6.1.4.1.

6.1.4.1.1 *Advantages of slab-column systems*—Slab-column systems generally provide a shallower depth. The system provides flexibility in column location, because some deviation from the horizontal axis or line of columns is allowed. In general, these systems do not need a false ceiling. Having a flat formwork decking simplifies construction. Reinforcement placement is easier because it does not require stirrups (except in waffle-slabs), thus allowing more efficient construction. Perforations, ducts, and shafts can be located in the central part of the slab panel.

6.1.4.1.2 *Disadvantages of slab-column systems*—The main disadvantage of these systems is vulnerability to punching shear failure. To account for this failure mode, slab depth may have to be increased, thus making the slab heavier. Using drop panels and column capitals help increase punching-shear resistance but complicate design and construction. Spans are shorter than for other systems, due to greater self-weight and less depth. As the span is increased, significant long-term deflections caused by the large permanent load are possible. These deflections can adversely affect walls and partitions, especially if they are masonry walls. Perforations, ducts, and shafts should not be located near columns because these openings reduce slab punching shear strength. The seismic performance of slab-column systems, when not laterally stiffened by reinforced concrete walls have, at times, not performed as well as other structural systems during seismic events.

6.1.4.2 *Flat plate*—A slab of uniform thickness supported by columns is called a flat plate (Fig. 6.1.4.2).

6.1.4.3 *Flat slabs*—To increase resistance to punching shear (Fig. 6.1.4.1) and to increase the overall flexural strength, the slab can be thickened around the columns. This system is called a flat-slab system.

The thicker, rectangular slab around the column is called a drop panel, as shown in Fig. 6.1.4.3a. When a drop panel is used, the punching shear strength at the column and at the edge of the drop panel needs to be checked.

Another option is to increase the contact area between the column and the slab by forming a column capital, as shown in Fig. 6.1.4.3b.

In some instances, both a capital and a drop panel are combined, as shown in Fig. 6.1.4.3c. Here, two potential zones of punching shear failure exist and both should be checked in design.

Drop panels should project below the slab at least one-fourth of the slab depth beyond the drop and should extend in each direction from the column centerline a distance not less than one-sixth the span length measured from center-to-center of supports in that direction (Fig. 6.1.4.3d).

6.1.4.4 *Waffle-slabs*—For longer spans, voids are formed in the bottom of a flat-plate, away from the columns. This system is called a waffle-slab, as shown in Fig. 6.1.4.4. The rib dimensions (joists) in waffle-slabs should comply with the minimum dimensions for joists given in 6.1.3.1.

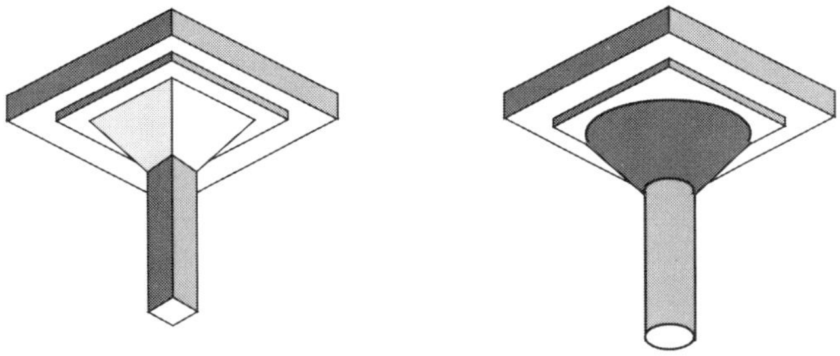

Fig. 6.1.4.3c—Column capital and drop panel.

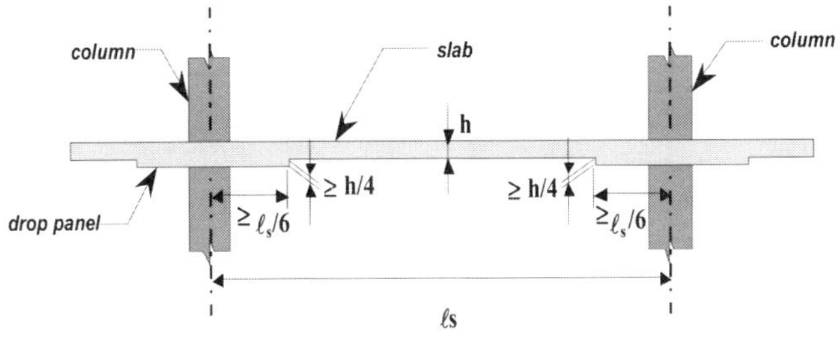

Fig. 6.1.4.3d—Minimum dimensions for drop panels.

This system is very similar to two-way joists. The main difference is that for two-way joists, beams are located in the column lines, whereas for waffle slabs, all members are joists and the voids surrounding the column are filled, thus forming a column capital. The solid capital should engage at least three joists in each direction for interior columns, and at least two joists parallel to the edge in edge and corner columns. Domes or pans for waffle-slabs can be permanent, made of concrete block, or removable when made of wood, fiberglass, or plastic, as described for the joist system in 6.1.3.1.

6.2—Selection of floor system

The licensed design professional should select a floor system from the systems covered by the guide as presented in 6.1. Several alternatives should be studied, and the final selection should consider:

(a) Magnitude of dead and live loads, including self-weight;

(b) Geometry of the structural plan layout, including span lengths in both plan directions and the ratio between them;

(c) Presence of cantilevers and their maximum span and direction;

(d) Occupancy type;

(e) Available concrete and reinforcing steel material strengths;

(f) Expected behavior of the slab system and the ability to comply with the serviceability and deflection criteria;

(g) Amount of materials for concrete, steel, and formwork that are required to build the floor system, which contains most of the materials needed to construct the building;

(h) Local preferences to simplify construction coordination;

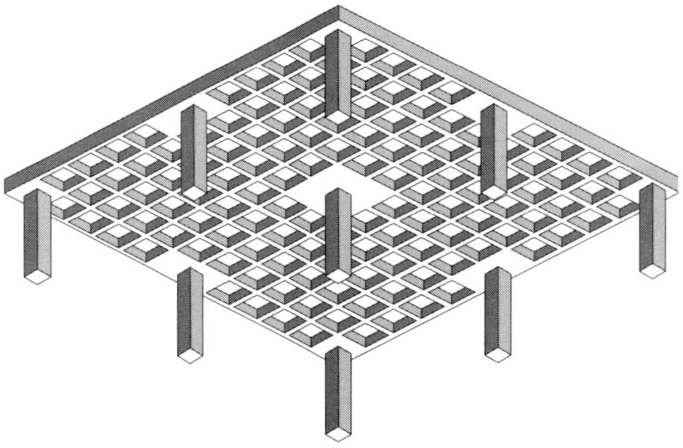

Fig. 6.1.4.4—Waffle-slab.

(i) Worker training and proficiency, as some systems require more training and proficiency than what local workers can comply with; and

(j) Relative cost of alternatives, with economic advantages evaluated only after providing adequate expected behavior and safety of the system.

6.3—Structural integrity

6.3.1 *General*—Sections 6.3.2 through 6.3.5 constitute the detailing concepts for structural integrity.

Structural integrity should improve redundancy and ductility of the structure as a whole for it to function in the event of damage to a major supporting member or an abnormal loading event by confining the damage to a relatively small area and maintaining overall stability.

6.3.2 *Perimeter beams in slab-and-girder and joist systems*—Beams should link all perimeter columns and reinforced concrete walls of the structure. These perimeter beams or girders should have a minimum area of continuous top and bottom longitudinal reinforcement and closed stirrups, as indicated in 8.5.2.1. This reinforcement should be lap spliced using the lap-splice length of 5.8.2.

6.3.3 *Nonperimeter beams and girders*—All beams and girders not covered in 6.3.2 should have closed stirrups and a minimum area of continuous bottom longitudinal reinforcement as indicated in 8.4 and 8.5. This longitudinal reinforcement should be lap spliced at or close to the supports using the lap splice length of 5.8.2.

6.3.4 *Joists*—In joists, at least one bottom bar should be continuous or spliced over the support using the lap-splice length of 5.8.2. At noncontinuous supports, at least one bottom bar should be terminated with a standard hook (8.4).

6.3.5 *Slab-column systems*—In slab-column systems, minimum continuous structural integrity reinforcement should be provided as indicated by Chapter 9.

6.4—One-way and two-way load paths

6.4.1 *General*—Load path of the floor system should be classified either as one-way or two-way. The actual load path, from the application point to the supports, depends on plan dimensions and stiffness of the supporting members.

Table 6.5.2.2— Minimum depth *h* for one-way solid slabs supporting deflection-insensitive nonstructural elements

Continuity across the supports	Minimum depth *h*
Simply supported	$\ell_s/20$
One end continuous	$\ell_s/24$
Both ends continuous	$\ell_s/28$
Cantilever	$\ell_s/10$

6.4.2 *One-way*—A slab should be considered one-way when it has:

(a) Two opposing free edges without vertical support, and has vertical support, such as beams or walls, in the other two opposing edges;

(b) A rectangular plan shape, has vertical support at all edges, and the longer slab span is greater than twice the shorter slab span; or

(c) Joists, except the distribution ribs, in only one direction.

6.4.3 *Two-way*—A slab should be considered two-way when:

(a) The slab panel has a rectangular plan shape and has vertical support, such as beams or walls, at all edges and the longer slab span is less than or equal to twice the shorter slab span; or

(b) In slab-column systems, the panel has a rectangular plan shape and the long slab span is less than or equal to twice the short slab span.

6.4.4 *Floor system load path*—An approximate load path (one-way or two-way) should be identified and used to assign tributary load to all slab-supporting members and determine preliminary dimensions of the slab and the supporting members. The load path and the tributary load should be verified as members are dimensioned and designed, and corrected as needed.

6.5—Minimum depth for floor system members

6.5.1 *General*—Minimum depth for floor system members given in 6.5.2 to 6.5.5 should be considered to meet the serviceability limit state, providing sufficient stiffness to avoid excessive member deflections caused by dead and live loads.

6.5.2 *Solid one-way slabs*

6.5.2.1 *Slab between joists*—Slabs should have a minimum depth of $\ell_s/12$, but should not be less than 1-1/2 in. (40 mm) when permanent concrete or clay filler blocks are used as forms, nor less than 2 in. (50 mm) in all other cases.

6.5.2.2 *Deflection-insensitive nonstructural elements*—Where the slab supports nonstructural elements built of materials that will not be damaged by large deflections, slab depth *h* should not be less than the values in Table 6.5.2.2, where the span length ℓ_s is the distance center-to-center of supports. Where the clear span is less than 10 ft (3m), ℓ_s is the clear span.

6.5.2.3 *Deflection-sensitive nonstructural elements*—Where the slab supports nonstructural elements built of materials that will be damaged by large deflections, slab depth *h* should not be less than the values in Table 6.5.2.3, where ℓ_s should be taken as the distance center-to-center of supports. Where the clear span is less than 10 ft (3 m), ℓ_s is the clear span.

Table 6.5.2.3—Minimum depth *h* for one-way solid slabs supporting deflection-sensitive nonstructural elements

Continuity across the supports	Minimum depth *h*
Simply supported	$\ell_s/14$
One end continuous	$\ell_s/16$
Both ends continuous	$\ell_s/19$
Cantilever	$\ell_s/7$

Table 6.5.3.1—Minimum depth *h* for girders, beams, and one-way joists supporting deflection-insensitive nonstructural elements

Continuity across the supports	Minimum depth *h*
Simply supported	$\ell_s/16$
One end continuous	$\ell_s/18.5$
Both ends continuous	$\ell_s/21$
Cantilever	$\ell_s/8$

6.5.3 *Girders, beams, and one-way joists*

6.5.3.1 *Deflection-insensitive nonstructural elements*—Where a girder, beam, or one-way joist supports nonstructural elements built of materials that will not be damaged by large deflections, *h* should not be less than the values in Table 6.5.3.1, where ℓ_s should be taken as the distance center-to-center of supports. For joists where the clear span is less than 10 ft (3 m), ℓ_s is the clear span.

6.5.3.2 *Deflection-sensitive nonstructural elements*—Where the girder, beam, or one-way joist supports nonstructural elements built of materials that will be damaged by large deflections, *h* should not be less than the values in Table 6.5.3.2, where ℓ_s should be taken as the distance center-to-center of supports. For joists where the clear span is less than 10 ft (3 m), ℓ_s is the clear span.

6.5.4 *Two-way slabs supported at all edges*—The minimum depth of two-way slabs, including two-way joist and waffle-on-beams systems, supported by girders, beams, or reinforced concrete walls on all panel edges, is given in Eq. (6.5.4). The supporting girders or beams should have a depth not less than three times the slab depth (7.9.1). Depth of solid slabs should be not less than 5 in. (125 mm) for ℓ_n greater than 10 ft (3 m), and should not be less than 4 in. (100 mm) for ℓ_n less than or equal to 10 ft (3 m).

$$h = \frac{\ell_n}{30 + 3\beta} \quad (6.5.4)$$

where ℓ_n is the clear span in the long direction, measured face-to-face of the supporting beams, and β is the ratio of longer clear span to shorter clear span of the slab panel.

6.5.5 *Slab-column systems*—For slabs without interior beams spanning between supports, including waffle-slabs, and having a ratio of longer to shorter span not greater than 2:

Table 6.5.3.2—Minimum depth *h* for girders, beams, and one-way joists supporting deflection-sensitive nonstructural elements

Continuity across the supports	Minimum depth h
Simply supported	$\ell_s/11$
One end continuous	$\ell_s/12$
Both ends continuous	$\ell_s/14$
Cantilever	$\ell_s/5$

(a) For slabs supporting nonstructural elements built of materials that will not be damaged by large deflections, minimum depth is given by Table 6.5.5a;

(b) For slabs supporting nonstructural elements built of materials that will be damaged by large deflections, minimum depth is given by Table 6.5.5b; and

(c) Minimum slab depth should not be less than the values given in (1) and (2) for solid slabs.

(1) Slabs without drop panels—6 in. (150 mm)

(2) Slabs with drop panels, where drop panel dimensions conform to 6.1.4.3—5 in. (125 mm);

Table 6.5.5a—Minimum depth of slab-column systems supporting deflection-insensitive nonstructural elements

Type of slab-column system	Location of panel	Minimum depth of slab, h
Without drop panels	Exterior	$\ell_n/30$
	Interior	$\ell_n/33$
With drop panels	Exterior	$\ell_n/33$
	Interior	$\ell_n/36$

Note: ℓ_n corresponds to the longer span.

Table 6.5.5b—Minimum depth of slab-column systems supporting deflection-sensitive nonstructural elements

Type of slab-column system	Location of panel	Minimum depth of slab, h
Without drop panels	Exterior	$\ell_n/22.5$
	Interior	$\ell_n/25$
With drop panels	Exterior	$\ell_n/25$
	Interior	$\ell_n/27$

Note: ℓ_n corresponds to the longer span.

6.6—Trial dimensions for floor system

Initial trial dimensions for floor system members should be defined using the minimum depth as indicated in Chapter 6. Trial dimensions should be assigned using h given in 6.5. Trial width b_w for a beam is one-half of h but not less than 8 in. (200 mm), and b_w for joists are the minimum width dimensions given in 6.1.3.1. These trial dimensions meet the serviceability limit state, and as the design proceeds, they should be modified as needed by the strength limit state. The self-weight calculated using the trial dimension should be corrected because modifications to the dimensions occur during the design process.

6.7—Floor finish

A floor finish should not be included as part of a structural member unless placed monolithically with the floor slab. It is permitted to consider all concrete and mortar floor finishes as part of the cover.

6.8—Ducts, shafts, openings, and embedded piping

6.8.1 *Ducts, shafts, and openings in slab systems*

6.8.1.1 *Slab-on-girder system*—In slab-on-girder systems, minor openings should not interrupt girders or beams. The total area of reinforcement for the slab without an opening should be maintained. Openings with plan dimensions greater than $\ell_n/4$ need beams on all edges.

6.8.1.2 *Joist construction*—In joist construction, minor openings should be located between joists. When an opening interrupts one or two joists, it should be surrounded by joists and distribution ribs. The joists at the edge of an opening should be continuous and proportioned to resist double the design vertical load. Distribution ribs at the edge of the opening should extend to a beam or girder on both sides of the opening. When more than two joists are interrupted, the opening should have beams on all sides.

6.8.1.3 *Openings in slab-column systems*—In Chapter 9, slab openings are permitted in some areas (9.3.9), provided the punching shear strength is not diminished and the total area of reinforcement needed for the panel without the opening is maintained.

6.8.2 *Embedded conduits and pipes*

6.8.2.1 *General*—Conduits and pipes of aluminum should not be embedded in structural members. Conduits and pipes of any material should not be embedded within a column.

6.8.2.2 *Conduits and pipes passing through girders, beams, and joists*—A conduit or pipe passing through a girder, beam, or joist should not be larger in outside diameter than $h/3$ when passing horizontally, nor greater than $b_w/3$ when passing vertically. Conduits or pipes should be located in plan no closer to the support face than $\ell_s/4$, or farther away than $\ell_s/3$. Conduits and pipes passing horizontally through the member should be located in the middle third of the member height. Conduits and pipes passing vertically through the member should be located in the middle third of the member width. Pipes and conduits should be spaced horizontally at least three diameters center-to-center (Fig. 6.8.2.2). Reinforcing bars should not be allowed to be cut or damaged due to pipe penetrations at any location unless approved by the licensed design professional.

6.8.2.3 *Conduits and pipes located longitudinally within girders, beams, and joists*—Conduits and pipes embedded longitudinally within girders, beams, or joists should not be larger in outside diameter than $b_w/3$ and should be located vertically within the middle third of the member depth. Conduits or pipes should not be spaced closer than three diameters center-to-center.

6.8.2.4 *Conduits and pipes embedded in slabs*—Conduits and pipes embedded within solid slabs should be placed between top and bottom reinforcement. Their outside diameter should not be greater than 2 in. (50 mm) or 25 percent of

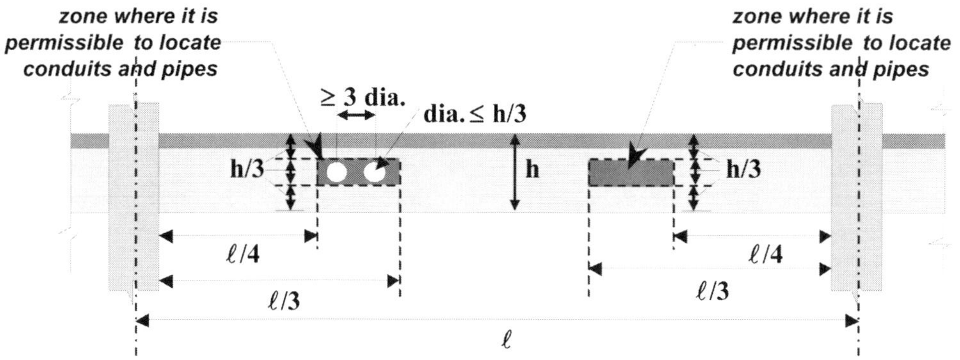

Fig. 6.8.2.2—Location of conduits and pipes passing horizontally through girders, beams, and joists.

the slab thickness and should be spaced no less than three diameters center-to-center.

6.8.2.5 *Pipes embedded in the top slab in joists*—Where conduits or pipes are embedded within the top slab in joists, depth of the top slab should be at least 1 in. (25 mm) greater than the total overall height of the conduits or pipes, and the concrete cover at any point should not be less than 1/2 in. (13 mm).

CHAPTER 7—SOLID SLABS SUPPORTED ON GIRDERS, BEAMS, JOISTS, OR REINFORCED CONCRETE WALLS

7.1—General

Design of one-way and two-way solid slabs supported at their edges by girders, beams, or reinforced concrete walls should be in accordance with Chapter 7. The top solid slabs that span between joists are also included.

7.2—Loads

7.2.1 *Loads to be included*—Design loads for solid slabs supported on girders, beams, joists, or reinforced concrete walls should be in accordance with Chapter 4. Gravity loads that should be included in the design are:

(a) Dead loads: Member self-weight, flat nonstructural elements, standing nonstructural elements, and any fixed equipment loads;

(b) Live loads; and

(c) Roof live load, rain load, and snow load should be used if the slab is part of the roof system.

7.2.2 *Dead load and live load*—The value of q_d for dead load should include slab self-weight and the weight of the flat and standing nonstructural elements as defined in 4.5.3. The value of q_ℓ for live load should be determined by 4.6. Where the slab is part of the roof system, roof live load given in 4.7, rain load in 4.8, and snow loads in 4.9 should be included, as appropriate.

7.2.3 *Factored design load*—The value of the factored design load q_u should be the greater value determined combining q_d and q_ℓ using the load factors and load combinations given in 4.2.

7.3—Reinforcement details

7.3.1 *General*—Reinforcement of slabs-on-girders should comply with 7.3.2 to 7.3.10.

7.3.2 *Minimum clear spacing between parallel bars in a layer*—In slabs, the minimum clear spacing between parallel bars in a layer should be the largest nominal bar diameter d_b, but not less than 1 in. (25 mm). Clear distance limitation between bars should apply also to the clear distance between a contact lap splice and adjacent splices or bars.

7.3.3 *Shrinkage and temperature reinforcement*

7.3.3.1 *Description*—Reinforcement for shrinkage and temperature effects normal to flexural reinforcement of the slab should be provided in slabs-on-girders where flexural reinforcement extends in one direction only.

7.3.3.2 *Location*—For floor slabs, shrinkage and temperature reinforcement should be placed on top of the positive flexural reinforcement perpendicular to it. For roof slabs, it should be placed under the negative moment reinforcement perpendicular to it to limit the size and spacing of cracking and thus reduce possible water infiltration.

7.3.3.3 *Maximum spacing*—In solid slabs, shrinkage and temperature reinforcement should be spaced no farther apart than four times the slab thickness, nor 14 in. (350 mm). (Refer to Fig. 7.3.3.3).

7.3.3.4 *Minimum area*—The minimum ratio of shrinkage and temperature reinforcement area to gross concrete area, $\rho_t = A_{st}/(bh)$, should be 0.0020.

7.3.3.5 *Splicing*—It is permitted to lap-splice shrinkage and temperature reinforcement at any location. Splice length should comply with 5.8.2.

7.3.3.6 *End anchorage*—At slab edges, shrinkage and temperature reinforcement should end in a standard hook. Where constrained by slab geometry, the end hook need not be placed vertically.

7.3.4 *Flexural reinforcement: General*

7.3.4.1 *Maximum spacing of flexural reinforcement in solid slabs*—In solid slabs, primary flexural reinforcement (Fig. 7.3.4.1) should be spaced no farther apart than three times the slab thickness, nor 12 in. (300 mm).

7.3.4.2 *Minimum area of tension flexural reinforcement*—The minimum area of tension flexural reinforcement in solid

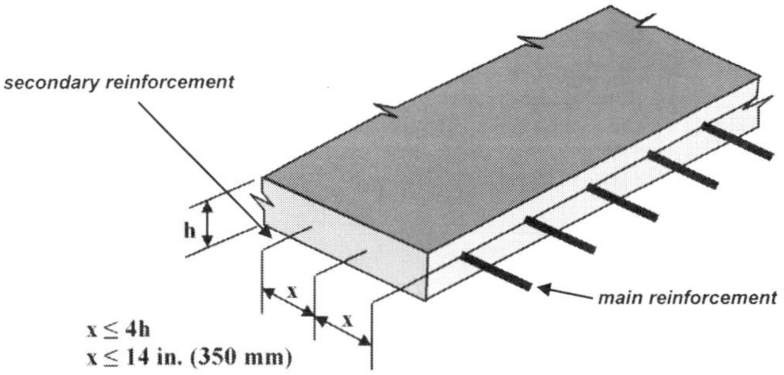

Fig. 7.3.3.3—Maximum spacing between shrinkage and temperature reinforcement in slabs.

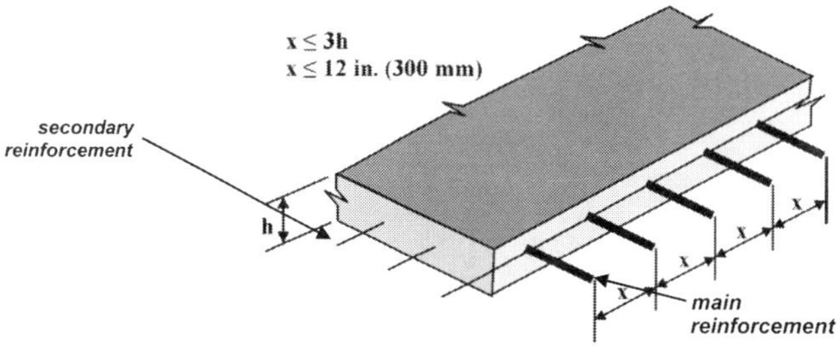

Fig. 7.3.4.1—Maximum spacing between flexural reinforcement in solid slabs.

slabs should be the area needed for shrinkage and temperature effects, $A_s \geq \rho_t bh$ (7.3.3.4 and Fig. 7.3.4.2).

7.3.4.3 *Maximum area of tension flexural reinforcement*—The tension flexural reinforcement ratio, $\rho = A_s/(bd)$, in solid slabs should not exceed ρ_{max} in Table 7.3.4.3. In solid slabs, compression flexural reinforcement should not be taken into account when computing design moment strength.

7.3.4.4 *Design moment strength*—For solid slabs, the design moment strength of 5.11.4 can be approximated using Eq. (7.3.4.4), with $\phi = 0.90$:

$$\phi M_n = \phi 0.85 A_s f_y d \tag{7.3.4.4}$$

7.3.4.5 *Obtaining flexural reinforcement ratio*—Flexural tension reinforcement ratio, $\rho = A_s/(bd)$, should be determined from Eq. (7.3.4.5):

$$\rho = \frac{A_s}{bd} = \frac{M_u}{\phi 0.85 f_y bd^2} \tag{7.3.4.5}$$

When ρ is less than ρ_{min}, as given in 7.3.4.2, increase A_s. When ρ is greater than ρ_{max}, as given in 7.3.4.3, increase slab thickness and correct the self-weight.

7.3.5 *Positive moment reinforcement*

7.3.5.1 *Description*—Positive moment reinforcement, as indicated in Chapter 7, should comply with 7.3.4, 7.3.5, and 7.5 to 7.9 for each slab type. The design moment strength of a section based on the provided area of positive moment reinforcement should be equal to or greater than the required flexural strength.

7.3.5.2 *Location*—Positive moment reinforcement should be provided parallel to the shorter span in one-way slabs and in both directions in two-way-slabs. Positive moment reinforcement should be located as close to the bottom surface of concrete as cover permits (5.4.1). In two-way systems, the shorter span positive moment reinforcement should be located below the longer span positive moment reinforcement because the positive moment in the shorter span is larger.

7.3.5.3 *Cutoff points*—No more than one-half of the positive moment reinforcement needed to obtain the required flexural

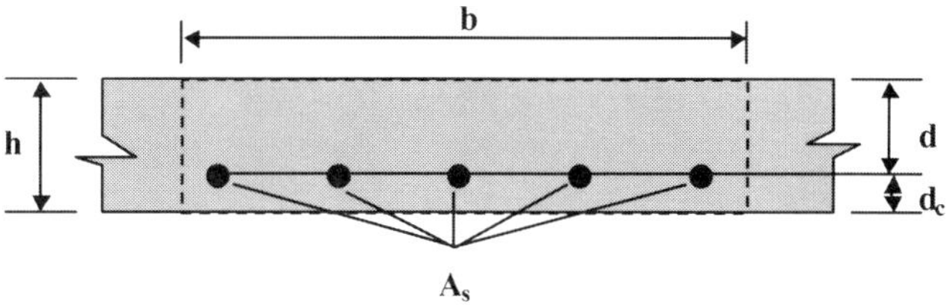

Fig. 7.3.4.2—Slab section.

Table 7.3.4.3—Maximum flexural reinforcement ratio ρ_{max} for solid slabs

		f_y, psi (MPa)	
		40,000 (280)	60,000 (420)
f_c', psi (MPa)	3000 (21)	0.0190	0.0100
	3500 (25)	0.0220	0.0125
	4000 (28)	0.0250	0.0140
	4500 (32)	0.0270	0.0160
	5000 (36)	0.0290	0.0170

Note: It should be permitted to interpolate for different values of f_y and f_c'.

strength at midspan may be cut off at the locations indicated in 7.7 to 7.9 for each slab type.

7.3.5.4 *Reinforcement splicing*—The remaining positive moment reinforcement may be lap spliced between the cutoff point and the opposite face of the support.

7.3.5.5 *Embedment at interior supports*—Positive moment reinforcement cut off at an interior support should be extended to the opposite face of the support.

7.3.5.6 *End anchorage of reinforcement*—Positive moment reinforcement perpendicular to a discontinuous edge should extend to the slab edge and should end with a standard hook.

7.3.6 *Negative moment reinforcement*

7.3.6.1 *Description*—Negative moment reinforcement should be provided in the amounts and lengths indicated in Chapter 7, and should comply with 7.3.4, 7.3.6, and 7.5 to 7.9 for each slab type. The design moment strength of a section based on the provided area of negative moment reinforcement should be equal to or greater than the required flexural strength.

7.3.6.2 *Location*—Negative moment reinforcement should be provided perpendicular to supporting girders, beams, and reinforced concrete walls. Negative moment reinforcement should be located as close to the top surface of concrete as

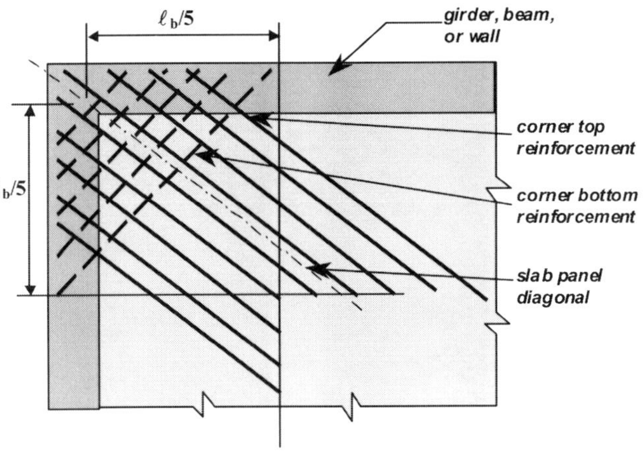

Fig. 7.3.8—Slab corner reinforcement.

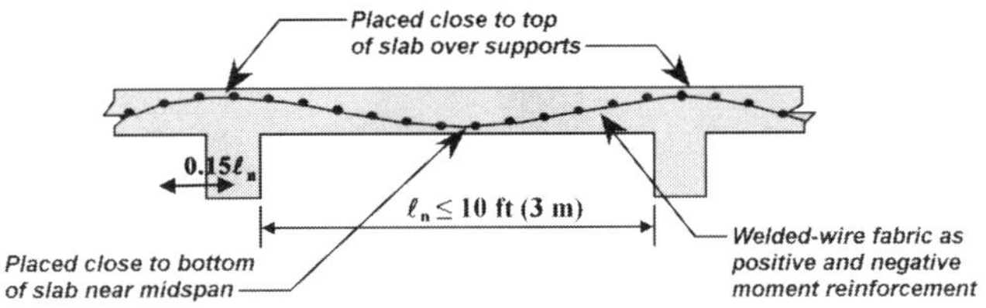

Fig. 7.3.9—Welded wire reinforcement in short one-way spans.

cover permits (5.4.1). In two-way systems, the shorter span negative moment reinforcement should be located above the longer span negative moment reinforcement because the negative moment in the shorter span is larger.

7.3.6.3 *Cutoff points*—Except for cantilevers, all negative moment reinforcement may be cut off at the locations indicated in 7.5 to 7.9 for each slab type. Where adjacent spans are unequal, negative moment reinforcement cutoff points should be based on the longer span.

7.3.6.4 *Reinforcement splicing*—Negative moment reinforcement should not be lap spliced between the cutoff point and the support.

7.3.6.5 *End anchorage of reinforcement*—Negative moment reinforcement perpendicular to a discontinuous edge should end in a standard hook at the edge, complying with 5.8.3. At the external edge of cantilevers, negative moment reinforcement perpendicular to the edge should end in a standard hook. Where constrained by slab geometry, the end hook need not be placed vertically.

7.3.7 *Shear reinforcement*—Procedures for design of shear or transverse reinforcement in solid slabs are beyond the scope of this guide, and the standards of 1.4 should be used if slab shear reinforcement is needed.

7.3.8 *Corner reinforcement*—Top and bottom corner reinforcement, in addition to other reinforcement, should be provided at exterior supported slab corners (for cantilevers, refer to 7.6) for a distance equal to one-fifth of the longer clear span of the slab panel (Fig. 7.3.8). The area of top and bottom corner reinforcement should be sufficient to resist a moment equal to the required positive flexural strength, per unit of width, in the slab panel, in accordance with 7.3.8.1 and 7.3.8.2.

7.3.8.1 *Top corner reinforcement*—Corner reinforcement parallel to the diagonal of the panel should be placed at the slab top, directly below the flexural reinforcement. Corner reinforcement should end with a standard hook at the support.

7.3.8.2 *Bottom corner reinforcement*—Corner reinforcement perpendicular to the diagonal of the panel should be placed at the slab bottom, directly above the flexural reinforcement. Corner reinforcement should end with a standard hook at the edge support.

7.3.9 *Welded wire reinforcement used in short one-way span slabs*—In one-way slabs with a clear span not exceeding 10 ft (3 m), welded wire reinforcement may act as both negative and positive moment reinforcement. It should be continuous, located near the slab top over the support and near the slab bottom at midspan, provided it either continues over or is anchored at the supports. The longitudinal reinforcement area should be adequate to resist the required positive and negative flexural strength (Fig. 7.3.9). The wire area in the perpendicular direction should comply with the area needed for shrinkage and temperature reinforcement.

7.3.10 *Value of* d_c *and* d *in solid slabs*—The distance from extreme tension fiber to centroid of tension reinforcement, d_c, should consider the concrete cover in 5.4, bar diameter, and reinforcement in the perpendicular direction placed between the reinforcement under study and the concrete surface.

The following values of d_c can be used to compute d as $d = h - d_c$. For one-way slabs and for reinforcement in the short direction in two-way slabs, $d_c = 1.6$ in. (40 mm) for interior exposure, and $d_c = 2.4$ in. (60 mm) for exterior exposure. For reinforcement in the longer direction of two-way slabs, $d_c = 2.2$ in. (55 mm) for interior exposure, and $d_c = 3$ in. (75 mm) for exterior exposure.

7.4—Shear strength

The design shear strength of solid slabs should be determined in accordance with 5.13 and 7.4.

7.4.1 *Required shear strength*—The factored shear V_u (required shear strength) caused by factored loads applied to the slab should be determined at the locations indicated in 7.5 to 7.9 for each slab type.

7.4.2 *Design shear strength*—The design shear strength ϕV_n for solid slabs should be based only on the concrete shear strength and should be equal to or greater than the required shear strength V_u, as shown in Eq. (7.4.2a) with $\phi = 0.75$ (Fig. 7.4.2).

$$\phi V_n = \phi V_c \geq V_u \quad (7.4.2a)$$

$$\phi V_c = \phi 2\sqrt{f_c'}bd \quad \left[\phi V_c = \phi 0.17\sqrt{f_c'}bd \text{ (SI)}\right] \quad (7.4.2b)$$

7.5—Slab between joists

7.5.1 *Dimensional limits*—A solid slab spanning between joists should comply with the minimum depth of 6.5.2.1.

Ducts, shafts, and slab openings should comply with 6.8. A top solid slab in joist construction should not cantilever past the edge joist (6.1.3.1).

7.5.2 *Required moment strength*—The factored moment M_u (required moment strength) for negative and positive moment in the slab between joists should be calculated using Eq. (7.5.2), where ℓ_n is the clear spacing between joists (Fig. 7.5.3).

$$M_u^+ = M_u^- = \frac{q_u \ell_n^2}{12} \tag{7.5.2}$$

7.5.3 *Reinforcement*—Flexural reinforcement ratio ρ perpendicular to the joist direction should be determined using M_u from Eq. (7.5.2) and using d as one-half the thickness of the top solid slab, and should not be less than ρ_{min} (Fig. 7.5.3). The slab reinforcement area parallel to the joist direction should meet that needed for shrinkage and temperature reinforcement.

7.5.4 *Shear strength*—The factored shear V_u (required shear strength) per unit slab width should be calculated using Eq. (7.5.4), where ℓ_n is the clear spacing between joists (Fig. 7.5.3).

$$V_u = \frac{q_u \ell_n}{2} \tag{7.5.4}$$

The value of d for computing design shear strength per unit slab width, ϕV_n, should be one-half the slab thickness.

7.5.5 *Calculation of slab reactions*—Slab reactions on the supporting joists, r_u, should be the values determined from Eq. (7.5.5), where V_u is the factored shear per unit slab width from 7.5.4, ℓ_s is the center-to-center joist spacing, and ℓ_n is the clear spacing between joists.

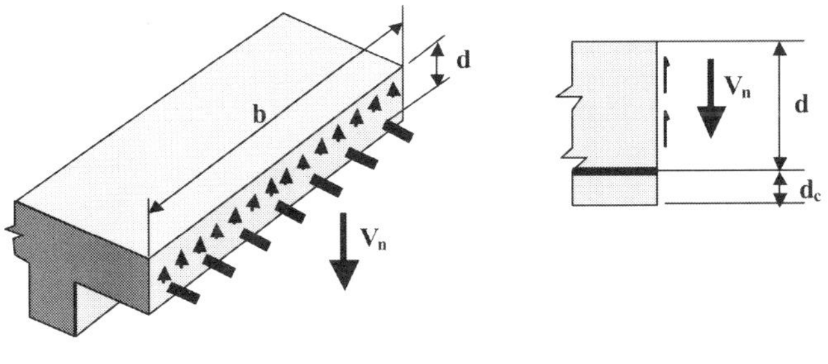

Fig. 7.4.2—Contribution of concrete to beam-action shear strength in concrete slabs.

$$r_u = \frac{2V_u \ell_s}{\ell_n} \tag{7.5.5}$$

7.6—Cantilevers of slabs supported on girders, beams, or walls

7.6.1 *Dimensional limits*—A cantilever slab spanning beyond the edge of girder, beam, or reinforced concrete wall should comply with the minimum depth of 6.5.2. The cantilever span should not exceed the limit given in 1.3. No slab openings for ducts or shafts should be located between the support and one-half the cantilever span. The slab may cantilever in two directions at corners, with the same limitations in each direction that apply for single cantilevers.

7.6.2 *Necessary moment strength*—The factored moment M_u (required moment strength) for a cantilever slab should be calculated assuming that one-half of the distributed factored load, q_u, acts as a concentrated load at the cantilever tip, and the other half acts as a uniformly distributed load over the full span, using Eq. (7.6.2) where ℓ_n is the clear span of the cantilever. Required negative moment strength for the cantilever should not be less than the strength required at the exterior support of the first interior span, or one-third of the required positive moment strength of the first interior span, in the same direction (Fig. 7.6.2).

$$M_u^- = \frac{3q_u \ell_n^2}{4} \tag{7.6.2}$$

7.6.3 *Reinforcement*

7.6.3.1 *Negative moment reinforcement*—Negative moment reinforcement ratio ρ in the direction of the cantilever should be based on M_u from Eq. (7.6.2). This reinforcement should extend beyond the first support not less than the distance indicated for negative moment reinforcement of the interior slab panel at the edge support, and not less than ℓ_d for the reinforcing bar used (Fig. 7.6.3.1).

7.6.3.2 *Positive moment reinforcement*—Positive moment reinforcement should be at least $A_{s,min}$. It is provided in the direction of the cantilever to reduce time-dependent deflections and to satisfy required moment strength (Fig. 7.6.3.1).

7.6.3.3 *Shrinkage and temperature reinforcement*—The reinforcement area parallel to the cantilever edge should not be less than that needed for shrinkage and temperature reinforcement (Fig. 7.6.3.1).

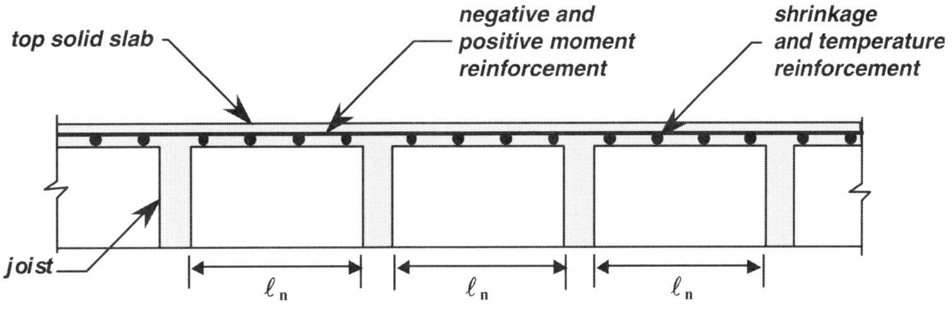

Fig. 7.5.3—Reinforcement of top solid slab that spans between joists.

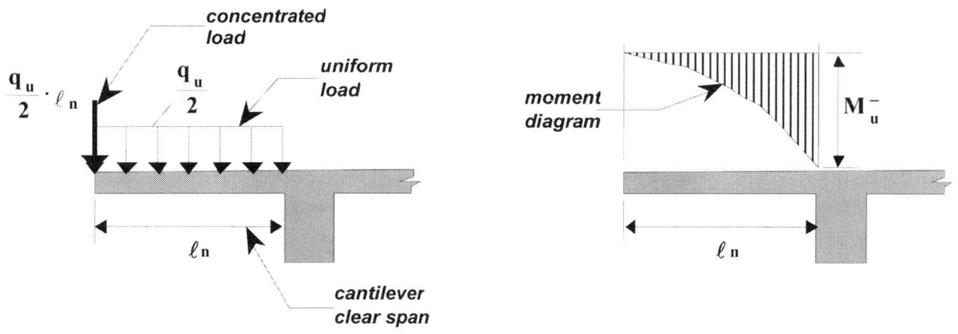

Fig. 7.6.2—Calculation of negative moment at slab cantilevers.

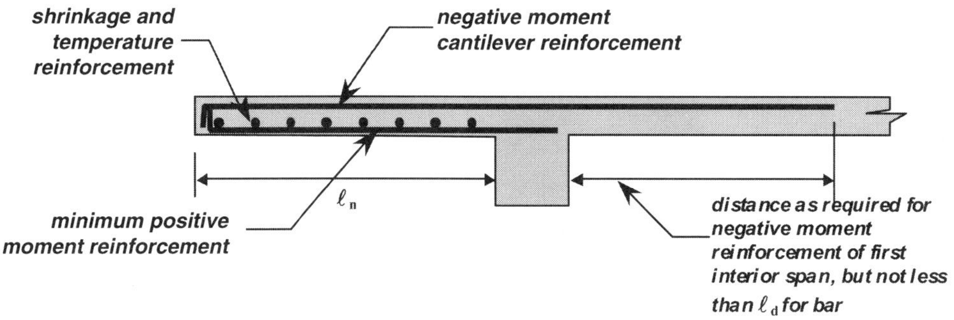

Fig. 7.6.3.1—Reinforcement for slab cantilevers.

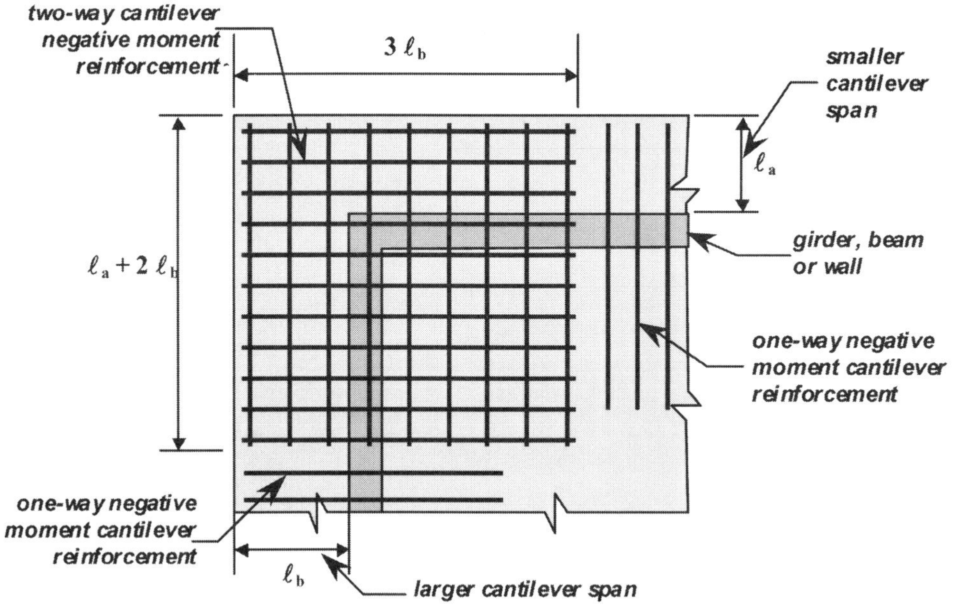

Fig. 7.6.3.4—Negative moment reinforcement in two-way slab cantilevers.

7.6.3.4 *Reinforcement of two-way cantilevers*—At corners where the slab cantilevers in two directions, negative moment reinforcement should be based on the longer cantilever. This reinforcement should be provided in both directions (Fig. 7.6.3.4). Minimum reinforcement length, measured from the corner, is equal to the cantilever clear span plus two times the longer cantilever, but not less than the distance indicated for the negative moment reinforcement of the first interior span plus the cantilever. Positive moment reinforcement as indicated by 7.6.3.2 should be placed in both directions.

7.6.4 *Shear strength*—The factored shear V_u (required shear strength) per unit slab width at the support of cantilever slabs should be calculated using Eq. (7.6.4), where ℓ_n is the clear span of the cantilever.

$$V_u = q_u \ell_n \tag{7.6.4}$$

For two-way cantilevers, the value of V_u per unit slab width should be taken as twice the value determined from Eq. (7.6.4) for the longer cantilever span.

7.6.5 *Calculation of support reactions*—Reactions at the supports, r_u, should be determined from Eq. (7.6.5), where V_u is the factored shear per unit slab width from 7.6.4; ℓ_s is the cantilever span measured from the support centerline; and ℓ_n is the cantilever clear span.

$$r_u = \frac{V_u \ell_s}{\ell_n} \qquad (7.6.5)$$

Where a two-way cantilever exists, r_u is determined from Eq. (7.6.4) for the larger cantilever span without using the factor of 2.

7.7—One-way single-span solid slabs spanning between girders, beams, or reinforced concrete walls

7.7.1 *Dimensional limits*—One-way single-span solid slabs should comply with the minimum depth of 6.5.2. In addition to 7.7, these slabs should comply with the general dimensional limits of 1.3 and the particular limits of 6.1.2 for slab-on-girder systems. Ducts, shafts, and slab openings should comply with 6.8.

7.7.2 *Required moment strength*—Positive and negative factored slab moment M_u (required moment strength) should be calculated using the equations given in Table 7.7.2.

7.7.3 *Primary flexural reinforcement*

7.7.3.1 *Positive moment reinforcement*—Positive moment reinforcement ratio ρ, in the direction of the span ℓ_n, should be based on M_u^+ determined from Eq. (7.7.2a). Where the slab is monolithic with the support, and the support has a depth at least three times greater than the slab depth, up to one-half of the positive moment reinforcement may be cut off at a distance equal to $\ell_n/8$ measured from the internal face of the support into the span (Fig. 7.7.3.1).

7.7.3.2 *Negative moment reinforcement*—Negative moment reinforcement ratio ρ, in the direction of the span ℓ_n, should be based on M_u^- determined from Eq. (7.7.2b). All negative moment reinforcement may be cut off at a distance equal to $\ell_n/4$ measured from the internal face of the support into the span. Refer to Fig. 7.7.3.1.

7.7.3.3 *Shrinkage and temperature reinforcement*—Reinforcement perpendicular to the span should not be less than that needed for shrinkage and temperature reinforcement (Fig. 7.7.3.1).

7.7.4 *Shear strength*—The factored shear V_u (required shear strength) per unit slab width for the single-span one-way slab should be calculated at the face of the supports using Eq. (7.7.4), where ℓ_n is the clear span (Fig. 7.7.3.1).

$$V_u = \frac{q_u \ell_n}{2} \qquad (7.7.4)$$

7.7.5 *Calculation of support reactions*—Slab reactions at the supports of one-way single-span slabs, r_u, should be the values determined from Eq. (7.7.5) plus the cantilever reaction where applicable. In Eq. (7.7.5), V_u is the factored shear per unit slab width from 7.7.4; ℓ_s is the center-to-center span of the slab; and ℓ_n is the clear span of the slab.

$$r_u = \frac{V_u \ell_s}{\ell_n} \qquad (7.7.5)$$

Table 7.7.2—Required moment strength for one-way, single-span slabs

Positive moment: $M_u^+ = \frac{q_u \ell_n^2}{8}$	(7.7.2a)
Negative moment at supports: $\frac{{}_{u}\;\;{}_{n}}{24}$	(7.7.2b)

7.8—One-way solid slabs supported on girders, beams, or walls, with two or more spans

7.8.1 *Dimensional limits*—In addition to 7.8, one-way solid slabs with two or more spans should comply with the minimum depth of 6.5.2, the general dimensional limits of 1.3, and the particular limits of 6.1.2 for slab-on-girder systems. Ducts, shafts, and slab openings should comply with 6.8.

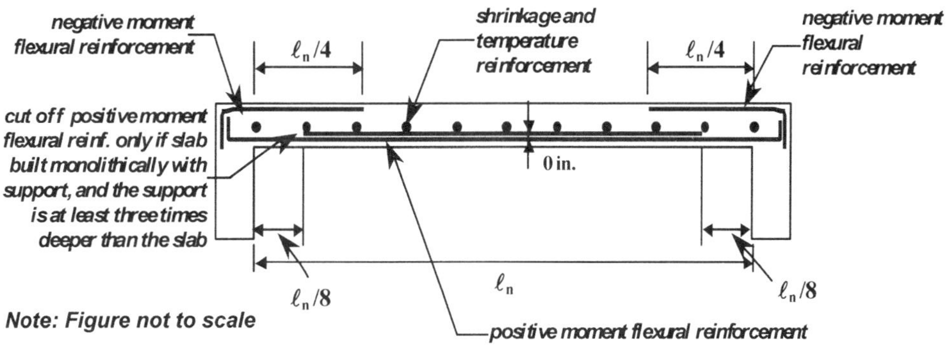

Fig. 7.7.3.1—Reinforcement for single-span one-way slabs.

Table 7.8.2—Required moment strength for one-way slabs with two or more spans

Positive moment at End spans:	
$M_u^+ = \frac{q_u \ell_n^2}{11}$	(7.8.2a)
Interior spans:	
$M_u^+ = \frac{q_u \ell_n^2}{16}$	(7.8.2b)
Negative moment at supports at Interior face of external support:	
$M_u^- = \frac{q_u \ell_n^2}{24}$	(7.8.2c)
Exterior face of first internal support, only two spans:	
$M_u^- = \frac{q_u \ell_n^2}{9}$	(7.8.2d)
Faces of internal supports, more than two spans:	
$M_u^- = \frac{q_u \ell_n^2}{10}$	(7.8.2e)
Faces of all supports for slabs with spans not exceeding 10 ft (3 m):	
$M_u^- = \frac{q_u \ell_n^2}{12}$	(7.8.2f)

The following restrictions should be in effect for slabs designed under 7.8:

(a) There are at least two spans;

(b) Span lengths should be approximately equal, and the shorter of two adjacent spans should not be smaller than 80 percent of the longer span (1.3);

(c) Loads are uniformly distributed;

(d) Unit live load q_ℓ does not exceeds three times unit dead load q_d; and

(e) For negative moment evaluation at internal supports, ℓ_n should correspond to the largest adjacent span.

7.8.2 *Required moment strength*—Positive and negative factored moment M_u (required moment strength) for the slab should be calculated using the equations in Table 7.8.2.

7.8.3 *Longitudinal flexural reinforcement*

7.8.3.1 *Positive moment reinforcement*—Positive moment reinforcement ratio ρ, in the direction of the span ℓ_n, should be based on M_u^+ determined from Eq. (7.8.2a) or (7.8.2b). At internal supports, at a distance $\ell_n/8$ measured from support face into the span, up to one-half of the positive moment reinforcement may be cut off (Fig. 7.8.3.1a and 7.8.3.1b).

7.8.3.2 *Negative moment reinforcement*—Negative moment reinforcement ratio ρ, in the direction of the span ℓ_n, should be based on M_u^- determined from Eq. (7.8.2c) to (7.8.2f). All negative moment reinforcement may be cut off at a distance $\ell_n/3$ from the internal supports, where ℓ_n should correspond to the longest adjacent span measured from the support face into the span. Because the negative moment at interior supports drops off quickly, the designer may want to provide two cutoff distances. In this case, the first cutoff distance is $\ell_n/5$ measured from the internal face of the support into the span.

At external supports, all negative moment reinforcement may be cut off (Fig. 7.8.3.1a and 7.8.3.1b) at a distance $\ell_n/4$ measured from the internal face of the support into the span.

7.8.3.3 *Shrinkage and temperature reinforcement*—Reinforcement perpendicular to the span should not be less than that needed for shrinkage and temperature reinforcement (Fig. 7.8.3.1a and 7.8.3.1b).

7.8.4 *Shear strength*—The factored shear V_u (required shear strength) per unit slab width should be calculated at the face of the support using the equations in Table 7.8.4, where ℓ_n is the clear span (Fig. 7.8.3.1a and 7.8.3.1b). The design shear strength ϕV_n per unit slab width should be calculated at the support face.

7.8.5 *Calculation of support reactions*—Slab support reactions r_u should be the values determined from Eq. (7.8.5), where V_u is the factored shear (required shear strength) per unit slab width from 7.8.4; ℓ_s is the center-to-center span; and ℓ_n is the clear span.

$$r_u = \frac{V_u \ell_s}{\ell_n} \quad (7.8.5)$$

The external support reaction is equal to the span reaction r_u, determined from Eq. (7.8.5) at the support, plus cantilever reaction, where applicable. Reaction at an internal support is the sum of reactions r_u, determined using Eq. (7.8.5) for both adjacent spans at that support.

7.9—Two-way solid slabs spanning between girders, beams, or reinforced concrete walls

7.9.1 *Dimensional limits*—Two-way solid slabs supported on all edges should comply with the minimum depth of 6.5.4. In addition to 7.9, two-way slabs should comply with the general dimensional limits of 1.3 and the particular limits of 6.1.2 for slab-on-girder systems. Ducts, shafts, and slab openings should comply with 6.8.

The following restrictions should be in effect to use the procedure described in 7.9:

(a) There are at least two spans;

(b) Span lengths should be approximately equal, and the shorter of two adjacent spans should not be less than 80 percent of the longer span (1.3);

(c) Supporting girders or beams should be monolithic with the slab and should have a total depth not less than three times the slab thickness;

(d) A slab with elevator and stair core openings is considered continuous when the floor slabs are supported on continuous beams around the opening, integrated into the core walls;

(e) Loads are uniformly distributed; and

(f) Unit live load q_ℓ does not exceed three times unit dead load q_d.

The slab panel should be divided, in both directions, into central and border regions. The central region should be the central half of the panel, and the border regions should be two one-quarter regions adjacent to both sides of the central region (Fig. 7.9.1).

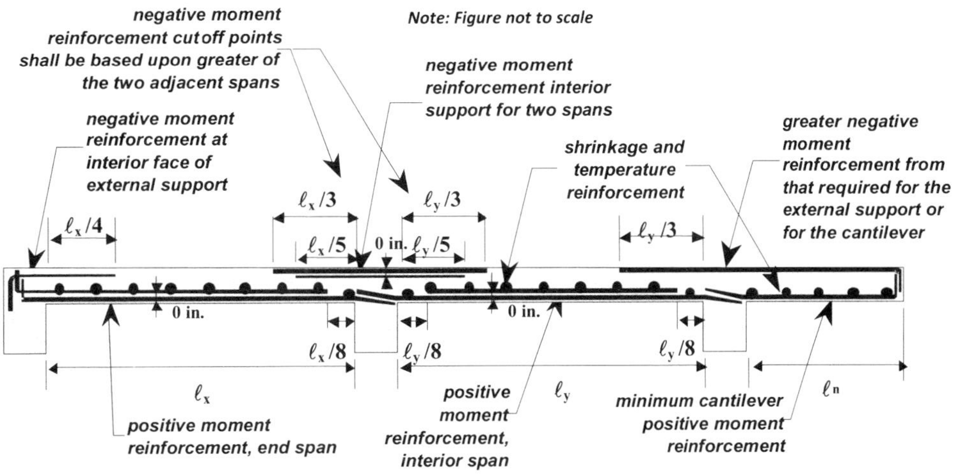

Fig. 7.8.3.1a—Reinforcement for two-span one-way slabs supported by girders, beams, or reinforced concrete walls.

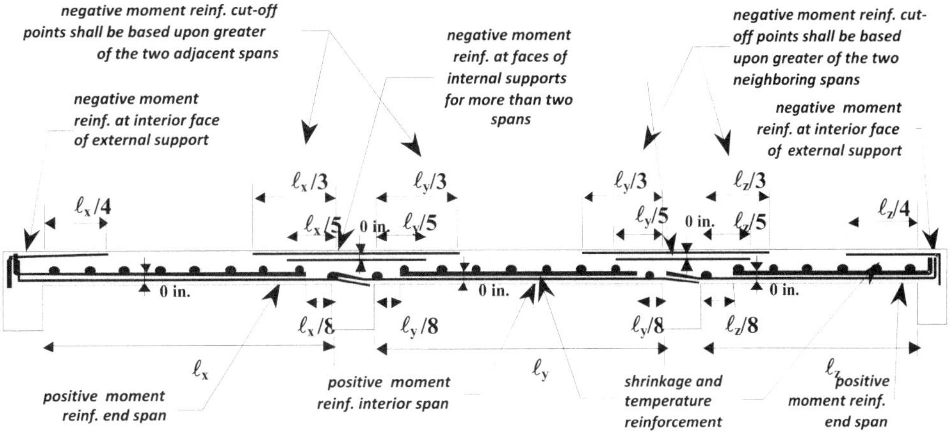

Fig. 7.8.3.1b—Reinforcement for one-way slabs supported by girders, beams, or reinforced concrete walls, with three or more spans.

7.9.2 *Required moment strength*—The factored positive and negative moment, M_u (required moment strength), for two-way solid slabs should be calculated using the procedure in this section. Negative and positive factored moments for the central region of the panel in each direction should be calculated using equations in Table 7.9.2a for interior panels, in Table 7.9.2b for edge panels with the shorter span at the edge, in Table 7.9.2c for edge panels with the longer span at the edge, and in Table 7.9.2d for corner panels. In each table, factored moments should be determined for the appropriate ratio of longer clear span to shorter clear span β and the corresponding edge continuity conditions. The negative moment at discontinuous edges should be taken as one-third of the positive moment in the same direction (Fig. 7.9.2a).

Moment values in the border of the central regions may be linearly decreased to one-third of the central region value at the panel edge, as shown in Fig. 7.9.2b for moments in the shorter direction and in Fig. 7.9.2c for moments in the longer direction.

7.9.3 *Longitudinal flexural reinforcement*

7.9.3.1 *Positive moment reinforcement*—In the central region of the slab panel, positive moment reinforcement should be based on the factored positive moment determined from the appropriate equations in Tables 7.9.2a to 7.9.2d. Positive moment steel ratio ρ for reinforcement parallel to the shorter span ℓ_a or the longer span ℓ_b should be determined using the corresponding value of M_a^+ or M_b^+. As much as one-half of the positive moment reinforcement at the center of the corresponding span may be cut off at a distance equal to $\ell_a/8$ or $\ell_b/8$, measured from the face of any interior support. Positive moment reinforcement perpendicular to a discontinuous edge may not be cut off.

Table 7.8.4—Required shear strength for one-way slabs with two or more spans

At exterior face of first interior support $$V_u = 1.15\frac{q_u \ell_n}{2}$$	(7.8.4a)
At faces of all other supports $$V_u = \frac{q_u \ell_n}{2}$$	(7.8.4b)

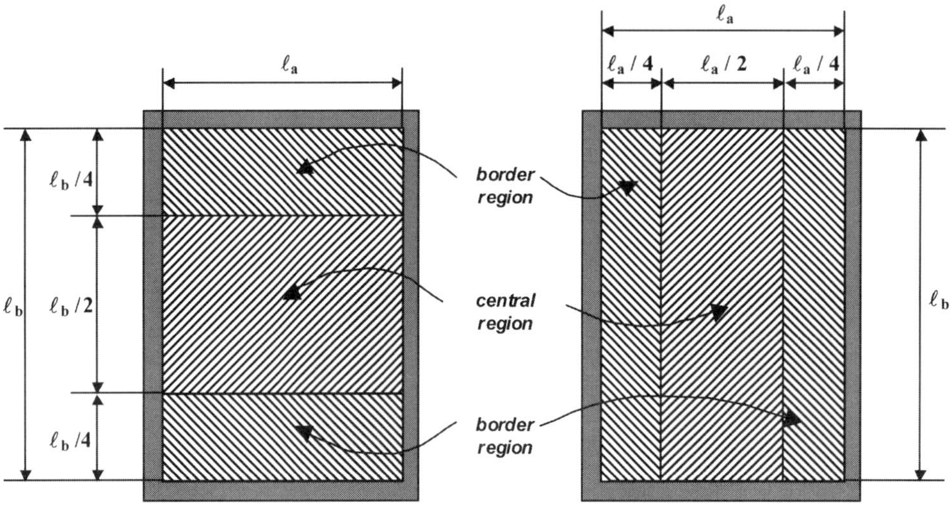

Fig. 7.9.1—Central and border regions for two-way slabs supported on girders, beams, or reinforced concrete walls.

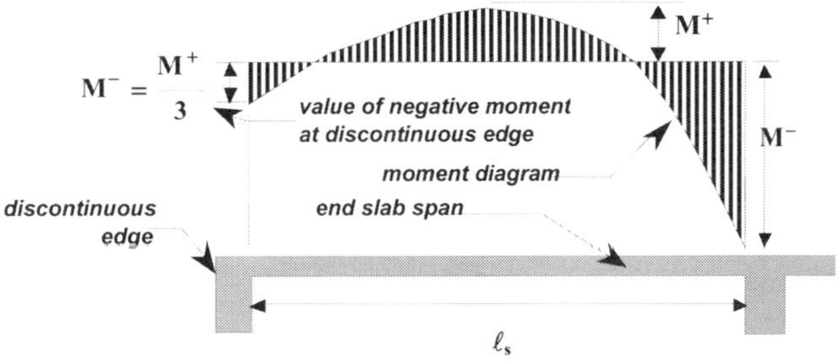

Fig. 7.9.2a—Negative moment at discontinuous edges of two-way solid slabs-on-girders.

Positive moment reinforcement of the border region may be decreased from that needed at the edge of the central region to one-third of this value at the panel edge, but not below the area needed for shrinkage and temperature (Fig. 7.9.3.1).

7.9.3.2 *Negative moment reinforcement*—At supporting edges of the slab panel's central region, the area of negative moment reinforcement should be based on the factored negative moment determined from appropriate equations of Tables 7.9.2a to 7.9.2d. The negative moment steel ratio ρ for reinforcement parallel to the shorter span ℓ_a or the longer span ℓ_b should be determined using the corresponding value of M_a^- or M_b^-. It is permitted to cut off up to one-half of the negative moment reinforcement area at the support at a distance equal to $\ell_a/5$ or $\ell_b/5$, measured from the face of any interior support. It is permitted to cut off all negative moment reinforcement at the corresponding span support at a distance equal to $\ell_a/3$ or $\ell_b/3$, measured from the face of any interior support. It is permitted to decrease gradually the negative moment reinforcement area at the central region from the edge of the central regions to one-third of this value at the panel edge, but not below the area needed for shrinkage and temperature (Fig. 7.9.3.1).

7.9.4 *Shear strength*—Factored shear per unit slab width, V_u, (required shear strength) at the faces of the supporting members should be determined using the load fractions α_a and α_b, spanning in the shorter and longer directions respectively, as given in Tables 7.9.2a to 7.9.2d for the corresponding panel edge conditions and panel span ratio β (Fig. 7.9.4a). The required shear strength should not be less than the shear caused by factored design load q_u acting on a tributary area bounded by 45-degree lines drawn from the corner and the centerline of the panel parallel to the longer span (Fig. 7.9.4b).

The factored shear per unit slab width, V_u, should not be less than the value determined from Eq. (7.9.4a) for the shorter-span supporting member and from Eq. (7.9.4b) for the longer-span supporting member.

$$V_u = \frac{\alpha_b q_u \ell_b}{2} \geq \frac{q_u \ell_a}{4} \qquad (7.9.4a)$$

$$V_u = \frac{\alpha_a q_u \ell_a}{2} \geq q_u \left[\frac{\ell_a}{2} - \frac{\ell_a^2}{4\ell_b}\right] \qquad (7.9.4b)$$

In determining the design shear strength ϕV_n, an effective depth d for the slab greater than or equal to the largest value from Eq. (7.9.4c), (7.9.4d), and (7.9.4e) should be used.

$$d \geq \frac{q_u \alpha_a \ell_a}{\phi 4\sqrt{f_c'}} \quad \left[d \geq \frac{3q_u \alpha_a \ell_a}{\phi\sqrt{f_c'}} \text{ (SI)}\right] \qquad (7.9.4c)$$

$$d \geq \frac{q_u \alpha_b \ell_b}{\phi 4\sqrt{f_c'}} \quad \left[d \geq \frac{3q_u \alpha_b \ell_b}{\phi\sqrt{f_c'}} \text{ (SI)}\right] \qquad (7.9.4d)$$

$$d \geq \frac{q_u \ell_a}{\phi 8\sqrt{f_c'}} \quad \left[d \geq \frac{3q_u \ell_a}{\phi 2\sqrt{f_c'}} \text{ (SI)}\right] \qquad (7.9.4e)$$

Table 7.9.2a—Interior panel of two-way slabs supported on girders, beams, or reinforced concrete walls

$\beta = \ell_b/\ell_a$	Shorter direction ℓ_a			Longer direction ℓ_b		
Panel span ratio	Negative moment	Positive moment	Load fraction	Negative moment	Positive moment	Load fraction
1.0	$M_a^- = \frac{q_u \ell_a^2}{22}$	$M_a^+ = \frac{q_u \ell_a^2}{42}$	$\alpha_a = 0.50$	$M_b^- = \frac{q_u \ell_b^2}{22}$	$M_b^+ = \frac{q_u \ell_b^2}{42}$	$\alpha_b = 0.50$
1.1	$M_a^- = \frac{q_u \ell_a^2}{18}$	$M_a^+ = \frac{q_u \ell_a^2}{35}$	$\alpha_a = 0.60$	$M_b^- = \frac{q_u \ell_b^2}{25}$	$M_b^+ = \frac{q_u \ell_b^2}{50}$	$\alpha_b = 0.40$
1.2	$M_a^- = \frac{q_u \ell_a^2}{16}$	$M_a^+ = \frac{q_u \ell_a^2}{30}$	$\alpha_a = 0.67$	$M_b^- = \frac{q_u \ell_b^2}{35}$	$M_b^+ = \frac{q_u \ell_b^2}{60}$	$\alpha_b = 0.33$
1.3	$M_a^- = \frac{q_u \ell_a^2}{15}$	$M_a^+ = \frac{q_u \ell_a^2}{27}$	$\alpha_a = 0.74$	$M_b^- = \frac{q_u \ell_b^2}{40}$	$M_b^+ = \frac{q_u \ell_b^2}{75}$	$\alpha_b = 0.26$
1.4	$M_a^- = \frac{q_u \ell_a^2}{14}$	$M_a^+ = \frac{q_u \ell_a^2}{25}$	$\alpha_a = 0.80$	$M_b^- = \frac{q_u \ell_b^2}{50}$	$M_b^+ = \frac{q_u \ell_b^2}{100}$	$\alpha_b = 0.20$
1.5	$M_a^- = \frac{q_u \ell_a^2}{13}$	$M_a^+ = \frac{q_u \ell_a^2}{23}$	$\alpha_a = 0.84$	$M_b^- = \frac{q_u \ell_b^2}{65}$	$M_b^+ = \frac{q_u \ell_b^2}{120}$	$\alpha_b = 0.16$
1.6	$M_a^- = \frac{q_u \ell_a^2}{13}$	$M_a^+ = \frac{q_u \ell_a^2}{22}$	$\alpha_a = 0.87$	$M_b^- = \frac{q_u \ell_b^2}{85}$	$M_b^+ = \frac{q_u \ell_b^2}{145}$	$\alpha_b = 0.13$
1.7	$M_a^- = \frac{q_u \ell_a^2}{12}$	$M_a^+ = \frac{q_u \ell_a^2}{21}$	$\alpha_a = 0.90$	$M_b^- = \frac{q_u \ell_b^2}{110}$	$M_b^+ = \frac{q_u \ell_b^2}{180}$	$\alpha_b = 0.10$
1.8	$M_a^- = \frac{q_u \ell_a^2}{12}$	$M_a^+ = \frac{q_u \ell_a^2}{20}$	$\alpha_a = 0.92$	$M_b^- = \frac{q_u \ell_b^2}{135}$	$M_b^+ = \frac{q_u \ell_b^2}{225}$	$\alpha_b = 0.08$
1.9	$M_a^- = \frac{q_u \ell_a^2}{12}$	$M_a^+ = \frac{q_u \ell_a^2}{20}$	$\alpha_a = 0.93$	$M_b^- = \frac{q_u \ell_b^2}{160}$	$M_b^+ = \frac{q_u \ell_b^2}{275}$	$\alpha_b = 0.07$
2.0	$M_a^- = \frac{q_u \ell_a^2}{11}$	$M_a^+ = \frac{q_u \ell_a^2}{18}$	$\alpha_a = 0.94$	$M_b^- = \frac{q_u \ell_b^2}{170}$	$M_b^+ = \frac{q_u \ell_b^2}{340}$	$\alpha_b = 0.06$
> 2.0	$M_a^- = \frac{q_u \ell_a^2}{10}$	$M_a^+ = \frac{q_u \ell_a^2}{16}$	$\alpha_a = 1.00$	Temperature and shrinkage reinforcement		$\alpha_b = 0.00$

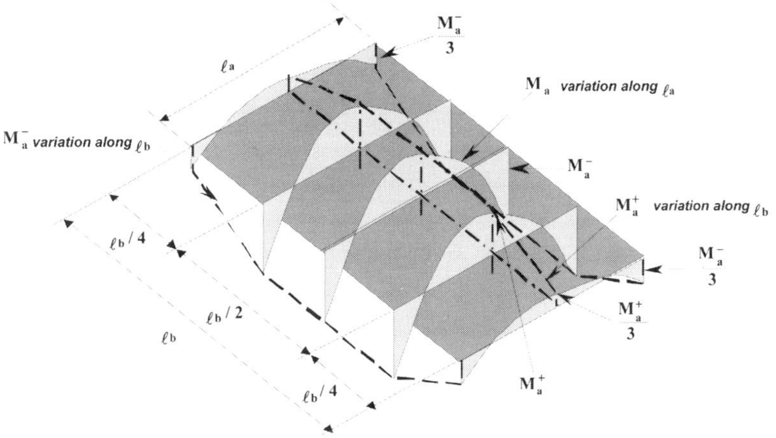

Fig. 7.9.2b—Variation of moment M_a *across the width of critical sections for design, two-way slabs supported on girders, beams, or reinforced concrete walls.*

Table 7.9.2b—Edge panel with ℓ_a parallel to edge of two-way slabs supported on girders, beams, or walls

$\beta = \ell_b/\ell_a$	Shorter direction ℓ_a			Longer direction ℓ_b		
Panel span ratio	Negative moment	Positive moment	Load fraction	Negative moment	Positive moment	Load fraction
1.0	$M_a^- = \frac{q_u \ell_a^2}{16}$	$M_a^+ = \frac{q_u \ell_a^2}{35}$	$\alpha_a = 0.67$	$M_b^- = \frac{q_u \ell_b^2}{33}$	$M_b^+ = \frac{q_u \ell_b^2}{40}$	$\alpha_b = 0.33$
1.1	$M_a^- = \frac{q_u \ell_a^2}{15}$	$M_a^+ = \frac{q_u \ell_a^2}{31}$	$\alpha_a = 0.74$	$M_b^- = \frac{q_u \ell_b^2}{35}$	$M_b^+ = \frac{q_u \ell_b^2}{50}$	$\alpha_b = 0.26$
1.2	$M_a^- = \frac{q_u \ell_a^2}{14}$	$M_a^+ = \frac{q_u \ell_a^2}{28}$	$\alpha_a = 0.80$	$M_b^- = \frac{q_u \ell_b^2}{50}$	$M_b^+ = \frac{q_u \ell_b^2}{65}$	$\alpha_b = 0.20$
1.3	$M_a^- = \frac{q_u \ell_a^2}{13}$	$M_a^+ = \frac{q_u \ell_a^2}{25}$	$\alpha_a = 0.85$	$M_b^- = \frac{q_u \ell_b^2}{70}$	$M_b^+ = \frac{q_u \ell_b^2}{85}$	$\alpha_b = 0.15$
1.4	$M_a^- = \frac{q_u \ell_a^2}{13}$	$M_a^+ = \frac{q_u \ell_a^2}{23}$	$\alpha_a = 0.88$	$M_b^- = \frac{q_u \ell_b^2}{90}$	$M_b^+ = \frac{q_u \ell_b^2}{110}$	$\alpha_b = 0.12$
1.5	$M_a^- = \frac{q_u \ell_a^2}{12}$	$M_a^+ = \frac{q_u \ell_a^2}{22}$	$\alpha_a = 0.91$	$M_b^- = \frac{q_u \ell_b^2}{115}$	$M_b^+ = \frac{q_u \ell_b^2}{135}$	$\alpha_b = 0.09$
1.6	$M_a^- = \frac{q_u \ell_a^2}{12}$	$M_a^+ = \frac{q_u \ell_a^2}{21}$	$\alpha_a = 0.93$	$M_b^- = \frac{q_u \ell_b^2}{135}$	$M_b^+ = \frac{q_u \ell_b^2}{160}$	$\alpha_b = 0.07$
1.7	$M_a^- = \frac{q_u \ell_a^2}{12}$	$M_a^+ = \frac{q_u \ell_a^2}{20}$	$\alpha_a = 0.94$	$M_b^- = \frac{q_u \ell_b^2}{165}$	$M_b^+ = \frac{q_u \ell_b^2}{185}$	$\alpha_b = 0.06$
1.8	$M_a^- = \frac{q_u \ell_a^2}{12}$	$M_a^+ = \frac{q_u \ell_a^2}{20}$	$\alpha_a = 0.95$	$M_b^- = \frac{q_u \ell_b^2}{200}$	$M_b^+ = \frac{q_u \ell_b^2}{220}$	$\alpha_b = 0.05$
1.9	$M_a^- = \frac{q_u \ell_a^2}{12}$	$M_a^+ = \frac{q_u \ell_a^2}{19}$	$\alpha_a = 0.96$	$M_b^- = \frac{q_u \ell_b^2}{250}$	$M_b^+ = \frac{q_u \ell_b^2}{270}$	$\alpha_b = 0.04$
2.0	$M_a^- = \frac{q_u \ell_a^2}{11}$	$M_a^+ = \frac{q_u \ell_a^2}{18}$	$\alpha_a = 0.97$	$M_b^- = \frac{q_u \ell_b^2}{330}$	$M_b^+ = \frac{q_u \ell_b^2}{340}$	$\alpha_b = 0.03$
> 2.0	$M \quad = \frac{u \quad a}{10}$	$M_a^+ = \frac{q_u \ell_a^2}{16}$	$\alpha_a = 1.00$	Temperature and shrinkage reinforcement		$\alpha_b = 0.00$

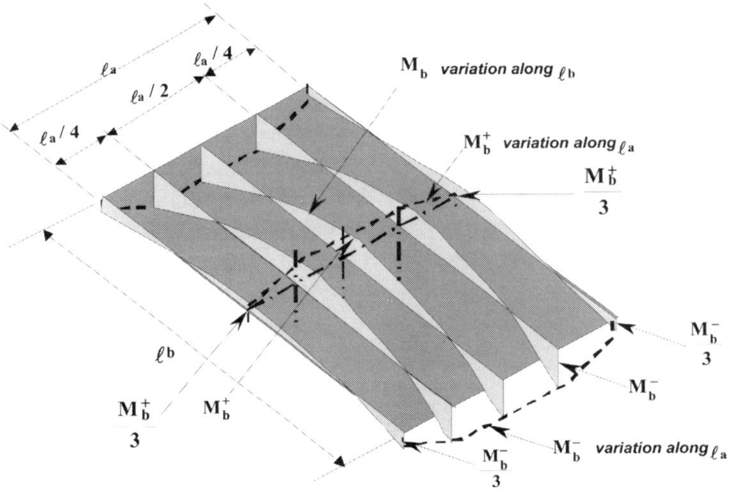

Fig. 7.9.2c—Variation of moment M_b *across the width of critical sections for design, two-way slabs supported on girders, beams, or reinforced concrete walls.*

Table 7.9.2c—Edge panel with ℓ_b parallel to edge of two-way slabs supported on girders, beams, or walls

$\beta = \ell_b/\ell_a$	Shorter direction ℓ_a			Longer direction ℓ_b		
Panel span ratio	Negative moment	Positive moment	Load fraction	Negative moment	Positive moment	Load fraction
1.0	$M_a^- = \frac{q_u\ell_a^2}{30}$	$M_a^+ = \frac{q_u\ell_a^2}{39}$	$\alpha_a = 0.33$	$M_b^- = \frac{q_u\ell_b^2}{16}$	$M_b^+ = \frac{q_u\ell_b^2}{35}$	$\alpha_b = 0.67$
1.1	$M_a^- = \frac{q_u\ell_a^2}{23}$	$M_a^+ = \frac{q_u\ell_a^2}{32}$	$\alpha_a = 0.42$	$M_b^- = \frac{q_u\ell_b^2}{19}$	$M_b^+ = \frac{q_u\ell_b^2}{40}$	$\alpha_b = 0.58$
1.2	$M_a^- = \frac{q_u\ell_a^2}{19}$	$M_a^+ = \frac{q_u\ell_a^2}{26}$	$\alpha_a = 0.51$	$M_b^- = \frac{q_u\ell_b^2}{22}$	$M_b^+ = \frac{q_u\ell_b^2}{50}$	$\alpha_b = 0.49$
1.3	$M_a^- = \frac{q_u\ell_a^2}{17}$	$M_a^+ = \frac{q_u\ell_a^2}{23}$	$\alpha_a = 0.59$	$M_b^- = \frac{q_u\ell_b^2}{27}$	$M_b^+ = \frac{q_u\ell_b^2}{60}$	$\alpha_b = 0.41$
1.4	$M_a^- = \frac{q_u\ell_a^2}{15}$	$M_a^+ = \frac{q_u\ell_a^2}{20}$	$\alpha_a = 0.66$	$M_b^- = \frac{q_u\ell_b^2}{32}$	$M_b^+ = \frac{q_u\ell_b^2}{70}$	$\alpha_b = 0.34$
1.5	$M_a^- = \frac{q_u\ell_a^2}{13}$	$M_a^+ = \frac{q_u\ell_a^2}{19}$	$\alpha_a = 0.72$	$M_b^- = \frac{q_u\ell_b^2}{40}$	$M_b^+ = \frac{q_u\ell_b^2}{85}$	$\alpha_b = 0.28$
1.6	$M_a^- = \frac{q_u\ell_a^2}{12}$	$M_a^+ = \frac{q_u\ell_a^2}{17}$	$\alpha_a = 0.77$	$M_b^- = \frac{q_u\ell_b^2}{50}$	$M_b^+ = \frac{q_u\ell_b^2}{100}$	$\alpha_b = 0.23$
1.7	$M_a^- = \frac{q_u\ell_a^2}{12}$	$M_a^+ = \frac{q_u\ell_a^2}{16}$	$\alpha_a = 0.81$	$M_b^- = \frac{q_u\ell_b^2}{60}$	$M_b^+ = \frac{q_u\ell_b^2}{125}$	$\alpha_b = 0.19$
1.8	$M_a^- = \frac{q_u\ell_a^2}{11}$	$M_a^+ = \frac{q_u\ell_a^2}{15}$	$\alpha_a = 0.85$	$M_b^- = \frac{q_u\ell_b^2}{70}$	$M_b^+ = \frac{q_u\ell_b^2}{150}$	$\alpha_b = 0.15$
1.9	$M_a^- = \frac{q_u\ell_a^2}{11}$	$M_a^+ = \frac{q_u\ell_a^2}{15}$	$\alpha_a = 0.88$	$M_b^- = \frac{q_u\ell_b^2}{85}$	$M_b^+ = \frac{q_u\ell_b^2}{175}$	$\alpha_b = 0.12$
2.0	$M_a^- = \frac{q_u\ell_a^2}{10}$	$M_a^+ = \frac{q_u\ell_a^2}{14}$	$\alpha_a = 0.92$	$M_b^- = \frac{q_u\ell_b^2}{100}$	$M_b^+ = \frac{q_u\ell_b^2}{200}$	$\alpha_b = 0.08$
> 2.0	$M_a^- = \frac{q_u\ell_a^2}{9}$	$M_a^+ = \frac{q_u\ell_a^2}{11}$	$\alpha_a = 1.00$	Temperature and shrinkage reinforcement		$\alpha_b = 0.00$

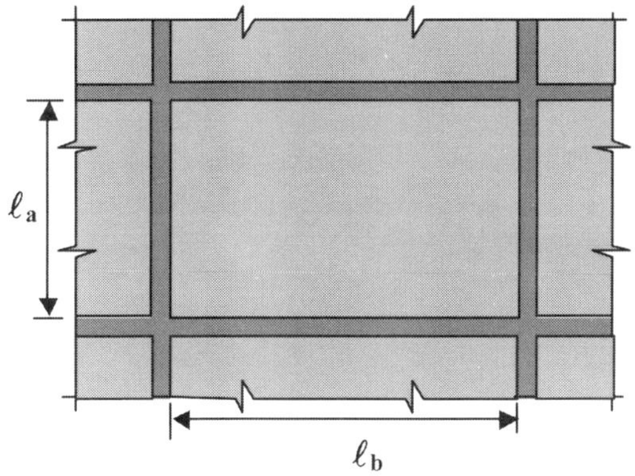

Fig. 7.9.2d—Interior panel of two-way slabs supported on girders, beams, or reinforced concrete walls.

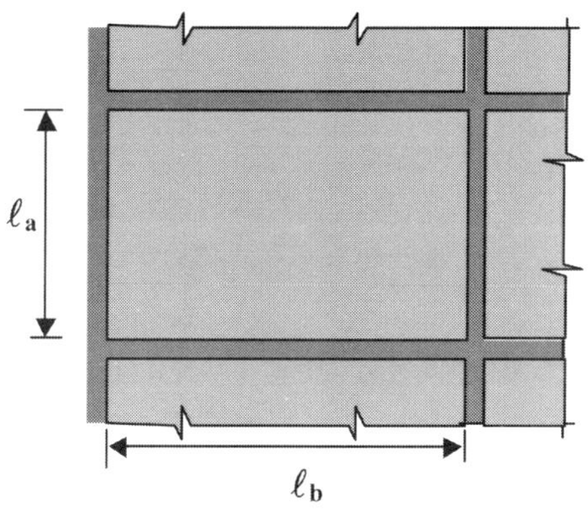

Fig. 7.9.2e—Edge panel with ℓ_a parallel to edge of two-way slabs supported on girders, beams, or walls.

Table 7.9.2d—Corner panel of two-way slabs supported on girders, beams, or reinforced concrete walls

$\beta = \ell_b/\ell_a$	Shorter direction ℓ_a			Longer direction ℓ_b		
Panel span ratio	Negative moment	Positive moment	Load fraction	Negative moment	Positive moment	Load fraction
1.0	$M_a^- = \frac{q_u \ell_a^2}{20}$	$M_a^+ = \frac{q_u \ell_a^2}{31}$	$\alpha_a = 0.50$	$M_b^- = \frac{q_u \ell_b^2}{20}$	$M_b^+ = \frac{q_u \ell_b^2}{31}$	$\alpha_b = 0.50$
1.1	$M_a^- = \frac{q_u \ell_a^2}{17}$	$M_a^+ = \frac{q_u \ell_a^2}{26}$	$\alpha_a = 0.59$	$M_b^- = \frac{q_u \ell_b^2}{25}$	$M_b^+ = \frac{q_u \ell_b^2}{38}$	$\alpha_b = 0.41$
1.2	$M_a^- = \frac{q_u \ell_a^2}{15}$	$M_a^+ = \frac{q_u \ell_a^2}{23}$	$\alpha_a = 0.67$	$M_b^- = \frac{q_u \ell_b^2}{30}$	$M_b^+ = \frac{q_u \ell_b^2}{45}$	$\alpha_b = 0.33$
1.3	$M_a^- = \frac{q_u \ell_a^2}{13}$	$M_a^+ = \frac{q_u \ell_a^2}{20}$	$\alpha_a = 0.74$	$M_b^- = \frac{q_u \ell_b^2}{40}$	$M_b^+ = \frac{q_u \ell_b^2}{55}$	$\alpha_b = 0.26$
1.4	$M_a^- = \frac{q_u \ell_a^2}{13}$	$M_a^+ = \frac{q_u \ell_a^2}{19}$	$\alpha_a = 0.80$	$M_b^- = \frac{q_u \ell_b^2}{50}$	$M_b^+ = \frac{q_u \ell_b^2}{70}$	$\alpha_b = 0.20$
1.5	$M_a^- = \frac{q_u \ell_a^2}{12}$	$M_a^+ = \frac{q_u \ell_a^2}{17}$	$\alpha_a = 0.84$	$M_b^- = \frac{q_u \ell_b^2}{60}$	$M_b^+ = \frac{q_u \ell_b^2}{85}$	$\alpha_b = 0.16$
1.6	$M_a^- = \frac{q_u \ell_a^2}{11}$	$M_a^+ = \frac{q_u \ell_a^2}{16}$	$\alpha_a = 0.87$	$M_b^- = \frac{q_u \ell_b^2}{75}$	$M_b^+ = \frac{q_u \ell_b^2}{100}$	$\alpha_b = 0.13$
1.7	$M_a^- = \frac{q_u \ell_a^2}{11}$	$M_a^+ = \frac{q_u \ell_a^2}{16}$	$\alpha_a = 0.90$	$M_b^- = \frac{q_u \ell_b^2}{100}$	$M_b^+ = \frac{q_u \ell_b^2}{125}$	$\alpha_b = 0.10$
1.8	$M_a^- = \frac{q_u \ell_a^2}{11}$	$M_a^+ = \frac{q_u \ell_a^2}{15}$	$\alpha_a = 0.92$	$M_b^- = \frac{q_u \ell_b^2}{120}$	$M_b^+ = \frac{q_u \ell_b^2}{150}$	$\alpha_b = 0.08$
1.9	$M_a^- = \frac{q_u \ell_a^2}{11}$	$M_a^+ = \frac{q_u \ell_a^2}{15}$	$\alpha_a = 0.94$	$M_b^- = \frac{q_u \ell_b^2}{145}$	$M_b^+ = \frac{q_u \ell_b^2}{175}$	$\alpha_b = 0.06$
2.0	$M_a^- = \frac{q_u \ell_a^2}{10}$	$M_a^+ = \frac{q_u \ell_a^2}{14}$	$\alpha_a = 0.96$	$M_b^- = \frac{q_u \ell_b^2}{165}$	$M_b^+ = \frac{q_u \ell_b^2}{200}$	$\alpha_b = 0.04$
> 2.0	$M_a^- = \frac{q_u \ell_a^2}{9}$	$M_a^+ = \frac{q_u \ell_a^2}{11}$	$\alpha_a = 1.00$	Temperature and shrinkage reinforcement		$\alpha_b = 0.00$

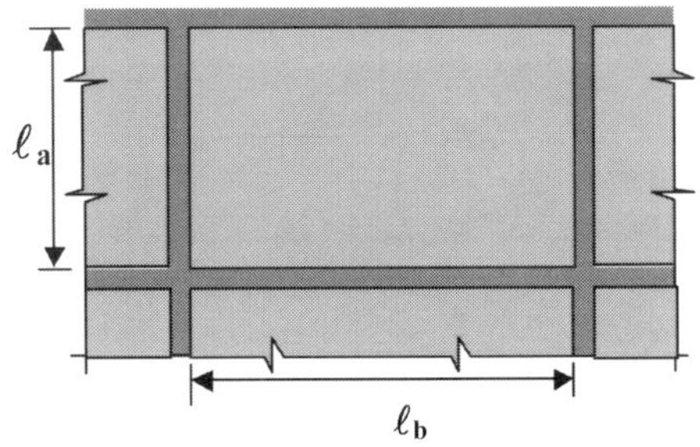

Fig. 7.9.2f—Edge panel with ℓ_b parallel to edge of two-way slabs supported on girders, beams, or walls.

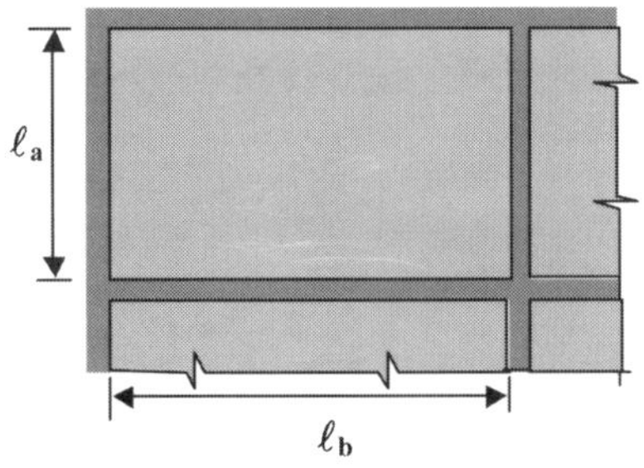

Fig. 7.9.2g—Corner panel of two-way slabs supported on girders, beams, or reinforced concrete walls.

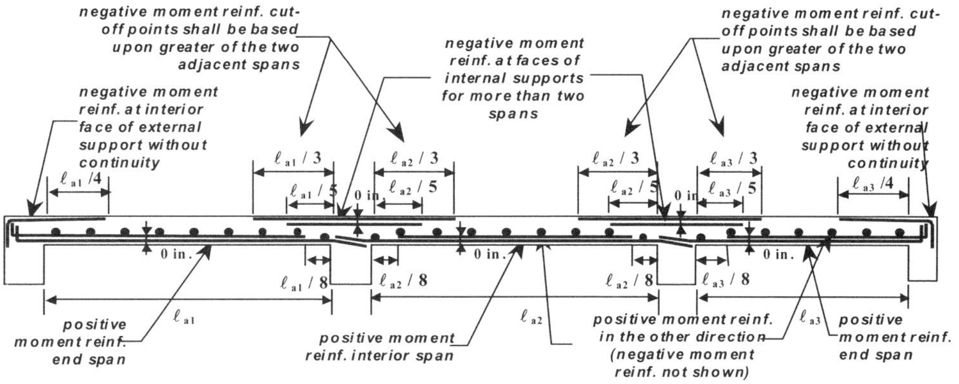

Fig. 7.9.3.1—Reinforcement for two-way slabs supported by girders, beams, or reinforced concrete walls.

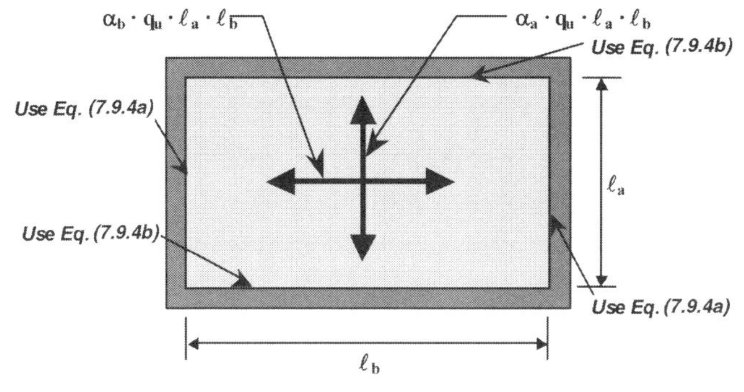

Fig. 7.9.4a—Fraction of the total load in the panel acting in each direction in two-way slabs supported on girders, beams, or reinforced concrete walls.

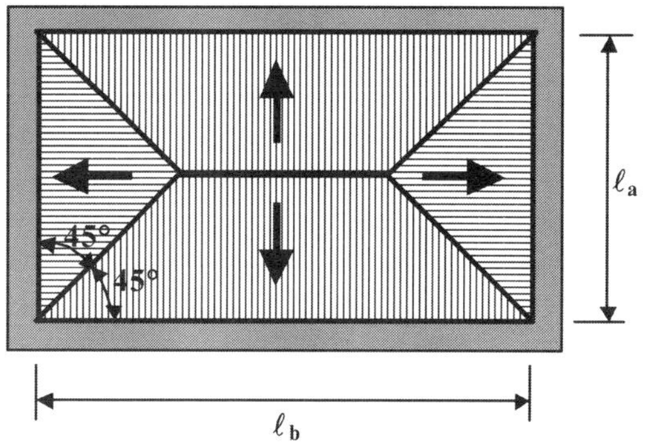

Fig. 7.9.4b—Tributary areas of minimum shear at the supports of two-way slabs supported on girders, beams, or reinforced concrete walls.

In Eq. (7.9.4c) to (7.9.4e), $\phi = 0.75$.

7.9.5 *Calculation of support reactions*—Support reactions from any panel of two-way slabs, r_u, in the shorter direction should be determined from Eq. (7.9.5a) and in the longer direction should be determined from Eq. (7.9.5b).

$$r_u = \frac{V_u \ell_s}{\ell_a} \tag{7.9.5a}$$

$$r_u = \frac{V_u \ell_s}{\ell_b} \tag{7.9.5b}$$

In Eq. (7.9.5a) and (7.9.5b), V_u is the factored shear from Eq. (7.9.4a) and (7.9.4b) respectively; ℓ_s is the center-to-center span in that direction; and ℓ_a and ℓ_b are the corresponding clear spans.

Reactions at the edge panel external supports should be equal to values from the panel r_u at the edge support, determined from Eq. (7.9.5a) or (7.9.5b), plus the cantilever reaction, where applicable. Reactions on internal supports should be the sum of reactions r_u, determined using either Eq. (7.9.5a) or (7.9.5b), for both adjacent spans at that support.

CHAPTER 8—GIRDERS, BEAMS, AND JOISTS

8.1—General

8.1.1 *Scope*—Girders, beams, and joists should be designed according to Chapter 8. The chapter applies to isolated beams, girders, beams, and joists that are part of a floor system and girders that are part of a moment-resisting frame supported on columns or concrete structural walls. Foundation beams (grade beams) should be designed following Chapter 14.

8.1.2 *The use of frame analysis*—Frame analysis that meets (a) through (g) may be used to obtain factored moments and shears as a substitute for values prescribed in Chapter 8:

(a) Analysis procedure should be based on established principles of structural mechanics;

(b) Procedure should consider equilibrium, deformation compatibility, general stability, and short- and long-term material properties;

(c) Analysis procedure should consider support flexibility and the interaction between flexure and torsion of supported and supporting members;

(d) Concrete modulus of elasticity should be taken as

$E_c = 57{,}000\sqrt{f'_c}$, psi $\left[E_c = 4700\sqrt{f'_c}, \text{MPa}\right]$.

(e) Reinforcing steel modulus of elasticity should be taken as E_s = 29,000,000 psi [E_s = 200,000 MPa].

(f) Reasonable assumptions should be used for computing relative flexural and torsional stiffness of structural members. The assumptions adopted should be consistent throughout analysis;

(g) Span length should be taken as the center-to-center distance between supports, but factored moments and shears may be obtained at faces of supports; and

(h) The arrangement of live load may be limited to load combinations of factored dead load on all spans with full factored live load on two adjacent spans and factored dead load on all spans with full factored live load on alternate spans.

8.2—Loads

8.2.1 *Loads to be included*—Loads for girders, beams, and joists should be established from Chapter 4. Gravity loads should be the sum of the reactions from supported members and loads applied directly on the member. Lateral loads should conform to Chapter 4.

8.2.1.1 *Reactions*—The gravity reactions from supported members should consider:

(a) Dead loads, including self-weight of supported members, loads due to flat and standing nonstructural elements, and loads from any fixed equipment carried by these supported members; and

(b) Live loads applied on supported members.

8.2.1.2 *Member loads*—Loads carried directly by beam, girder, or joist should include:

(a) Dead loads, including member self-weight, and flat and standing nonstructural elements, and fixed equipment loads, applied directly on the member; and

(b) Live loads applied directly on the member.

8.2.2 *Dead and live loads carried by the member*

8.2.2.1 *Uniformly distributed loads*—The following should be used to obtain values of w_d for dead load and w_ℓ for live load:

(a) w_d should include the self-weight per member unit length;

(b) w_d should include the weight of the flat and standing nonstructural elements directly supported, as defined in 4.5.3, per member unit length;

(c) For girders, beams, or joists located at the edge of floor slabs, w_d should include distributed load due to the weight of façades and building enclosure elements, as indicated by 4.5.3.2;

(d) w_ℓ should include live load directly applied on the member, as indicated by 4.6; and

(e) For a roof member, the directly applied roof live loads given in 4.7, rain loads in 4.8, and snow loads in 4.9 should be included as appropriate.

8.2.2.2 *Concentrated loads*—The following should be used to obtain values of p_d for dead load and p_ℓ for live load for all locations where concentrated loads are applied to the member:

(a) p_d should include the concentrated loads produced by the weight of the horizontal and vertical nonstructural elements directly supported by the member, as defined in 4.5.3;

(b) For girders, beams, or joists located at the edge of floor slabs, p_d should include concentrated loads due to the weight of elements of the façade and the building enclosure; and

(c) p_ℓ should include any concentrated live load directly applied on the member.

8.2.3 *Factored load*

8.2.3.1 *Factored reactions*:

(a) For uniformly distributed reactions carried by the member, the largest value of r_u should be determined; and

(b) For concentrated loads carried by the member, the largest value of R_u should be determined for all concentrated load locations.

8.2.3.2 *Factored member loads*:

(a) For uniformly distributed loads carried directly by the member, w_u should be the greatest value determined by combining w_d and w_ℓ using load factors and load combinations specified in 4.2; and

(b) For all concentrated loads carried directly by the member, p_u should be the greatest value determined by combining p_d and p_ℓ using load factors and load combinations specified in 4.2 for each concentrated load.

8.2.3.3 *Factored total load*:

(a) Factored uniformly distributed total load W_u should be the sum of w_u, determined from 8.2.3.1, and r_u, determined from 8.2.3.2; and

(b) Factored concentrated total load P_u should be the sum of p_u, determined from 8.2.3.1, and R_u, determined from 8.2.3.2.

8.3—Reinforcement types

8.3.1 *General*—Reinforcement of girders, beams, and joists should comply with Chapter 8 and should be divided into longitudinal and transverse reinforcement. Longitudinal and transverse reinforcement in girders, beams, and joists have applications listed in Tables 8.3.1a and 8.3.1b and should comply with the sections listed. Figure 8.3.1 illustrates the principal types of reinforcement.

8.3.2 *Seismic*—In girders and beams supported directly on columns and reinforced concrete walls that are part of a moment-resisting frame in a dual system located in seismic zones, reinforcement should also comply with Chapter 11. Joists and beams not part of a frame are exempt from Chapter 11.

8.3.3 *Torsion*—When girders and beams are subjected to torsional moments, a minimum amount of transverse reinforcement as specified in 5.13.6 should be provided.

8.4—Longitudinal reinforcement

8.4.1 *General*—Section 8.4 should apply to all types of longitudinal reinforcement used in girders, beams, and joists.

8.4.2 *Minimum spacing*

8.4.2.1 *Longitudinal bars in a layer*—In girders, beams, and joists, the minimum clear spacing between parallel longitudinal bars in a layer should be the nominal bar diameter d_b of the largest bar but not less than 1 in. (25 mm) (Fig. 8.4.2.1). Clear distance limitation between bars should apply also to the clear distance between a contact lap splice and adjacent splices or bars.

8.4.2.2 *Parallel layers*—In girders, beams, and joists where parallel longitudinal bars are placed in two or more layers, bars in the upper layer should be placed directly

Table 8.3.1a—Types of longitudinal reinforcement in girders, beams, and joists

Longitudinal reinforcement	Flexural	Positive moment reinforcement	Section 8.4.14
		Negative moment reinforcement	Section 8.4.15
		Structural integrity	Section 6.3
	Other	Skin reinforcement	Section 8.4.12
		Reinforcement for support of stirrups	Section 8.4.15.6
		T-beam flange reinforcement	Section 8.4.11
		Longitudinal torsion reinforcement	Section 5.13.6

Table 8.3.1b—Types of transverse reinforcement in girders, beams, and joists

Transverse reinforcement	Stirrups	Shear reinforcement	Section 8.5.4
		Transverse support of compression reinforcement	Section 8.4.9.5
		Hanger reinforcement	Section 8.5.5
		Transverse torsion reinforcement	Section 5.13.6
	Hoops	Confinement in seismic zones	Section 11.1.2.3

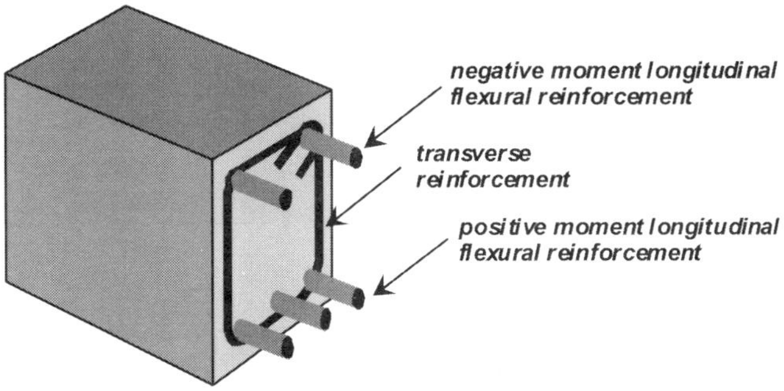

Fig. 8.3.1—Main types of reinforcement for girders, beams, and joists.

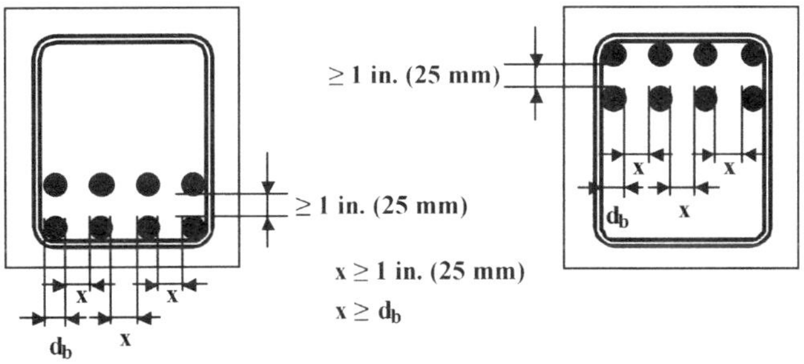

Fig. 8.4.2.1—Minimum clear spacing between bars in a layer and clear distance between layers of reinforcement.

above bars in the lower layer with clear distance between layers of at least 1 in. (25 mm) (Fig. 8.4.2.1).

8.4.3 *Maximum number of bars in a layer*—The maximum number of longitudinal bars in a layer should be computed considering longitudinal and transverse bar diameters, concrete cover (5.4), maximum nominal coarse aggregate size, and minimum clear spacing between bars (8.4.2). When these calculations are not performed, Sections 8.4.3.1 to 8.4.3.3 may be used.

8.4.3.1 *Girders and beams with* $b_w \geq$ *12 in. (300 mm)*—For girders and beams whose width b_w is equal to or greater than 12 in. (300 mm), the maximum number of bars in a layer may be determined using Eq. (8.4.3.1) (Table 8.4.3.2).

Table 8.4.3.2—Maximum number of longitudinal bars in a layer for girders and beams

Beam or girder web width b_w, in. (mm)	Maximum number of longitudinal bars
b_w < 8 in. (200 mm)	Section not permitted
8 in. (200 mm) ≤ b_w < 10 in. (250 mm)	Two bars
10 in. (250 mm) ≤ b_w < 12 in. (300 mm)	Three bars
12 in. (300 mm) ≤ b_w	$\leq \frac{b_w}{2} - 3 \left(\leq \frac{b_w}{50} - 3 \text{ (SI)}\right)$ bars

maximum number of bars in a layer

$$\leq \frac{b_w}{2} - 3 \quad \left(\leq \frac{b_w}{50} - 3 \text{ (SI)}\right) \qquad (8.4.3.1)$$

where b_w is in inches (mm).

8.4.3.2 *Girders and beams with* b_w < *12 in. (300 mm)*—Up to three longitudinal bars may be used in girders and beams whose width b_w is equal to or greater than 10 in. (250 mm) and less than 12 in. (300 mm). Two longitudinal bars should be used for b_w less than 10 in. (250 mm) (Table 8.4.3.2).

8.4.3.3 *Joists*—In joists with web widths b_w less than or equal to 6 in. (150 mm), either one longitudinal bar or a two-bar bundle in contact, located one above the other, is permitted. For web widths greater than 6 in. (150 mm) and less than 8 in. (200 mm), either one longitudinal bar or two bars in a single layer are permitted, but bundling of reinforcement is not permitted. For web widths equal to or greater than 8 in. (200 mm), the maximum number of bars in a single layer should be one more than those allowed for in Table 8.4.3.2.

8.4.4 *Minimum number of bars in a layer*—To minimize the number and width of flexural cracks at points of maximum moment, a large number of smaller-diameter bars should be provided rather than a small number of large-diameter bars. For joists, the minimum number of longitudinal bars should be one. For girders and beams with b_w less than 12 in. (300 mm),

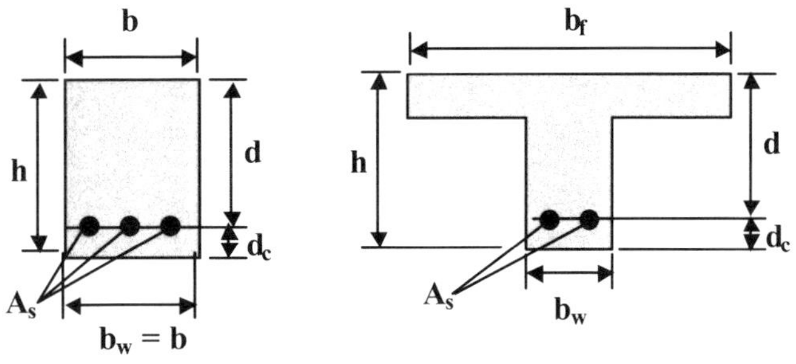

Fig. 8.4.5a—Rectangular section and T-shaped section with flange in compression.

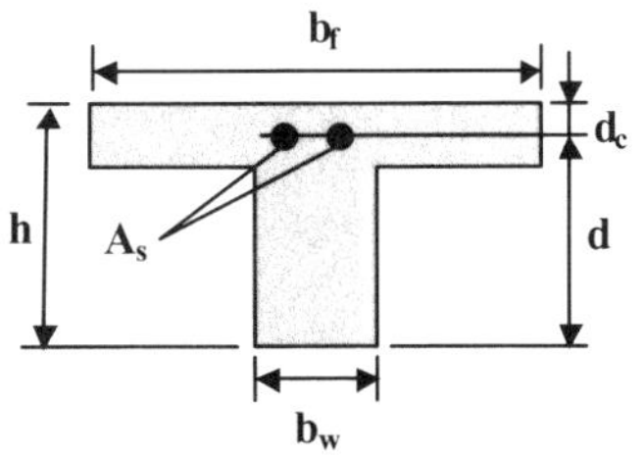

Fig. 8.4.5b—T-shaped section with flange in tension.

the minimum number of longitudinal bars should be two. For girders and beams whose width b_w is equal to or greater than 12 in. (300 mm), 8.4.4.1 and 8.4.4.2 should be met at sections of maximum positive and negative moment.

8.4.4.1 *Exterior exposure*—The minimum number of longitudinal bars in a layer for girders and beams exposed to earth or weather should be the value given by Eq. (8.4.4.1)

minimum number of bars in a layer

$$\leq \frac{b_w}{4} \left(\leq \frac{b_w}{100} \text{ (SI)} \right) \tag{8.4.4.1}$$

where b_w is in inches (mm).

8.4.4.2 *Interior exposure*—The minimum number of longitudinal bars in a layer for girders and beams not exposed to earth or weather should be the value given by Eq. (8.4.4.2)

minimum number of bars in a layer

$$\leq \frac{b_w}{8} \left(\leq \frac{b_w}{200} \text{ (SI)} \right) \tag{8.4.4.2}$$

where b_w is in inches (mm).

8.4.5 *Minimum area of flexural tension reinforcement*—Where Chapter 8 indicated that tension flexural reinforcement is needed, $A_{s,min}$ should be computed by (a) or (b), where ρ_{min} is the value given in Table 8.4.5:

(a) For rectangular sections and for T-shaped sections where the flange is in compression (Fig. 8.4.5a):

$$A_{s,min} = \rho_{min} d b_w \tag{8.4.5a}$$

Table 8.4.5—Minimum flexural reinforcement ratio ρ_{min} for girders, beams, and joists

	f_y,* psi (MPa)	
	40,000 (280)	**60,000 (420)**
ρ_{min}	0.0050	0.0033

*It is permitted to interpolate for different values of f_y.

Table 8.4.6—Maximum flexural reinforcement ratio ρ_{max} for girders, beams, and joists

		f_y,* psi (MPa)	
		40,000 (280)	**60,000 (420)**
f_c', psi (MPa)	3000 (21)	0.0280	0.0160
	3500 (24)	0.0325	0.0190
	4000 (28)	0.0370	0.0210
	4500 (31)	0.0400	0.0230
	5000 (35)	0.0435	0.0250

*It is permitted to interpolate for different values of f_y and f_c'.

(b) For cantilevers having a T-section where the flange is in tension (Fig. 8.4.5b), $A_{s,min}$ should be the smaller value computed from Eq. (8.4.5b) and (8.4.5c):

$$A_{s,min} = 2\rho_{min} d b_w \tag{8.4.5b}$$

$$A_{s,min} = \rho_{min} d b_f \tag{8.4.5c}$$

Refer to Section 8.4.10 for b_f calculation for T-section.

8.4.6 *Maximum flexural reinforcement ratio*—The ratio of tension flexural reinforcement, ρ, should be computed by (a) and (b), and should not exceed the values of ρ_{max} as given in Table 8.4.6:

(a) In girders, beams, and joists having only tension flexural reinforcement:

$$\rho = \frac{A_s}{bd} \leq \rho_{max} \tag{8.4.6a}$$

(b) In girders, beams, and joists having tension and compression flexural reinforcement (Fig. 8.4.6):

$$\rho - \rho' = \frac{A_s - A_s'}{bd} \leq \rho_{max} \tag{8.4.6b}$$

8.4.7 *Minimum design moment strength*—The section design moment strength ϕM_n should be equal to or greater than the required moment strength M_u.

$$\phi M_n \geq M_u \tag{8.4.7}$$

8.4.8 *Design moment strength for rectangular sections with tension reinforcement only*

8.4.8.1 *Design moment strength*—For girders, beams, and joists, the design moment strength of 5.11.4 may be approximated using Eq. (8.4.8.1) with ϕ = 0.90 (Fig. 8.4.8.1):

$$\phi M_n = \phi 0.85 A_s f_y d \tag{8.4.8.1}$$

8.4.8.2 *Flexural tension reinforcement area*—The flexural tension reinforcement area is approximately computed from Eq. (8.4.8.2).

$$A_s \geq \frac{M_u}{\phi 0.85 f_y d} \tag{8.4.8.2}$$

The flexural reinforcement ratio, $\rho = A_s/bd$, should be computed. When ρ is less than ρ_{min} as established in 8.4.5, A_s should be increased. If ρ is greater than ρ_{max} as established in 8.4.6, member dimensions should be increased, making the appropriate correction for self-weight, or the use of compression reinforcement should be investigated.

A note on employment of U.S. customary units. Computed moments generally are in units of lb·ft because they are determined from concentrated loads in lb, distributed loads in lb/ft, and spans in ft. They should be converted to lb·in (12 lb·in. = 1·lb·ft), for use with f_y in psi, d and b in in., and A_s in in.^{2}

A note on employment of SI units. The computed moments generally are in units of kN·m because they are determined from concentrated loads in kN, distributed loads in kN/m, and spans in m. They shall be converted to N·mm (1 kN·m = 10^6 × N·mm), for use with f_y in MPa (1 MPa = 1 N/mm^2), d and b in mm, and A_s in mm^2.

8.4.9 *Compression reinforcement in girders, beams, and joists*

8.4.9.1 *Tension reinforcement less than maximum*—For moment strength calculations, compression reinforcement is not necessary when the tension reinforcement ratio ρ is less than ρ_{max}.

8.4.9.2 *Shallow doubly reinforced sections*—For moment strength calculations, compression reinforcement is not necessary when the value of ratio d/d' is less than the values given in Table 8.4.9.2.

8.4.9.3 *Design moment strength with compression reinforcement*—For moment strength calculations with compression reinforcement, use Eq. (8.4.9.3) (Fig. 8.4.9.3):

$$\phi M_n = \phi\left[0.85\left(A_s - A_s'\right) f_y d + A_s' f_y \left(d - d'\right)\right] \tag{8.4.9.3}$$

where $\phi = 0.90$.

Equation (8.4.9.3) assumes the compression reinforcement yields, which should be verified.

8.4.9.4 *Flexural tension and compression reinforcement area*—The area of flexural tension reinforcement, A_s, and compression reinforcement, A_s', should be computed using M_u as follows

$$A_s' = \frac{M_u}{\phi f_y\left(d - d'\right)} - \frac{0.85\rho_{max} b d^2}{\left(d - d'\right)} \tag{8.4.9.4a}$$

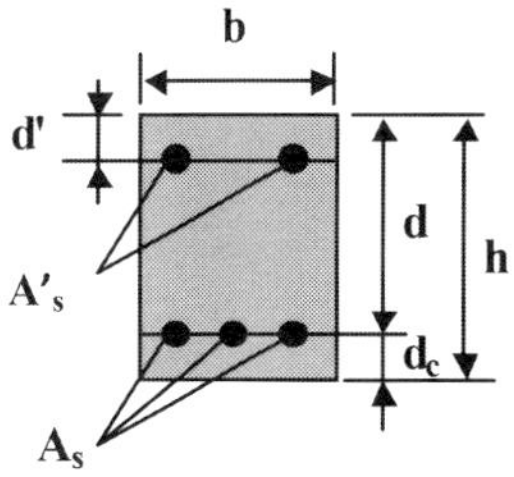

Fig. 8.4.6—Section with tension and compression reinforcement.

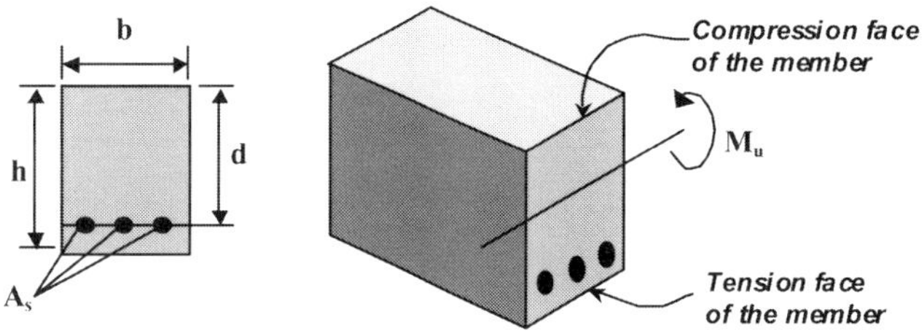

Fig. 8.4.8.1—Design dimensions for moment strength: tension reinforcement only.

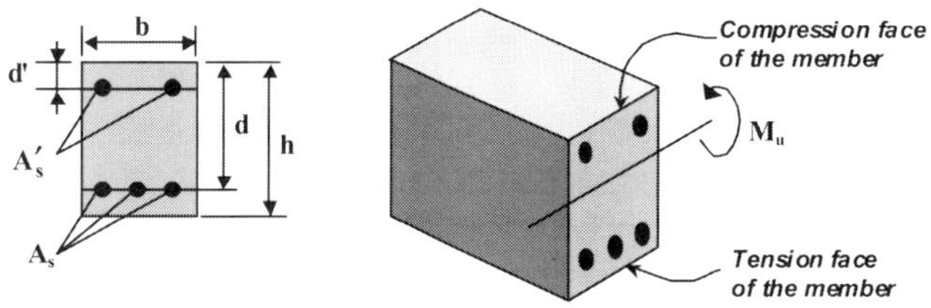

Fig. 8.4.9.3—Design dimensions for moment strength: sections with compression reinforcement.

Table 8.4.9.2 —Minimum values of *d/d'* for compression reinforcement to be effective

f_y, psi (MPa)	40,000 (280)	60,000 (420)
$d/d' \geq$	4	7

$$A_s = A_s' + \rho_{max} bd \tag{8.4.9.4b}$$

The ratio ρ_{max} should be determined from 8.4.6 when the minimum d/d' is met. After reinforcing bars are selected, ρ_{max}, as given in Eq. (8.4.6b), should be checked.

8.4.9.5 *Transverse reinforcement where flexural compression reinforcement is present*—Longitudinal flexural compression reinforcement should be enclosed by ties or stirrups satisfying the size and spacing limitations of column ties in 10.4.3.2. Such ties or stirrups should be provided throughout the distance where compression reinforcement is needed.

8.4.10 *T-beam effect*—Where a beam is monolithic with a slab and a moment induces compression in the slab, the slab may be assumed to act as a beam flange, and the design should comply with 8.4.10.1 to 8.4.10.5.

8.4.10.1 *Effective flange width for beams with slab on both sides*—The effective flange width b should not exceed the least of (a), (b), and (c) (Fig. 8.4.10.1):

(a) One-quarter the beam span length;

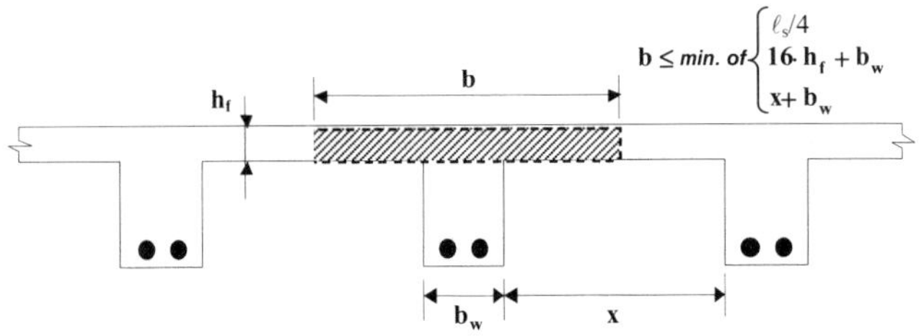

Fig. 8.4.10.1—Effective flange width for T-beams with slab on both sides.

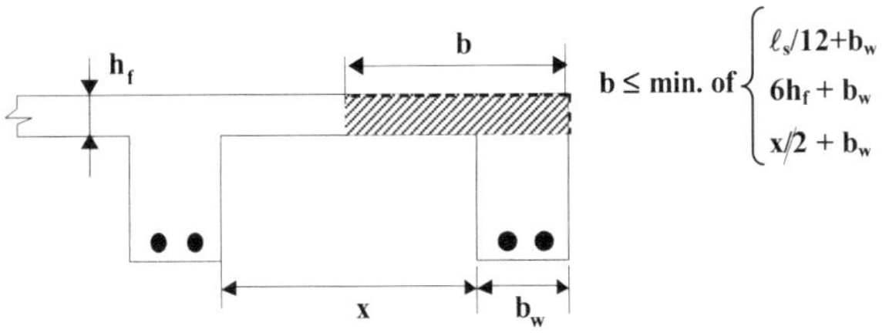

Fig. 8.4.10.2—Effective flange width for T-beams with slab on one side only

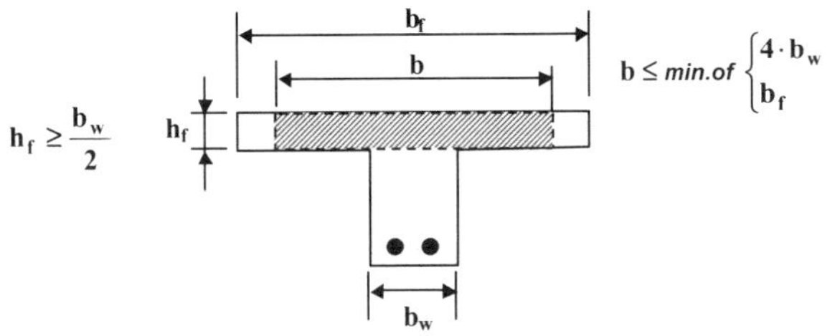

Fig. 8.4.10.3—Effective flange width for isolated T-beams.

(b) Sixteen times the slab thickness h_f plus the web thickness b_w; or

(c) The clear distance between webs plus the web thickness b_w.

8.4.10.2 *Effective flange width for beams with slab on one side only*—The effective flange width b should not exceed the least of (a), (b), and (c) (Fig. 8.4.10.2):

(a) One-twelfth the beam span length plus the web thickness b_w;

(b) Six times the flange thickness h_f plus the web thickness b_w; or

(c) One-half the clear distance to the next web plus the web thickness b_w.

8.4.10.3 *Isolated T-beams*— Flange thickness h_f in isolated T-beams should be at least one-half of the web thickness b_w, and the effective flange width b should not exceed the smaller of $4b_w$ and b_f (Fig. 8.4.10.3).

8.4.10.4 *Design moment strength of T-beams*—The design moment strength should be calculated using 8.4.8 for sections where the flange is in compression (Fig. 8.4.10.4), and the depth of the equivalent uniform stress block a lies within the flange thickness h_f, as computed by Eq. (8.4.10.4).

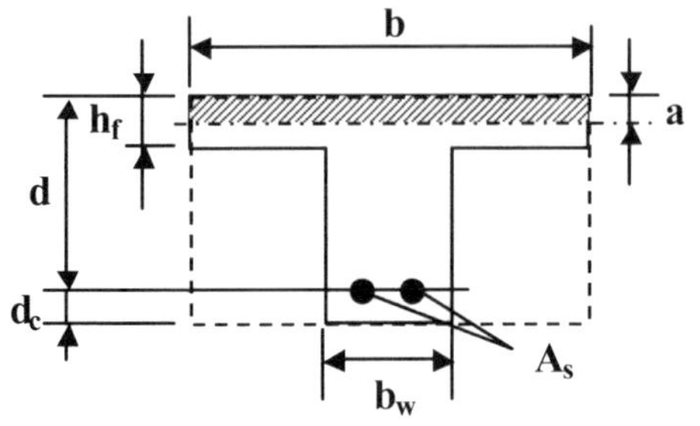

Fig. 8.4.10.4—Effective cross section for moment strength calculation of T-beams.

$$h_f \geq a \text{ and } a = \frac{A_s f_y}{0.85 f'_c b} \tag{8.4.10.4}$$

8.4.10.5 *Flexural tension reinforcement ratio*—When the value ρ given by Eq. (8.4.10.5) is not exceeded, the flexural reinforcement ratio, $\rho = A_s/(bd)$ for T-beams, should be computed using A_s from Eq. (8.4.8.2).

$$\rho \leq \frac{0.85 f'_c h_f}{f_y d} \tag{8.4.10.5}$$

Where the value of ρ is smaller than ρ_{min} from 8.4.5, A_s should be increased. Where the value of ρ is greater than ρ_{max} from 8.4.6, member dimensions should be increased, correcting the self-weight.

8.4.11 *Reinforcement in T-beam flanges*—The minimum T-beam flange reinforcement of 8.4.11.1 and 8.4.11.2 should apply to girders and beams but not to joist construction. Flange reinforcement should not be less than needed by the slab system.

8.4.11.1 *Distribution of negative moment reinforcement*—Where T-beam flanges are in tension, negative moment beam reinforcement should be distributed over a width equal to the smaller of the effective flange width as defined in 8.4.10, or one-tenth of the beam span. The portion of the effective flange width that exceeds one-tenth of the beam span should have a minimum of shrinkage and temperature reinforcement (7.3.3) in the beam direction. Refer to Fig. 8.4.11.1. The restrictions of 8.4.3 should not apply to this item.

8.4.11.2 *Transverse flange reinforcement*—In the slab, reinforcement perpendicular to the beam should resist a factored negative moment computed assuming the flange width acts as a cantilever supported by the beam (Fig. 8.4.11.1) and for isolated T-beams the full width of the overhanging flange. This reinforcement should also comply with 7.3.6.

8.4.12 *Skin reinforcement*—Where the height h of a girder, beam, or joist exceeds 36 in. (900 mm), longitudinal skin reinforcement should be provided along both side faces of the member for a vertical distance equal to $h/2$ nearest the flexural tension reinforcement. Vertical spacing s_{sk} between bars should not exceed the least value of Eq. (8.4.12), $d/6$, and 12 in. (300 mm) (Fig. 8.4.12).

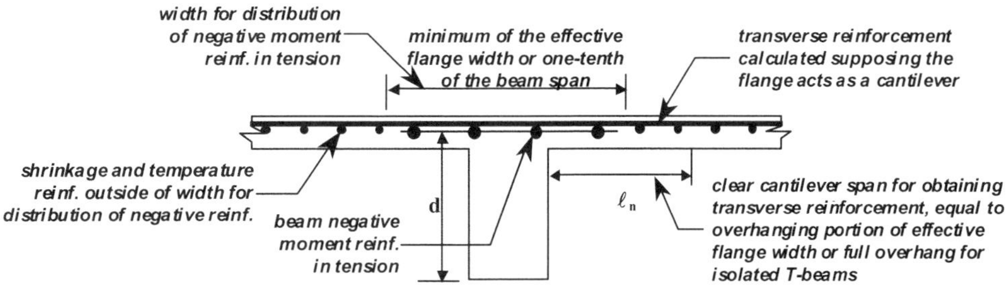

Fig. 8.4.11.1—Reinforcement in T-beam flanges.

$$s_{sk} = \frac{900{,}000}{f_y} - 2.5c_c \le \frac{720{,}000}{f_y}$$

$$\left[s_{sk} = \frac{159{,}600}{f_y} - 2.5c_c \le \frac{126{,}000}{f_y} \text{ (SI)} \right] \tag{8.4.12}$$

8.4.13 *Value of* d_c *and* d *to use in girders, beams, and joists*—Calculation of d_c, the distance from extreme tension fiber to centroid of tension reinforcement, should consider concrete cover from 5.4, bar diameters, and reinforcement layers.

The following values of d_c can be used to compute d as $d = h - d_c$, for cases where only one reinforcement layer is used. For girders and beams, d_c = 2.4 in. (60 mm) for interior exposure, and d_c = 2.8 in. (70 mm) for exterior exposure. For joists, d_c = 2 in. (50 mm) for interior exposure, and d_c = 2.4 in. (60 mm) for exterior exposure.

8.4.14 *Positive moment reinforcement*

8.4.14.1 *Description*—Positive moment reinforcement should be provided in the girder, beam, or joist section, as indicated in Chapter 8, and should comply with 8.4 and the particular limitations for each member type in 8.6 or 8.7.

8.4.14.2 *Location*—Positive moment reinforcement should be placed longitudinally in the girder, beam, or joist. Positive moment reinforcement should be located as close to the bottom surface of the girder, beams, or joist as practicable following the concrete cover of 5.4. Where girders, beams, or joists support other girders, beams, or joists, the positive moment reinforcement of the supported member should be placed above the positive moment reinforcement of the supporting member.

8.4.14.3 *Cutoff amount*—No more than one-half the positive moment reinforcement at midspan may be cut off at the locations indicated in 8.6.5 or 8.7.5.

8.4.14.4 *Reinforcement splicing*—The remaining positive moment reinforcement from 8.4.14.3 may be lap spliced between the cutoff point and the opposite face of the support.

8.4.14.5 *Embedment at interior supports*—Positive moment reinforcement terminated at an interior support should be continued to the opposite face of the support plus the lap splice distance of 5.8.2.

8.4.14.6 *End anchorage of reinforcement*—At the end of the girder, beam, or joist, the positive moment reinforcement should extend to the edge and end in a standard hook.

8.4.15 *Negative moment reinforcement*

8.4.15.1 *Description*—Negative moment reinforcement should be provided in the girder, beam, or joist section, at

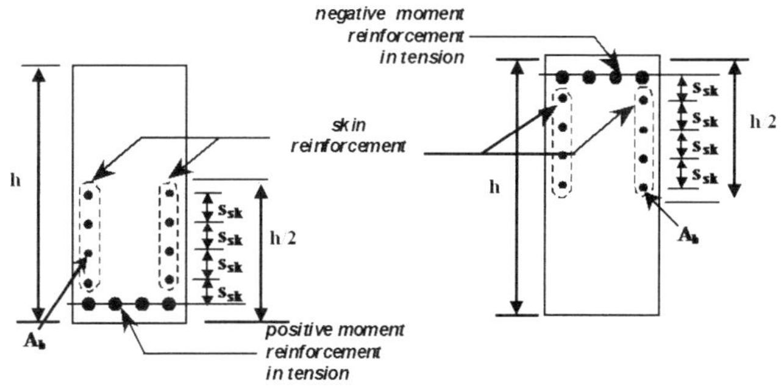

Fig. 8.4.12—Skin reinforcement for girders, beams, and joists with h > *36 in. (900 mm).*

edges and supports as indicated in Chapter 8, and should comply with 8.4 and the particular provisions of 8.6 or 8.7.

8.4.15.2 *Location*—Negative moment reinforcement should be provided at all supports and located as close to the upper surface of the girder, beam, or joist as practicable following concrete cover of 5.4. At supports where girders or beams intersect, the negative moment reinforcement of the member with the longer span should be located on top.

8.4.15.3 *Cutoff amount*—Negative moment reinforcement at the locations indicated in 8.6.5 or 8.7.5 may be cut off, except cantilever negative moment reinforcement is not allowed to be cut off. Where adjacent spans are unequal, negative moment reinforcement cutoff points should be based on the longer span.

8.4.15.4 *Reinforcement splicing*—Negative moment reinforcement between cutoff point and the support should not be lap spliced.

8.4.15.5 *End anchorage*—Negative moment reinforcement at the end of a girder, beam, or joist should end in a standard hook at the far edge of the supporting girder, beam, column, or reinforced concrete wall, complying with anchorage distance described by 5.8.3. At the external edge of cantilevers, negative moment reinforcement should end in a standard hook.

8.4.15.6 *Stirrup support*—In areas where no negative reinforcement is needed, top bars should be provided for attachment and anchorage of stirrups. The diameter of these top bars should be equal to or greater than the stirrup bar diameter. Minimum lap length of these top bars should be 6 in. (150 mm).

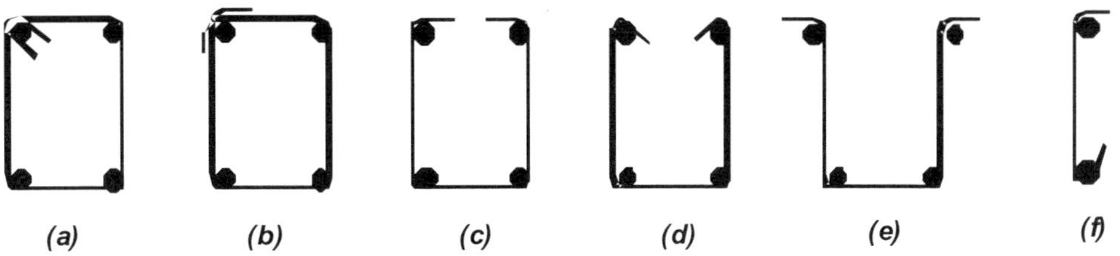

Fig. 8.5.2a—Typical stirrup shapes for girders and beams.

Fig. 8.5.2b—Typical stirrup shape for joists, in addition to Fig. 8.5.2a.

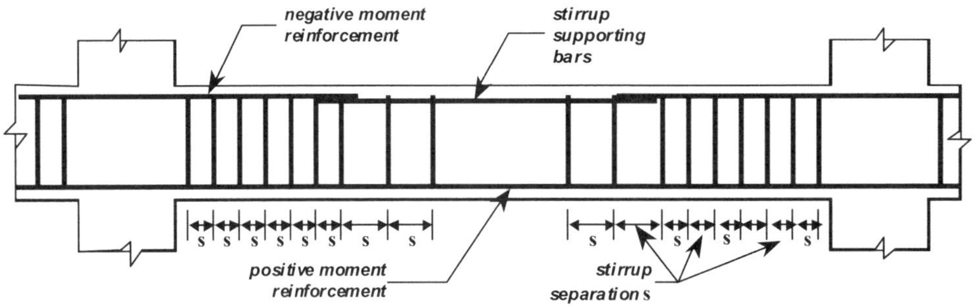

Fig. 8.5.3—Typical stirrup spacing along the girder, beam, or joist.

8.5—Transverse reinforcement

8.5.1 *Description*—Transverse reinforcement for girders, beams, and joists should consist of stirrups that enclose the longitudinal reinforcement and are placed perpendicular to the longitudinal axis of the member at varying intervals.

The main functions for transverse reinforcement in girders, beams, and joists are:

(a) Contribute to member shear strength;

(b) Provide lateral support for longitudinal reinforcement subjected to compression stresses;

(c) Act as hanger reinforcement in girders, supporting beams, and joists;

(d) Contribute to member torsion strength; and

(e) Provide confinement to the concrete in seismic zones at selected locations within the member.

8.5.2 *Stirrup shape*—A stirrup should consist of single or multiple vertical legs. Each vertical leg should engage a longitudinal bar either by bending around it when the stirrup continues or by using a standard stirrup hook (5.6) to surround the longitudinal bar at the stirrup end (Fig. 8.5.2a and 8.5.2b).

8.5.2.1 *Permitted stirrup shape for girders and beams*—All stirrups in girders and beams should be closed stirrups with 135-degree hooks, as shown in Fig. 8.5.2a(a). The other stirrup shapes are in common use but are shown to clarify that they are not to be used with this guide. In seismic areas, the stirrup shape is further limited.

8.5.2.2 *Permitted stirrup shape for joists*—All stirrup types shown in Fig. 8.5.2a and 8.5.2b may be used in joists.

8.5.2.3 *Minimum clear spacing between stirrups legs*—In girders, beams, and joists, the minimum clear space between stirrups or parallel legs in a stirrup should be 1 in. (25 mm).

8.5.2.4 *Stirrup support*—Stirrups should be attached to longitudinal bars so that stirrups do not displace during concrete placement (8.4.15.6).

8.5.2.5 *Stirrup leg splicing*—Stirrup bars should not be lap spliced.

8.5.3 *Location of transverse reinforcement*—Stirrup spacing intervals s should comply with 8.5.4.5 (Fig. 8.5.3).

8.5.4 *Contribution of transverse reinforcement to shear strength*

8.5.4.1 *General*—Beam-action shear accompanies flexural moments and occurs in girders, beams, and joists along their length, and is of greater magnitude in the vicinity of supports and concentrated loads.

8.5.4.2 *Design shear strength*—Design shear strength ϕV_n of a girder, beam, or joist section should be computed following the procedure in 5.13.4 for beam-action shear as:

$$\phi V_n = \phi\left(V_c + V_s\right) \quad (8.5.4.2)$$

where ϕV_c is the concrete contribution to the design shear strength; ϕV_s is the shear reinforcement contribution to the design shear strength; and $\phi = 0.75$.

8.5.4.3 *Contribution of concrete to shear strength*—At each location to be investigated (Fig. 8.5.4.3), the concrete contribution of the web of the girder, beam, or joist should be taken into account (Fig. 8.5.3) and should be computed using Eq. (8.5.4.3), with $\phi = 0.75$.

$$\phi V_c = \phi 2\sqrt{f'_c} b_w d \quad \left[\phi V_c = \phi 0.17\sqrt{f'_c} b_w d \text{ (SI)}\right] \quad (8.5.4.3)$$

Table 8.5.4.5—Shear reinforcement in girders, beams, and joists: maximum spacing *s*

Value of factored required shear strength V_u	Limiting value of ϕV_s	Minimum area of shear reinforcement A_v within a distance s	Maximum spacing s
$\frac{\phi V_c}{2} > V_u$		Not needed	
$\phi V_c > V_u \geq \frac{\phi V_c}{2}$		$A_v = 0.75\sqrt{f'_c}\frac{b_w s}{f_{yt}}$ $\left[A_v = 0.062\sqrt{f'_c}\frac{b_w s}{f_{yt}} \text{ (SI)}\right]$	$s \leq \text{min. of} \begin{cases} d/2 \\ 24 \text{ in. (600 mm)} \end{cases}$
$V_u \geq \phi V_c$	$2\phi V_c > \phi V_s$	$A_v = \frac{(V_u - \phi V_c)s}{\phi f_{yt} d}$	$s \leq \text{min. of} \begin{cases} d/2 \\ 24 \text{ in. (600 mm)} \\ A_v f_{yt}/(50 b_w) \\ \left[A_v f_{yt}/(0.35 b_w) \text{ (SI)}\right] \end{cases}$
	$4\phi V_c > \phi V_s > 2\phi V_c$	$A_v = \frac{(V_u - \phi V_c)s}{\phi f_{yt} d}$	$s \leq \text{min. of} \begin{cases} d/4 \\ 12 \text{ in. (300 mm)} \\ A_v f_{yt}/(50 b_w) \\ \left[A_v f_{yt}/(0.35 b_w) \text{ (SI)}\right] \end{cases}$
	$\phi V_s \geq 4\phi V_c$	Not permitted	

8.5.4.4 *Contribution of transverse reinforcement to shear strength*—For reinforcement perpendicular to the member axis, its contribution to design shear strength should be:

$$\phi V_s = \phi \frac{A_v f_{yt} d}{s} \qquad (8.5.4.4a)$$

where A_v is the area of shear reinforcement perpendicular to the member axis (the stirrup bar area A_b multiplied by the number of vertical stirrup legs) within a distance s; f_{yt} is the yield strength of the shear reinforcement steel; and $\phi = 0.75$.

Reinforcement contribution to the design shear strength should not be taken greater than:

$$\phi V_s \leq \phi 8\sqrt{f'_c} b_w d = 4\phi V_c \quad \left[\phi V_s \leq \phi 0.66\sqrt{f'_c} b_w d = 4\phi V_c \text{ (SI)}\right] \qquad (8.5.4.4b)$$

8.5.4.5 *Design of shear reinforcement*—Shear reinforcement in girders, beams, and joists should be provided using stirrups perpendicular to the member axis with a maximum spacing s, measured along member axis:

(a) Where the factored shear V_u is less than $\phi V_c/2$, the use of shear reinforcement may be waived;

(b) Where V_u exceeds $\phi V_c/2$ and is less than ϕV_c, a minimum area of shear reinforcement should be provided as specified by Eq. (8.5.4.5). The spacing s along the member axis should not exceed the smaller of $d/2$ and 24 in. (600 mm) (Fig. 8.5.4.5).

$$A_v = 0.75\sqrt{f'_c}\frac{b_w s}{f_{yt}} \quad \left[A_v = 0.062\sqrt{f'_c}\frac{b_w s}{f_{yt}} \text{ (SI)}\right] \qquad (8.5.4.5)$$

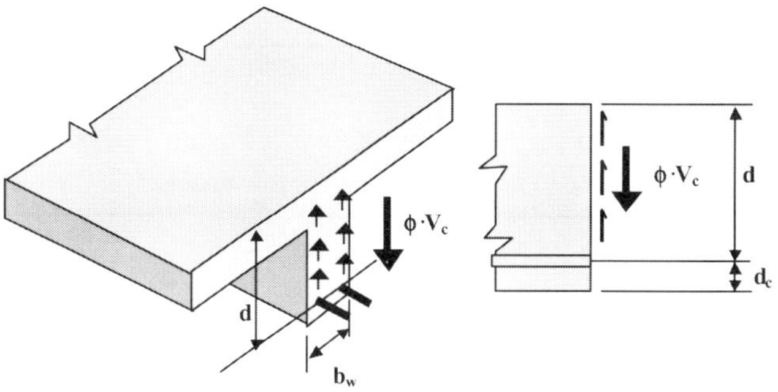

Fig. 8.5.4.3—Contribution of concrete to beam-action shear strength in girders, beams, and joists.

where A_v is A_b multiplied by the number of stirrup legs; and

(c) Where V_u exceeds ϕV_c, the difference $(V_u - \phi V_c)$ should be provided for by shear reinforcement, using Eq. (8.5.4.3) and (8.5.4.4a), and the limitations (1) through (4) should apply (Table 8.5.4.5):

(1) Minimum shear reinforcement area should not be less than that determined using Eq. (8.5.4.5);

(2) Where the value of ϕV_s, calculated using Eq. (8.5.4.4a), is less than $2\phi V_c$, the spacing limits of (b) should be used;

(3) Where the value of ϕV_s, calculated using Eq. (8.5.4.4a), is greater than $2\phi V_c$, spacing limits should be half the values specified in (b); and

(4) The value of ϕV_s, calculated using Eq. (8.5.4.4a), should not be taken greater than $4\phi V_c$.

8.5.4.6 *Shear diagram*—The value of V_u at the support face should be determined in conformance with 8.6 or 8.7. A diagram showing the shear variation within the span should be constructed, with the value of V_u at the left support face taken as positive. The shear from this point proceeding to the right should be decreased at a rate equal to:

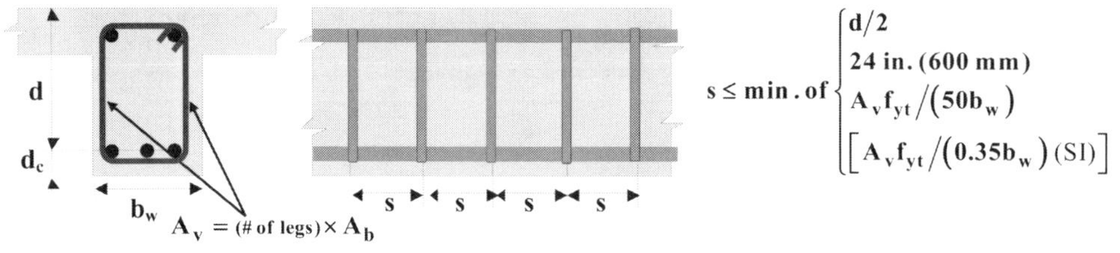

Fig. 8.5.4.5—Minimum shear reinforcement in girders, beams, and joists when ($\phi V_c/2 \le V_u < \phi V_c$).

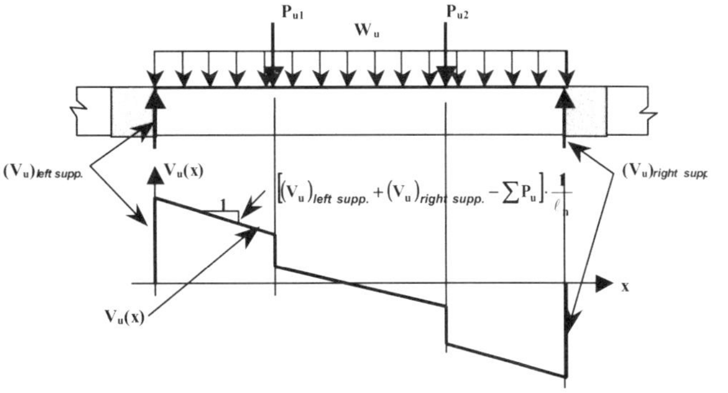

Fig. 8.5.4.6—Calculation of the shear diagram of a girder, beam, or joist.

$$\frac{\left[\left(V_u\right)_{\text{left supp.}} + \left(V_u\right)_{\text{right supp.}} - \Sigma P_u\right]}{\ell_n} \qquad (8.5.4.6)$$

where ΣP_u corresponds to the sum of all factored concentrated loads on the span. At the point where a concentrated load is applied, the value of P_u should be subtracted from the value of shear at the left of the load point. For beams with point loads, proceeding to the right, at the face of the right support, the negative value of V_u is reached (Fig. 8.5.4.6). At all sections within the span, the value of ϕV_n, as determined from Eq. (8.5.4.2), should be equal to or greater than the absolute value of $V_u(x)$, as shown in Fig. 8.5.4.6.

Limits for ϕV_n, as defined in Table 8.5.4.5, should be marked in the shear diagram, and stirrup spacing s should be defined for different regions within the shear diagram. The first stirrup should be placed not further than $s/2$ from the face of the supporting member, with s being the stirrup spacing at the support. The minimum stirrup spacing should comply with 8.5.2.3. If the computed s is less than 2 in. (50 mm), using stirrups with more vertical legs or a larger bar diameter should be investigated.

8.5.5 *Hanger stirrups*—Where a beam is supported by a girder of similar depth, hanger reinforcement should be provided in the joint. The reaction from the supported beam tends to push down the bottom of the supporting girder. This reaction should be resisted by hanger reinforcement in the form of closed stirrups placed in both members. Hanger stirrups are in addition to stirrups needed for shear (Fig. 8.5.5) and should comply with 8.5.5.1 and 8.5.5.2.

8.5.5.1 *Hanger stirrup area:*

(a) Provide hanger stirrups where V_u from the supported beam at the interface is equal to or greater than $\phi 3\sqrt{f'_c}b_w d$ $\left[\phi 0.25\sqrt{f'_c}b_w d\right]$, where $\phi = 0.75$;

(b) Provide hanger stirrups where h_b is equal to or less than one-half the total depth of the supporting girder, where h_b is the vertical dimension from the bottom of the supporting girder to the bottom of the supported beam (Fig. 8.5.5);

(c) The area of hanger reinforcement, A_i, should be determined from Eq. (8.5.5.1):

$$A_i \ge \frac{\left[1 - \left(h_b/h_g\right)\right]V_u}{\phi f_{yt}} \qquad (8.5.5.1)$$

where V_u is the beam factored shear at the support face; A_i is the total area of hanger stirrups; h_g is the girder height; f_{yt} is the stirrup specified yield strength; and $\phi = 0.75$.

8.5.5.2 *Hanger stirrup placement*—At least two-thirds of A_i should be evenly distributed within the supported beam width b_w, plus h_b at each side. The remaining area of hanger stirrups, not more than one-third of A_i, should be evenly distributed within $d/4$ from the supporting girder face, where d is the supported beam effective depth. Beam bottom longitudinal bars should be placed above the girder bottom longitudinal bars.

8.6—Joists and beams supported by girders

8.6.1 *General*—Section 8.6 applies to joists and beams, monolithic with and supported by girders. Two-way joist systems or waffle-on-beams systems as described in 6.1.3.3 should also comply with 8.6. Waffle-slab systems without beams spanning between columns as described in 6.1.4.4 should be designed using Chapter 9 for slab-column systems.

8.6.2 *Dimensional limits*

8.6.2.1 *Joists*—In addition to Chapter 8, joists should comply with the dimensional limits of 1.3 and the restrictions of 6.1.3.1. Ducts, shafts, and openings should comply with 6.8. Minimum depth should comply with 6.5.3 for one-way joists and 6.5.4 for two-way joists.

8.6.2.2 *Beams*—In addition to Chapter 8, beams supported by girders should comply with the dimensional limits of 1.3 and the restrictions of 6.1.2. Ducts, shafts, and openings should comply with 6.8. Minimum depth should comply with 6.5.3. Beam web width b_w should not be less than 8 in.

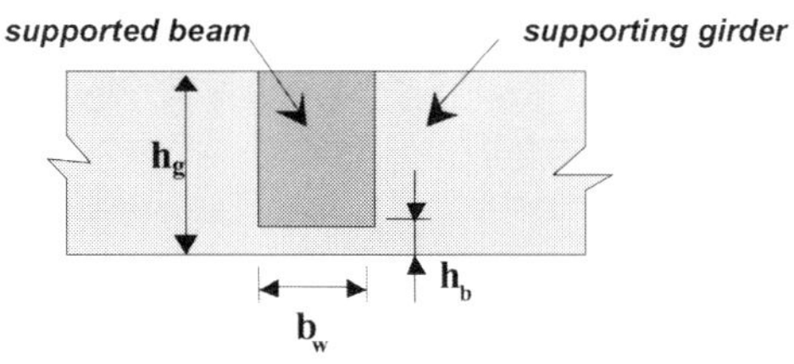

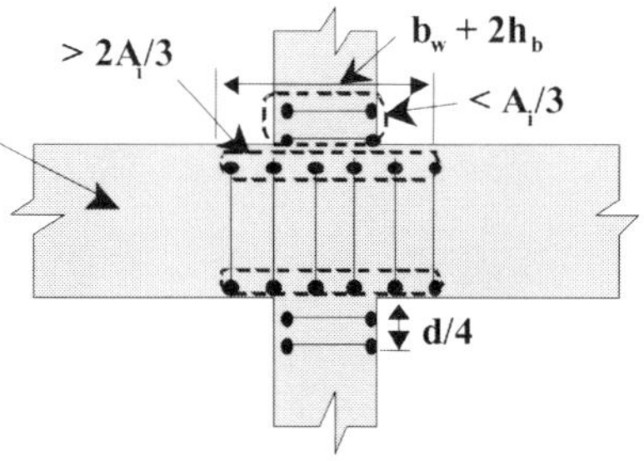

Fig. 8.5.5—Hanger reinforcement.

(200 mm). Maximum spacing between lateral supports of isolated beams should be 50 times the least width *b* of the compression flange.

8.6.2.3 *Cantilevers*—All cantilevers of joists or beams should be continuous with at least one interior span. A double cantilever without an interior span is not permitted.

8.6.3 *Required moment strength*

8.6.3.1 *Cantilevers*—The factored negative moment M_u (required moment strength) for beam and joist cantilevers supported by girders, beams, or reinforced concrete walls should be calculated using Eq. (8.6.3.1), assuming:

(a) One-half of the distributed factored load W_u acts as a concentrated load at the cantilever tip along with all concentrated loads that act on the cantilever span ΣP_u; and

(b) One-half W_u acts as uniformly distributed load over the full span.

$$M_u^- = \frac{3W_u\ell_n^2}{4} + \ell_n \Sigma P_u \qquad (8.6.3.1)$$

The cantilever required negative moment strength at the support should equal or exceed the maximum negative factored moment at the first interior support and one-third the positive factored moment of the first interior span.

8.6.3.2 *Single-span joists and beams supported by beams, girders, or reinforced concrete walls*—Factored positive and negative moment M_u (required moment strength) for single-span beams and single-span one-way joists should be calculated using Table 8.6.3.2, where ΣP_u is the sum of all factored concentrated loads that act on the span.

8.6.3.3 *Multi-span joists and beams supported by beams, girders, or walls*—Factored positive and negative moment M_u (required moment strength) for beams and one-way joists, with two or more spans, supported by beams, girders, or reinforced concrete walls, should be calculated using Table 8.6.3.3, where ΣP_u is the sum of all factored concentrated loads that act on the span.

8.6.3.4 *Use of frame analysis*—Frame analysis, which meets 8.1.2, may be used to determine factored moments and shears as a substitute for values in 8.6.3.1 to 8.6.3.3.

8.6.3.5 *Two-way joists supported by beams, girders, or walls*—Required moment strength for two-way joists supported by beams, girders, or structural walls may be obtained using 7.9.1 and 7.9.2, ignoring the minimum depth of the supporting beams or girders as given by 7.9.1(c) and 6.1.3.3.

Table 8.6.3.2—Factored moment for single-span beams and joists

Positive moment:	
$M_u^+ = \frac{W_u\ell_n^2}{8} + \frac{\ell_n}{4}\Sigma P_u$	(8.6.3.2a)
Negative moment at supports:	
$M_u^- = \frac{W_u\ell_n^2}{24} + \frac{\ell_n}{16}\Sigma P_u$	(8.6.3.2b)

8.6.4 *Required shear strength*

8.6.4.1 *Cantilevers of joists and beams supported by beams, girders, or walls*—Factored shear V_u at the cantilever support should be calculated using Eq. (8.6.4.1):

$$V_u = W_u\ell_n + \Sigma P_u \qquad (8.6.4.1)$$

where ΣP_u is the sum of all factored concentrated loads that act on the span.

8.6.4.2 *Single-span joists and beams supported by beams, girders, or walls*—Factored shear V_u for single-span beams and single-span one-way joists should be calculated using Eq. (8.6.4.2):

$$V_u = \frac{W_u\ell_n}{2} + 0.8\Sigma P_u \qquad (8.6.4.2)$$

where ΣP_u is the sum of all factored concentrated loads that act on the span.

8.6.4.3 *Joists and beams supported by beams, girders, or walls, with two or more spans*—Factored shear V_u for beams and one-way joists with two or more spans supported by beams, girders, or structural walls should be calculated using equations given in Table 8.6.4.3, where ΣP_u corresponds to the sum of all factored concentrated loads on the span.

8.6.4.4 *Use of frame analysis*—Frame analysis meeting 8.1.2 may be used to determine factored shear as a substitute for the values determined from 8.6.4.1 to 8.6.4.3.

8.6.4.5 *Two-way joists supported by beams, girders, or walls*—Factored shear for two-way joists supported

Table 8.6.3.3—Factored moment for beams and one-way joists with two or more spans

Positive moment at	
End spans: $M_u^+ = \frac{W_u \ell_n^2}{11} + \frac{\ell_n}{9}\Sigma P_u$	(8.6.3.3a)
Interior spans: $M_u^+ = \frac{W_u \ell_n^2}{16} + \frac{\ell_n}{5}\Sigma P_u$	(8.6.3.3b)
Negative moment at	
Interior face of external support: $M_u^- = \frac{W_u \ell_n^2}{24} + \frac{\ell_n}{16}\Sigma P_u$	(8.6.3.3c)
Exterior face of first internal support, only two spans: $M_u^- = \frac{W_u \ell_n^2}{9} + \frac{\ell_n}{6}\Sigma P_u$	(8.6.3.3d)
Faces of internal supports, more than two spans: $M_u^- = \frac{W_u \ell_n^2}{10} + \frac{\ell_n}{7}\Sigma P_u$	(8.6.3.3e)
Faces of all supports for joists with spans not exceeding 10 ft (3 m): $M_u^- = \frac{W_u \ell_n^2}{12} + \frac{\ell_n}{8}\Sigma P_u$	(8.6.3.3f)

Table 8.6.4.3—Factored shear for beams and one-way joists with two or more spans at

Exterior face of first interior support: $V_u = 1.15\frac{W_u \ell_n}{2} + 0.8\Sigma P_u$	(8.6.4.3a)
Faces of all other supports: $V_u = \frac{W_u \ell_n}{2} + 0.75\Sigma P_u$	(8.6.4.3b)

by beams, girders, or structural walls may be determined using 7.9.1 and 7.9.4, ignoring the minimum depth of the supporting beams or girders indicated in 7.9.1(c) and 6.1.3.3.

8.6.5 *Reinforcement*

8.6.5.1 *Positive moment reinforcement*—The positive moment reinforcement area should be determined for the calculated value of M_u^+. When a slab is present in the upper part of the section or when the beam or joist is T-shaped, the T-beam effect may be used. Positive moment reinforcement should comply with 8.4.14. At internal supports, at a distance equal to $\ell_n/8$ measured from the face of the supports into the span, up to one-half the positive moment reinforcement may be cut off if there are no concentrated loads within that distance (Fig. 8.6.5.1). For single-span beams and joists, positive moment reinforcement should not be cut off.

8.6.5.2 *Negative moment reinforcement*—The negative moment reinforcement area should be determined for the larger value of M_u^- calculated for both sides of the support. This reinforcement should comply with 8.4.15. When a slab is present in the upper part of the section or when the beam or joist is T-shaped, negative moment reinforcement should comply with 8.4.11.1. At a distance equal to $\ell_n/4$ for external supports and $\ell_n/3$ for internal supports, measured from the internal face of the support into the span, all negative moment reinforcement may be cut off (Fig. 8.6.5.1). Negative moment reinforcement should not be cut off in cantilevers.

8.6.5.3 *Transverse reinforcement*—Values of V_u at the right and left support faces should be determined by the appropriate equation from 8.6.4. The transverse reinforcement should comply with 8.5.

8.6.6 *Reactions on beams and girders*

8.6.6.1 *One-way joists*—Factored reaction on the joist system support may be considered as uniformly distributed. Factored reaction on the supports, r_u, per unit length, should be the value determined from Eq. (8.6.6.1) plus the uniformly distributed reaction from any cantilever spanning from that support.

$$r_u = \frac{V_u \ell_s}{s_j \ell_n} \quad (8.6.6.1)$$

where V_u is the factored shear from 8.6.4; ℓ_s is the center-to-center span of the joist; ℓ_n is the clear span of the joist; and s_j is the center-to-center spacing between parallel joists.

8.6.6.2 *Two-way joists supported by beams, girders, or walls*—Factored reactions for two-way joists supported by beams, girders, or structural walls may be calculated using 7.9.1 and 7.9.5, ignoring minimum depth of the supporting beams or girders given in 7.9.1(c) and 6.1.3.3.

8.6.6.3 *Beams*—Factored reactions on the supports, R_u, should be the values determined from Eq. (8.6.6.3) plus the factored reaction from any cantilever spanning from that support.

$$R_u = \frac{V_u \ell_s}{\ell_n} \quad (8.6.6.3)$$

where V_u is the factored shear from 8.6.4; ℓ_s is the center-to-center span; and ℓ_n is the clear span of the beam.

8.7—Girders that are part of a frame

8.7.1 *General*—Section 8.7 applies to girders that are part of a moment-resisting frame where the girders are monolithic and are directly supported by columns or reinforced concrete walls.

8.7.2 *Dimensional limits*

8.7.2.1 *General*—In addition to Chapter 8, girders that are part of a frame should comply with the dimensional limits set forth in 1.3. Embedded conduits and pipes should comply with 6.8.

8.7.2.2 *Girder depth and width*—The girder should be prismatic without haunches, brackets, or corbels. The height h should comply with the minimum depth of 6.5.3. Clear span of the member should not be less than four times its height h. The width-to-height ratio $b_w h$ should not be less than 0.3. The width b_w should not be less than 8 in. (200 mm), nor exceed the corresponding width of the supporting column plus $3/4h$ on each side of the supporting column (Fig. 8.7.2.2).

8.7.2.3 *Girders supported by reinforced concrete walls*—Girders, supported by a reinforced concrete wall located in the plane of the frame, should continue along the full horizontal wall length. Girder width should not be less than wall thickness. Where girders are supported by walls perpendicular to the longitudinal axis of the girder, a beam should run

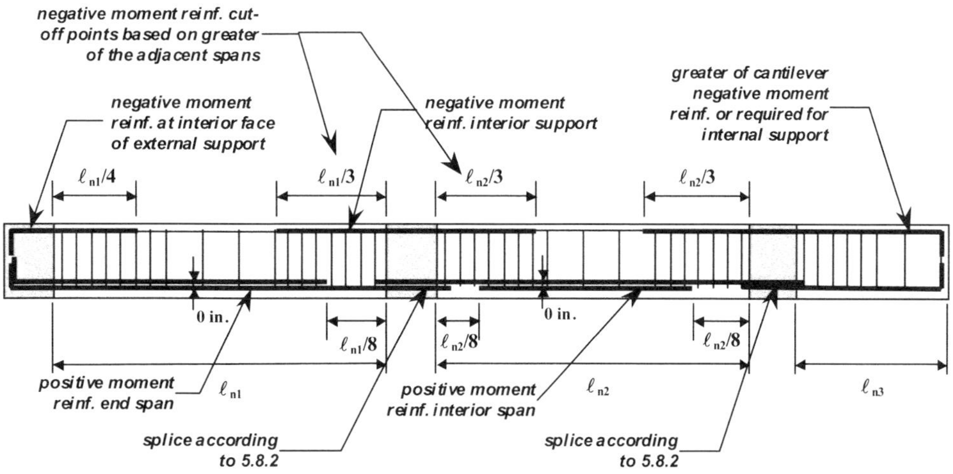

Fig. 8.6.5.1—Reinforcement for beams and joists supported by beams or girders.

along the full horizontal wall length at the same level and have the same height as the girder. Beam width should not be less than wall thickness or 8 in. (200 mm). Vertical reinforcement of the wall should pass through the girder or beam as indicated in Chapter 12.

8.7.2.4 *Lateral support*—For girders not continuously laterally supported by the floor slab or secondary beams, the clear distance between lateral supports should not exceed 50 times the least width b of compression flange or face.

8.7.2.5 *Restrictions*—The following restrictions should be in effect for girders of frames designed under 8.7:

(a) There are two or more spans;

(b) Spans are approximately equal, with the shorter of two adjacent spans greater than or equal to 80 percent of the larger span (1.3);

(c) Loads are uniformly distributed and adjustments for concentrated loads are performed;

(d) Unfactored unit live load w_ℓ does not exceed three times unfactored unit dead load w_d; and

(e) Girders should not have a slope exceeding 15 degrees.

8.7.3 *Required moment strength*

8.7.3.1 *Factored positive and negative moment*—Factored positive and negative moment M_u (required moment strength) for girders and beams that are part of a frame whose vertical members are columns and concrete structural walls should be calculated using equations in Table 8.7.3.1, where ΣP_u is the sum of all factored concentrated loads that act on the span.

8.7.3.2 *Girders parallel to one-way joist systems*—For girders parallel to joists, the assumed tributary width for the calculation of girder factored loads should be twice the joist spacing plus the girder width.

8.7.3.3 *Use of frame analysis*—Frame analysis meeting 8.1.2 may be used to determine factored moments and shears as a substitute for values computed from 8.7.3.1 and 8.7.4.1.

8.7.4 *Required shear strength*

8.7.4.1 *Factored shear*—Girder V_u should be calculated at support faces using equations of Table 8.7.4.1, where ΣP_u is the sum of all factored concentrated loads that act on the span. Refer to 8.7.3.2 for assumed tributary width.

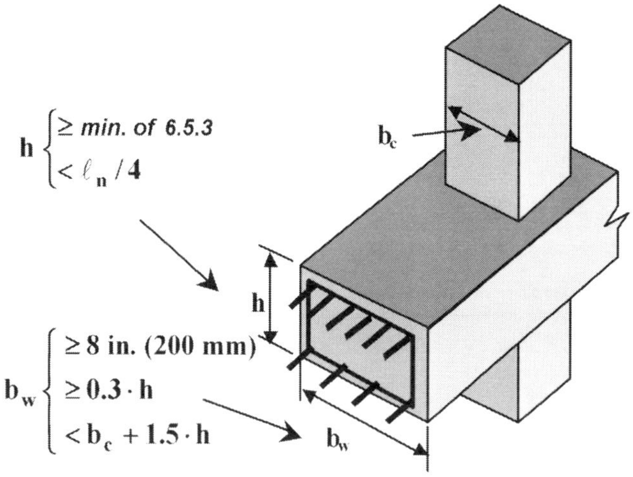

Fig. 8.7.2.2—Limits on girder depth and width.

8.7.4.2 *Use of frame analysis*—Frame analysis, meeting 8.1.2, may be used to determine factored shear as a substitute for values computed from 8.7.4.1.

8.7.5 *Reinforcement*

8.7.5.1 *Positive moment reinforcement*—The positive moment reinforcement area should be determined using the calculated value of M_u^+. If a slab exists in the upper part of the girder, the T-beam effect may be taken into account. Positive moment reinforcement should comply with 8.4.14. At internal supports, at a distance equal to $\ell_n/8$ measured from the support face into the span, up to one-half of the positive moment reinforcement may be cut off if there are no concentrated loads within that distance (Fig. 8.7.5.1a and 8.7.5.1b).

8.7.5.2 *Negative moment reinforcement*—The negative moment reinforcement area should be determined using the larger value of M_u^- computed at both sides of the support. This reinforcement should comply with 8.4.15. When a slab is present in the upper part of the section or when the girder is T-shaped, negative moment reinforcement should comply with 8.4.11.1. For perimeter beams, as indicated in 6.3.2, one-quarter of the negative moment reinforcement should

Table 8.7.3.1—Factored moment for girders of frames

Positive moment at	
End spans: $M_u^+ = \frac{W_u \ell_n^2}{14} + \frac{\ell_n}{6} \Sigma P_u$	(8.7.3.2a)
Interior spans: $M_u^+ = \frac{W_u \ell_n^2}{16} + \frac{\ell_n}{7} \Sigma P_u$	(8.7.3.2b)
Negative moment at supports at	
Interior face of external column or perpendicular structural wall: $M_u^- = \frac{W_u \ell_n^2}{16} + \frac{\ell_n}{10} \Sigma P_u$	(8.7.3.2c)
Exterior face of first internal column or perpendicular structural wall, only two spans: $M_u^- = \frac{W_u \ell_n^2}{9} + \frac{\ell_n}{6} \Sigma P_u$	(8.7.3.2d)
Faces of internal columns or perpendicular structural walls, more than two spans: $M_u^- = \frac{W_u \ell_n^2}{10} + \frac{\ell_n}{6.5} \Sigma P_u$	(8.7.3.2e)
Faces of structural walls parallel to the plane of the frame: $M_u^- = \frac{W_u \ell_n^2}{12} + \frac{\ell_n}{7} \Sigma P_u$	(8.7.3.2f)
Support of girder cantilevers: $M_u^- = \frac{3W_u \ell_n^2}{4} + \ell_n \Sigma P_u$	(8.7.3.2g)

Table 8.7.4.1—Factored shear for girders of frames

At exterior face of first interior column: $V_u = 1.15\frac{W_u \ell_n}{2} + 0.8\Sigma P_u$	(8.7.4.1a)
At faces of all other columns: $V_u = \frac{W_u \ell_n}{2} + 0.75\Sigma P_u$	(8.7.4.1b)
At supports of girder cantilevers: $V_u = W_u \ell_n + \Sigma P_u$	(8.7.4.1c)

either be continuous or spliced at midspan (Fig. 8.7.5.1b). Negative moment reinforcement in cantilevers should not be cut off. For other beams, all negative moment reinforcement may be cut off at a distance equal to $\ell_n/3$ for internal supports, and at a distance equal to $\ell_n/4$ for external supports, measured from the support face into the span (Fig. 8.7.5.1a).

8.7.5.3 *Transverse reinforcement*—Values of V_u at the right and left faces of the supports should be determined using the appropriate equation from 8.7.4. Transverse reinforcement should meet 8.5.

8.7.5.4 *Hanger stirrups*—Where a beam is supported by a girder of similar depth, the use of hanger stirrups as indicated by 8.5.5 should be investigated.

8.7.6 *Reactions on columns and reinforced concrete walls*

8.7.6.1 *Vertical reactions at columns and walls*—Factored reactions at the supports, R_u, should be the values determined from Eq. (8.7.6.1) plus the factored reaction from any cantilever spanning from that support.

$$R_u = \frac{V_u \ell_s}{\ell_n} \quad (8.7.6.1)$$

where V_u is the factored shear from 8.7.4; ℓ_s is the center-to-center span; and ℓ_n is the clear span of the girder.

8.7.6.2 *Unbalanced moment from vertical loading*—Moment reaction on columns should be evaluated using the unbalanced factored moment ΔM_u, caused by the factored vertical loads on girders that span from the column in the plane of the frame. The unbalanced moment should be distributed to the column above and below the girder in proportion to their relative stiffness. To calculate the unbalanced moment, (a), (b), and (c) should apply, or a frame analysis complying with 8.1.2 should be performed:

(a) The unbalanced moment ΔM_u is the largest difference in girder factored negative moments at the column when load cases in (b) and (c) are evaluated;

(b) In the first case (Case A of Fig. 8.7.6.2), the entire girder should be loaded with the factored dead load and alternate odd spans should be loaded with the factored live load; and

(c) In the second case (Case B of Fig. 8.7.6.2), the entire girder should be loaded with the factored dead load, and the alternate even spans should be loaded with the factored live load.

8.7.6.3 *Unbalanced moment distribution*—To distribute the unbalanced moment to columns or walls above and below the girder, (a) applies to roof joints, and (b) and (c) apply to joints with columns' top and bottom, or a frame analysis complying with 8.1.2 should be performed:

(a) For roof joints (Types D, E, and F of Fig. 8.7.6.3), the column factored-moment should correspond to ΔM_u;

(b) For joints with columns top and bottom (Types A, B, and C of Fig. 8.7.6.3), the unbalanced moment should be distributed to the column, or wall, above using Eq. (8.7.6.3a):

$$(M_u)_{up} = \Delta M_u \frac{(I_c/h_{pi})_{up}}{(I_c/h_{pi})_{up} + (I_c/h_{pi})_{down}} \quad (8.7.6.3a)$$

(c) For joints with columns top and bottom (Types A, B, and C of Fig. 8.7.6.3), the unbalanced moment should be distributed to the column, or wall, below using Eq. (8.7.6.3b):

$$(M_u)_{down} = \Delta M_u \frac{(I_c/h_{pi})_{down}}{(I_c/h_{pi})_{up} + (I_c/h_{pi})_{down}} \quad (8.7.6.3b)$$

(d) In Eq. (8.7.6.3a) and (8.7.6.3b), I_c should be evaluated using Eq. (8.7.6.3c) (Fig. 8.7.6.3).

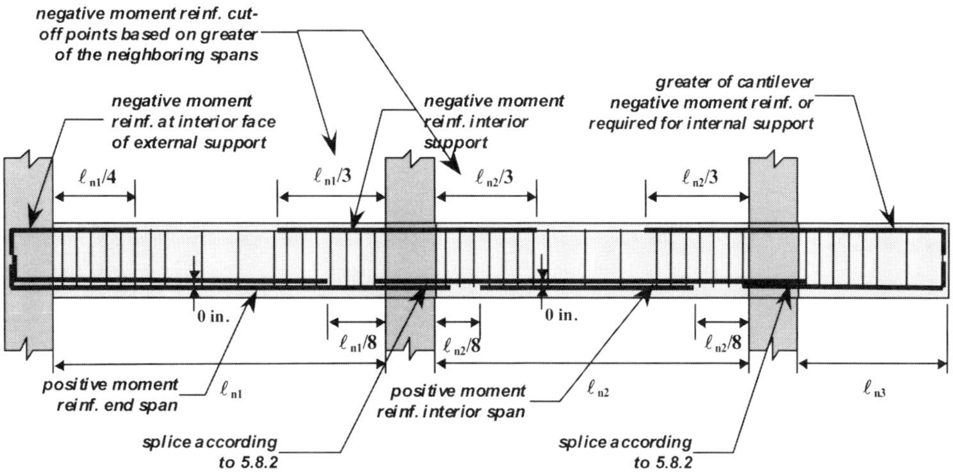

Fig. 8.7.5.1a—Reinforcement in girders that are part of a moment-resisting frame supported by columns or reinforced concrete walls.

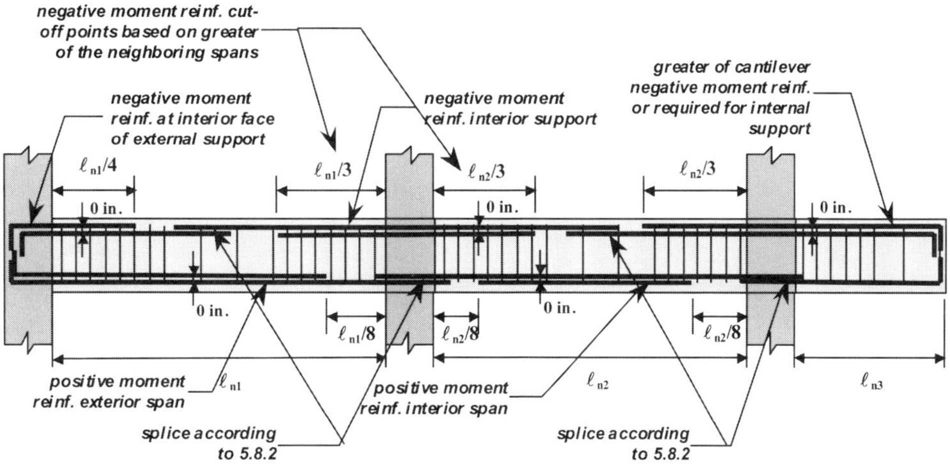

Fig. 8.7.5.1b—Reinforcement in girders that are part of the perimeter frame.

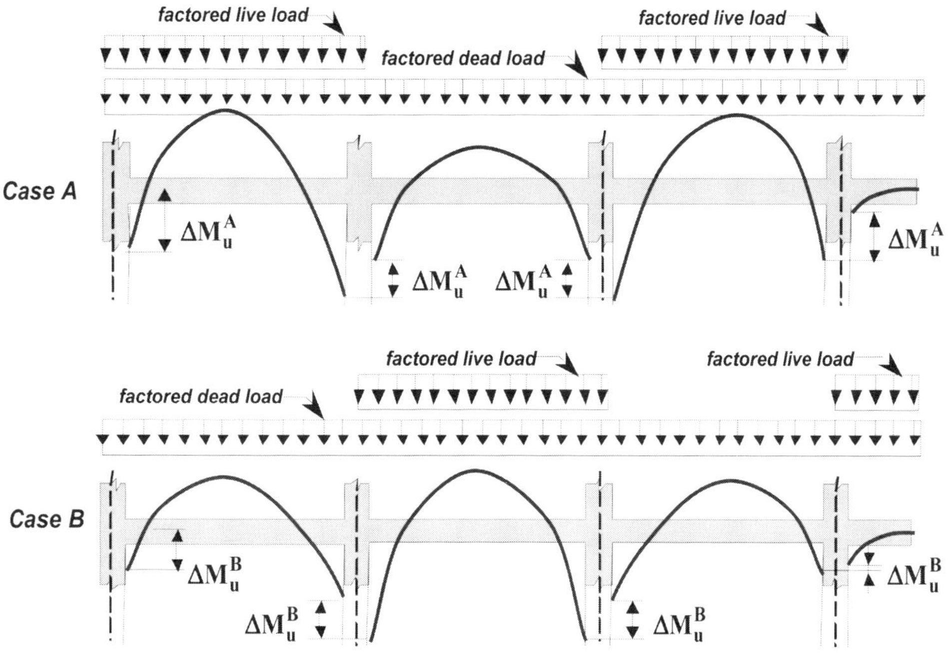

Fig. 8.7.6.2—Girder unbalanced moment to be transferred to columns.

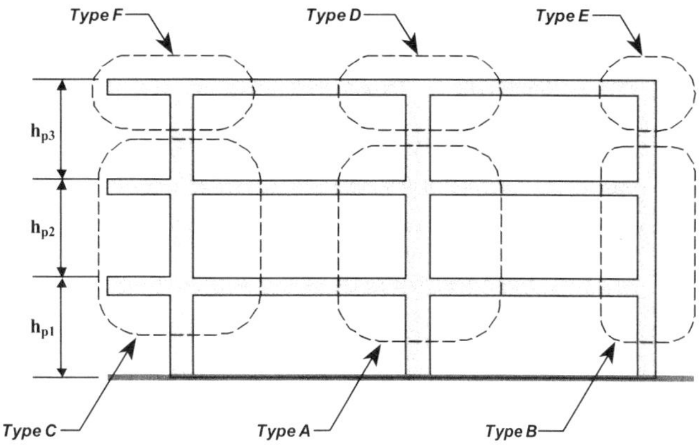

Fig. 8.7.6.3—Joint types for determination of column moments.

$$I_c = \frac{b_c h_c^3}{12} \quad (8.7.6.3c)$$

where b_c is the column dimension, or wall section in the direction perpendicular to the girder span; h_c is the column dimension, or wall section in the direction parallel to the girder span; and h_{pi} is the column or wall story height.

CHAPTER 9—SLAB-COLUMN SYSTEMS

9.1—General

Slabs in slab-column systems as described in 6.1.4 should be designed using Chapter 9. The design of waffle-slabs as described in 6.1.4.4 is included.

9.2—Loads

9.2.1 *Loads to be included*—Service loads for slabs that are part of slab-column systems should be established from Chapter 4. Gravity loads that should be included in the design are:

(a) Dead loads: Slab self-weight, flat nonstructural elements, standing nonstructural elements, and fixed equipment loads, if any;

(b) Live loads; and

(c) Where the slab is part of the roof system, appropriate values of roof live load, rain load, and snow load should be used.

9.2.2 *Dead load and live load*—Values of q_d for dead load should include slab self-weight and the weight of the horizontal and vertical nonstructural elements as defined in 4.5.3. The value q_ℓ for live load should be determined by 4.6. Where the slab is part of the roof system, roof live load given in 4.7, rain load in 4.8, and snow loads in 4.9 should be included, as appropriate.

9.2.3 *Factored loads*—Factored load q_u should be the largest value determined using load factors and load combinations in 4.2.

9.3—Dimensional limits

9.3.1 *General*—In addition to Chapter 9, the slab in slab-column systems should comply with the dimensional limits in 1.3, and the restrictions of 6.1.4 for slab-column systems.

9.3.2 *Column strip*—A column strip is a design strip with a width on each side of a column centerline equal to $\ell_2/4$ or $\ell_1/4$, whichever is less (Fig. 9.3.2). Where the strip is adjacent and parallel to an edge, the column-strip width on the external side should be the distance from the column centerline to the edge, without exceeding the width of the internal side. For waffle-slabs, the column strip comprises all joists that engage the column capital. Refer to 6.1.4.4 for the minimum number of joists that should engage the capital.

9.3.3 *Middle strip*—A middle strip is a design strip bounded by two column strips (Fig. 9.3.2).

9.3.4 *Definition of a panel*—A panel is bounded by column or wall centerlines on all sides (Fig. 9.3.2).

9.3.5 *Minimum slab thickness*—Minimum slab thickness to meet the serviceability limit state of slab-column systems should conform to 6.5.5.

9.3.6 *Restriction on column dimensions*—In slab-column systems, the ratio of long-to-short cross-sectional column dimensions should not exceed 2.

9.3.7 *Support dimensions and clear span*—For a slab system supported by columns or walls, the clear span ℓ_n should extend from face-to-face of columns, capitals, brackets, or walls.

9.3.8 *Other restrictions*—Slab-column systems should comply with (a) through (f):

(a) There should be a minimum of three continuous spans in each direction;

(b) Panels should be rectangular, with a ratio of longer to shorter span center-to-center of supports within a panel not greater than 2;

(c) Successive span lengths measured center-to-center of supports in each direction should not be smaller than 80 percent of the adjacent larger span, except in elevator and stair cores (1.3.6);

(d) Columns may be offset by a maximum of 10 percent of the span (in direction of offset) from either axis between centerlines of successive columns;

(e) All loads should be due to gravity only and be uniformly distributed over an entire panel; and

(f) Unfactored live load q_ℓ should not exceed twice the unfactored dead load q_d.

9.3.9 *Ducts, shafts, and openings*

9.3.9.1 *General.* Ducts, shafts, and openings in the slab should comply with 9.3.9.2, 9.3.9.3, and 9.3.9.4.

9.3.9.2 *Openings in middle strips*—Openings of any size are permitted in the area common to intersecting middle strips, provided total area of reinforcement in the panel without the opening is maintained.

9.3.9.3 *Openings in column strips*—In the area common to intersecting column strips, not more than one-eighth the column strip width in either span should be interrupted by openings. An area of reinforcement equivalent to that interrupted by an opening should be added on the sides of the opening. When an opening is located in the column strip or at a distance less than 10 times the slab thickness from the support face, or within the capital zone in waffle-systems, the critical section perimeter b_o defined in 9.5.4.2 should be modified by removing that part of the perimeter enclosed by

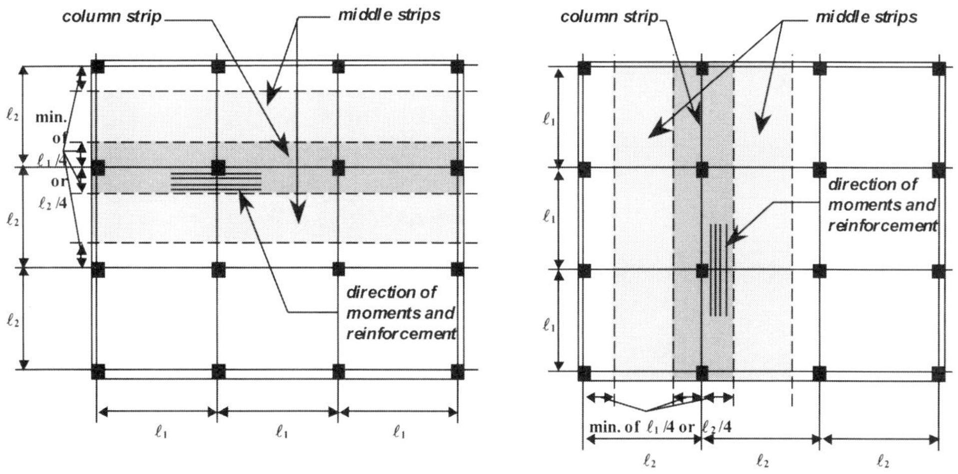

Fig. 9.3.2—Column and middle strips.

straight lines projecting from the centroid of the support and tangent to the boundaries of the opening (Fig. 9.3.9.3).

9.3.9.4 *Openings in zones common to column and middle strips*—In the area common to one column strip and one middle strip, not more than one-quarter of the reinforcement in either strip should be interrupted by openings. An area of reinforcement equivalent to that interrupted by an opening should be added on the sides of the opening.

9.3.10 *Drop panels*—Drop panels should comply with the dimensional limits of 6.1.4.4 (Fig. 6.1.4.3d). In slab reinforcement calculations, the drop panel thickness below the slab should not be assumed greater than one-quarter the distance from drop panel edge to edge of column or column capital.

9.3.11 *Waffle-slab systems*—Joists, slabs spanning between joists, and capitals of two-way waffle-slab systems should comply with the dimensional limits of 6.1.3 and 6.1.4.4.

9.4—Reinforcement details

9.4.1 *General*—Slab reinforcement of slab-column systems should be of the types described and should comply with 9.4, in addition to (a) through (f) (Fig 9.4.1):

(a) Slab reinforcement area in each direction should be determined from moments at critical sections, but should not be less than shrinkage and temperature reinforcement in 7.3.3;

(b) Flexural reinforcement should comply with 7.3.4;

(c) Slab reinforcement in waffle-slabs over cellular spaces should comply with 7.5;

(d) Joists in waffle-slab systems should comply with 8.4, flexural reinforcement, and 8.5, transverse reinforcement;

(e) Reinforcement in joists of waffle-slabs should meet 8.6.5 for joists located in the middle strip and of 8.7.5 for joists located in the column strip; and

(f) Reinforcement in waffle-slabs need not comply with 9.4.2 to 9.4.4.

9.4.2 *Positive moment reinforcement*

9.4.2.1 *Description*—Positive moment reinforcement should be provided in the lower part of the slab section, as indicated in Chapter 9, and should comply with 9.4.2.

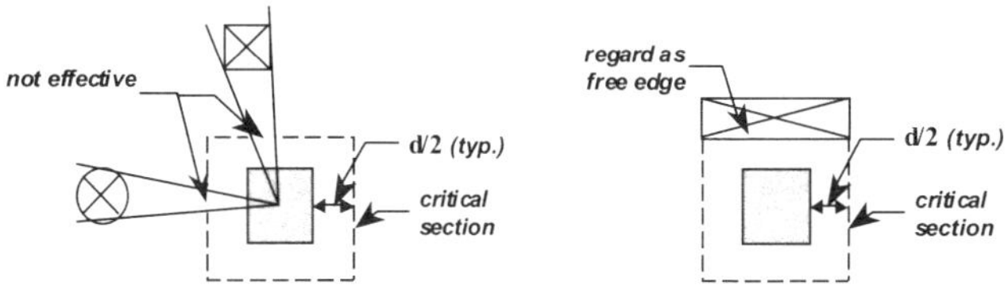

Fig. 9.3.9.3—Effect of openings in the slab.

9.4.2.2 *Location*—Positive moment reinforcement should be provided in both directions. Positive moment reinforcement should be located as close to the bottom surface of the slab as practicable following 5.4.1. Short-span positive moment flexural reinforcement should be located below the long-span positive moment flexural reinforcement.

9.4.2.3 *Minimum reinforcement area*—Positive moment reinforcement should have an area at least equal to the area determined by 7.3.3.4.

9.4.2.4 *Maximum reinforcement area*—Positive moment reinforcement area should not exceed values in 7.3.4.3.

9.4.2.5 *Maximum reinforcement spacing*—Positive moment reinforcement at critical sections should not be spaced further apart than twice the slab thickness.

9.4.2.6 *Cutoff points*—All bottom bars within the column strip, in each direction, should be continuous or spliced complying with 5.8.2 at locations shown in Fig. 9.4.2.6. At least two column strip bottom bars in each direction should pass within the column core. In the middle strip at interior supports, up to one-half of the positive moment reinforcement needed to resist the corresponding factored positive moment at midspan may be cut off at the locations indicated in Fig. 9.4.2.6.

9.4.2.7 *Reinforcement splicing*—Continuous positive moment reinforcement may be lap-spliced between the cutoff point and the opposite face of the support.

9.4.2.8 *Embedment at interior supports*—Positive moment reinforcement cut off at an interior support should extend to the opposite face of the support.

9.4.2.9 *End anchorage of reinforcement*—Positive moment reinforcement perpendicular to a discontinuous

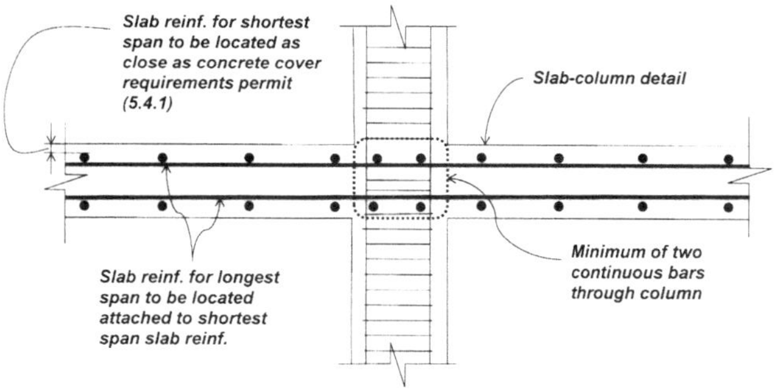

Fig. 9.4.1—Slab reinforcement detail.

edge should extend to the slab edge and should end in a standard hook.

9.4.3 *Negative moment reinforcement*

9.4.3.1 *Description*—Negative moment reinforcement should be provided in the amounts and lengths conforming to Chapter 9 and should comply with 9.4.3.

9.4.3.2 *Location*—Negative moment reinforcement should be provided in both directions at supports. Negative moment flexural reinforcement should be located as close to the upper surface of the slab as practicable following the concrete cover of 5.4.1. Short-span negative moment flexural reinforcement should be located above long-span negative moment flexural reinforcement.

9.4.3.3 *Minimum reinforcement area*—Negative moment reinforcement should have an area at least equal to the area determined by 7.3.3.4.

9.4.3.4 *Maximum reinforcement area*—Negative moment reinforcement area should not exceed the values in 7.3.4.3.

9.4.3.5 *Maximum reinforcement spacing*—Negative moment reinforcement at critical sections should not be spaced further apart than twice the slab thickness

9.4.3.6 *Cutoff points*—All negative moment reinforcement, except for cantilevers, may be cut off at the locations indicated in Fig. 9.4.2.6. Where adjacent spans are unequal, negative moment flexural reinforcement cutoff points should be based on the longer span.

9.4.3.7 *Reinforcement splicing*—It is not permitted to lap splice negative moment reinforcement between the cutoff point and the support.

9.4.3.8 *End anchorage of reinforcement*—Negative moment reinforcement perpendicular to a discontinuous edge should be anchored with a standard hook into the edge. At the external edge of cantilevers, negative moment flexural reinforcement perpendicular to the edge should end in a standard hook. Where constrained by geometry, the end hook need not be placed vertically.

9.4.4 *Shear reinforcement*—Design procedures for solid slabs do not include the use of shear reinforcement in slabs. Procedures for design of shear reinforcement in solid slabs are beyond the scope of this guide, and the standards of 1.4 should be used for design in that case. The use of shear reinforcement for beam-action shear in joists of waffle-slabs is permitted.

9.4.5 *Values of* d_c *and* d *to use in slab-column systems*—Calculation of the distance from extreme tension fiber to centroid of tension reinforcement, d_c, should consider concrete cover from 5.4, bar diameters, and the existence of reinforcement in the perpendicular direction placed between the reinforcement under consideration and the concrete surface.

The following values of d_c could be used for computing d as $d = h - d_c$. For reinforcement in the long direction of the slab panel, d_c = 2.2 in. (55 mm) for interior exposure, and d_c = 3 in. (75 mm) for exterior exposure. For reinforcement in the short direction of the slab panel, d_c = 1.6 in. (40 mm) for interior exposure, and d_c = 2.4 in. (60 mm) for exterior exposure. For joists that are part of waffle-slabs, d_c = 2 in. (50 mm) for interior exposure, and d_c = 2.4 in. (60 mm) for exterior exposure. Refer to 9.3.10 for slab zones where drop panels exist.

9.4.6 *Slab reinforcement in seismic zones*—A slab-column system should not be used in seismic zones unless the structural system is stiffened by reinforced concrete structural walls acting in both principal directions. In seismic zones, slab reinforcement should also comply with Chapter 11.

9.5—Shear strength

9.5.1 *General*—In slab-column systems, calculation of the design strength of slabs subjected to shear loads should be performed using 9.5. Two types of shear force effects occur in the vicinity of supports and concentrated loads:

(a) Punching-shear or two-way action shear; and

(b) Beam-action shear that accompanies flexural moments.

9.5.2 *Required shear strength*—Factored shear V_u (required shear strength) in the slab due to factored loads applied to the structure should be determined from 9.6.2 and 9.7.2.

9.5.3 *Design shear strength*—The design shear strength at the member section, ϕV_n, should be equal to or greater than the factored shear V_u, as shown in Eq. (9.5.3) with $\phi = 0.75$.

$$\phi V_n = \phi V_c \geq V_u \tag{9.5.3}$$

9.5.4 *Two-way action shear (punching shear) in solid slabs and footings*

9.5.4.1 *General*—Shear strength for two-way action shear, or punching-shear, should be investigated at edges of columns, concentrated loads, and supports, and at changes of thickness such as edges of capitals and drop panels.

9.5.4.2 *Definition of critical section for two-way shear*—Each of the critical sections investigated should be located so that the perimeter b_o is as defined in Fig. 9.5.4.2 but need not approach closer than a distance $d/2$ to edges or corners of columns and to changes in slab thickness, such as edges of capitals or drop panels.

9.5.4.3 *Two-way action shear strength*—Shear strength should be the value determined from Eq. (9.5.4.3) with ϕ = 0.75:

$$\phi V_n = \phi V_c = \phi \lambda_{ps} 4\sqrt{f'_c} b_o d$$
$$\left(\phi V_n = \phi V_c = \phi \lambda_{ps} \frac{\sqrt{f'_c}}{3} b_o d \text{ (SI)} \right) \tag{9.5.4.3}$$

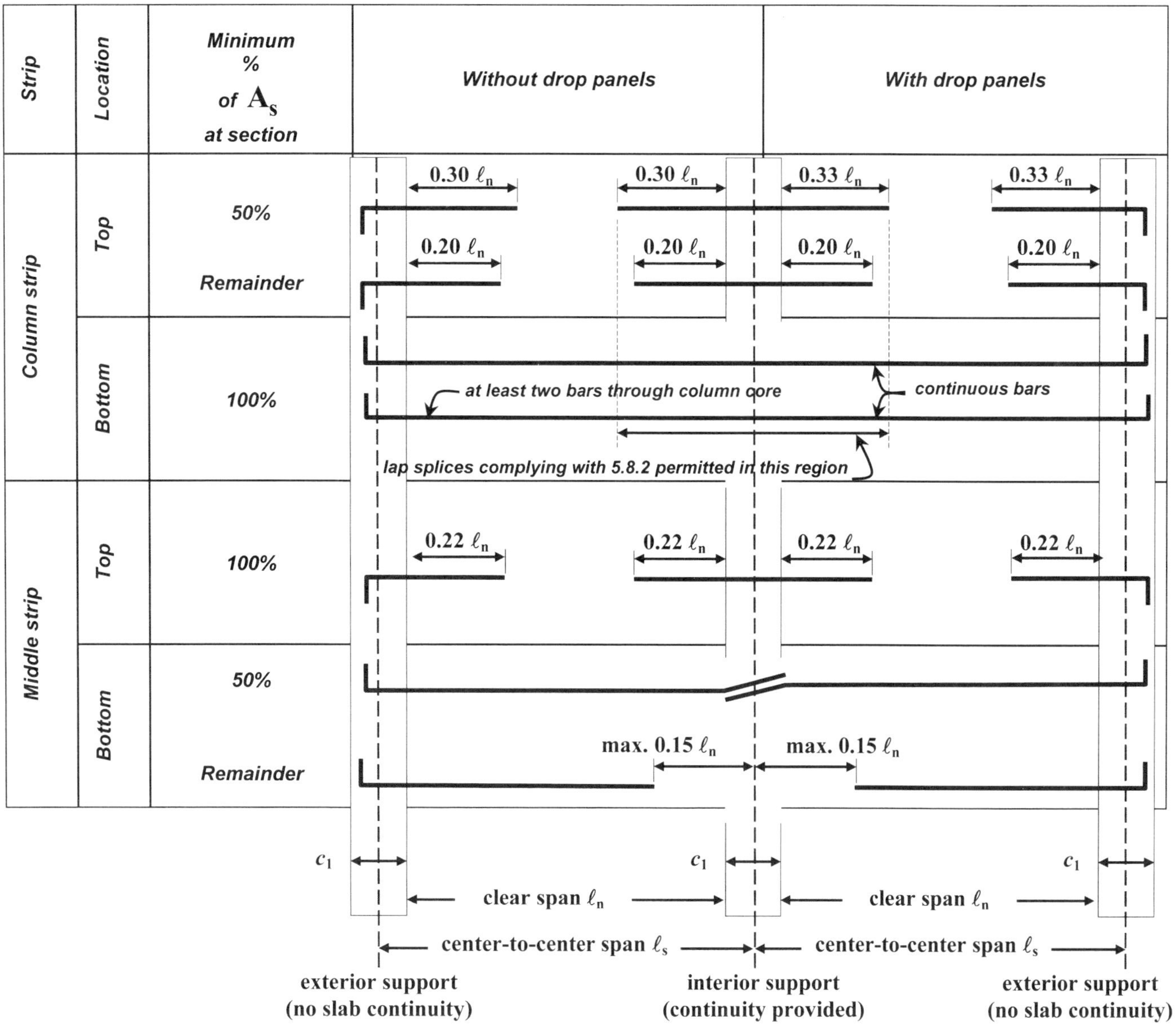

Fig. 9.4.2.6—Minimum length of slab reinforcement in slab-column systems.

where λ_{ps} should be taken as:
1.0 for $b_o/d \leq 20$
0.75 for $20 < b_o/d \leq 40$
0.5 for $40 < b_o/d$

9.5.4.4 *Effects of moment transfer to the slab-column joint*—Transfer of unbalanced moment defined in 9.8.1.7 increases the punching shear acting on the slab. The increase in punching shear due to moment transfer about any principal axis of the support should be disregarded where the ratio of the unbalanced factored moment to the factored shear does not exceed $0.2d$, where d is the slab effective depth. Where this ratio is exceeded, (a), (b), and (c) should be met:

(a) For corner column connections transferring moments in both principal directions and for edge column connections transferring unbalanced moments only perpendicular to the slab edge, the connection should be assumed to have adequate punching shear strength when the factored shear force V_u does not exceed $0.75\phi V_c$, with ϕV_c determined from Eq. (9.5.4.3);

(b) For interior and edge column connections transferring moments simultaneously in the two principal directions, a

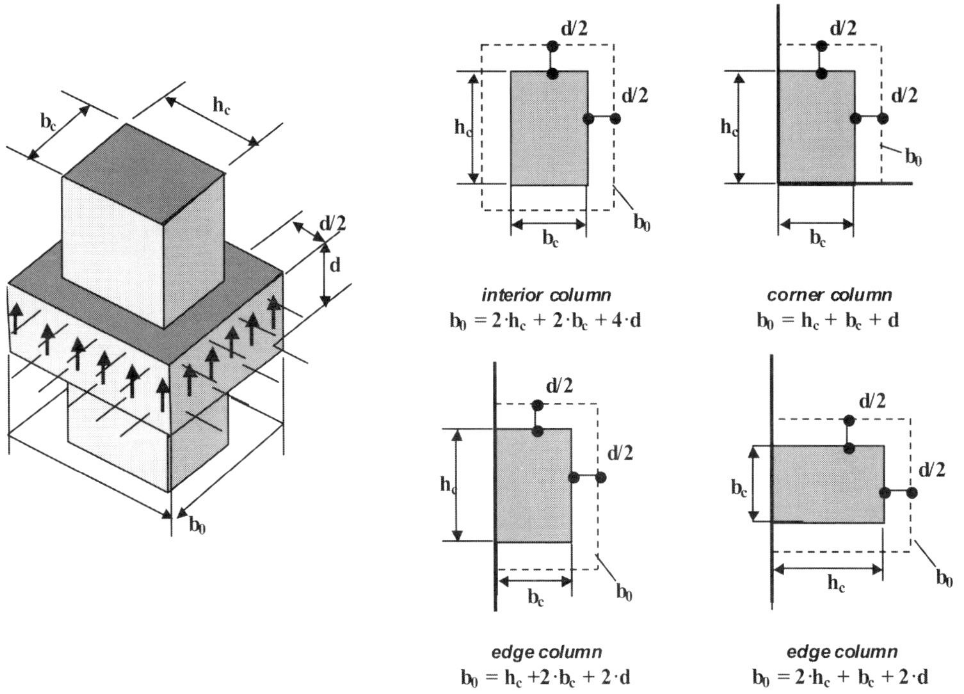

Fig. 9.5.4.2—Definition of b_o.

factored shear due to the unbalanced moment, ΔV_u, is added to the factored shear V_u. ΔV_u should be computed by Eq. (9.5.4.4) using the unbalanced moments in both directions:

$$\Delta V_u = \alpha_{sh}\left(\frac{\Delta M_{u1} + \Delta M_{u2}}{b_o}\right) \quad (9.5.4.4)$$

Interior and edge column connections should be assumed to have adequate punching shear strength when the total factored shear, $(V_u + \Delta V_u)$, does not exceed ϕV_c determined from Eq. (9.5.4.3), with ΔV_u from Eq. (9.5.4.4) computed using $\alpha_{hs} = 5$ for interior column connections and $\alpha_{sh} = 3.5$ for edge column connections.

For edge column connections, moments perpendicular to the slab edge may be taken equal to zero in Eq. (9.5.4.4) if V_u does not exceed $0.75\phi V_c$, with ϕV_c determined from Eq. (9.5.4.3); and

(c) The details of 9.8.1.8 for the longitudinal reinforcement resisting the unbalanced moment should be met.

9.5.5 *Beam-action shear*—Sections between the support face and a distance d from the support should be designed for the factored shear V_u computed at d. Shear strength at each critical location to be investigated should be computed using Eq. (9.5.5) with $\phi = 0.75$ (Fig. 9.5.5).

$$\phi V_c = \phi 2\sqrt{f_c'}bd$$
$$\left(\phi V_c = \phi\left[\frac{\sqrt{f_c'}}{6}\right]bd \text{ (SI)}\right) \quad (9.5.5)$$

9.6—Minimum slab thickness as required by punching shear

9.6.1 *Required punching-shear strength*—Factored punching-shear V_u should be determined as the value of the total factored load q_u multiplied by the area bounded by panel centerlines around the column, less the area defined by the critical shear perimeter (Fig. 9.6.1).

9.6.2 *Minimum slab thickness*—Punching-shear strength should not be less than the value determined from Eq. (9.5.4.3) modified for unbalanced moment transfer as indicated by 9.5.4.4. The slab effective depth should be calculated from Eq. (9.6.2a) to (9.6.2d), with h_c for the long dimension and b_c for the short column dimension and $\phi = 0.75$. If drop panels are used, both h_c and b_c apply to the plan drop panel dimensions. In waffle-slab systems, calculations for punching shear are needed only in the critical zone within the capital.

For interior columns:

$$d \geq \sqrt{\left(\frac{h_c + b_c}{4}\right)^2 + \frac{V_u}{\phi 6\sqrt{f_c'}}} - \left(\frac{h_c + b_c}{4}\right)$$
$$\left(d \geq \sqrt{\left(\frac{h_c + b_c}{4}\right)^2 + \frac{2V_u}{\phi\sqrt{f_c'}}} - \left(\frac{h_c + b_c}{4}\right) \text{ (SI)}\right) \quad (9.6.2a)$$

For edge columns with h_c parallel to edge:

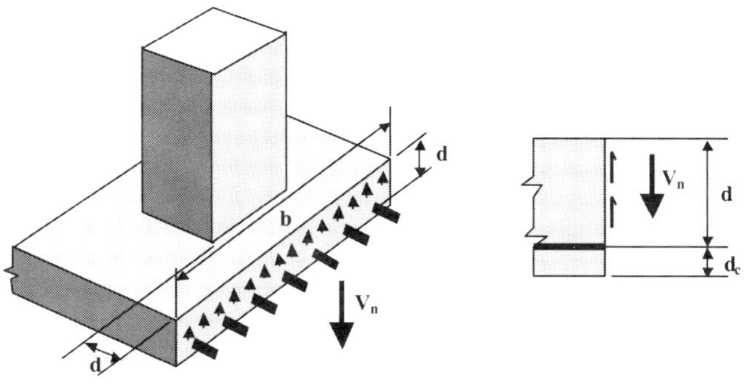

Fig. 9.5.5—Beam-action shear strength in solid slabs.

$$d \geq \sqrt{\left(\frac{h_c + 2b_c}{4}\right)^2 + \frac{V_u}{\phi 6\sqrt{f_c'}}} - \left(\frac{h_c + 2b_c}{4}\right) \quad (9.6.2b)$$

$$\left(d \geq \sqrt{\left(\frac{h_c + 2b_c}{4}\right)^2 + \frac{2V_u}{\phi\sqrt{f_c'}}} - \left(\frac{h_c + 2b_c}{4}\right) \text{ (SI)}\right)$$

For edge columns with b_c parallel to edge:

$$d \geq \sqrt{\left(\frac{2h_c + b_c}{4}\right)^2 + \frac{V_u}{\phi 6\sqrt{f_c'}}} - \left(\frac{2h_c + b_c}{4}\right) \quad (9.6.2c)$$

$$\left(d \geq \sqrt{\left(\frac{2h_c + b_c}{4}\right)^2 + \frac{2V_u}{\phi\sqrt{f_c'}}} - \left(\frac{2h_c + b_c}{4}\right) \text{ (SI)}\right)$$

For corner columns:

$$d \geq \sqrt{\left(\frac{h_c + b_c}{2}\right)^2 + \frac{V_u}{\phi 3\sqrt{f_c'}}} - \left(\frac{h_c + b_c}{2}\right) \quad (9.6.2d)$$

$$\left(d \geq \sqrt{\left(\frac{h_c + b_c}{2}\right)^2 + \frac{4V_u}{\phi\sqrt{f_c'}}} - \left(\frac{h_c + b_c}{2}\right) \text{ (SI)}\right)$$

9.7—Minimum slab thickness as required by beam action

9.7.1 *General*—The design for beam-action shear strength of a slab section should be evaluated by 9.5 before designing the slab for flexure. Shear strength for beam-action shear should be investigated at a distance *d* from the face of columns, concentrated loads, and supports.

9.7.2 *Required shear strength*—Factored beam-action shear V_u should be determined as the total factored load q_u multiplied by the area bounded by panel centerlines of the column and a line at a distance *d* from the face of the column (Fig. 9.7.2).

9.7.3 *Minimum slab thickness as required by beam-action shear*

9.7.3.1 *Solid slabs*—Minimum effective depth *d* as related to beam-action shear, should be determined by combining Eq. (9.5.3) with Eq. (9.5.4.3), and solving for *d*:

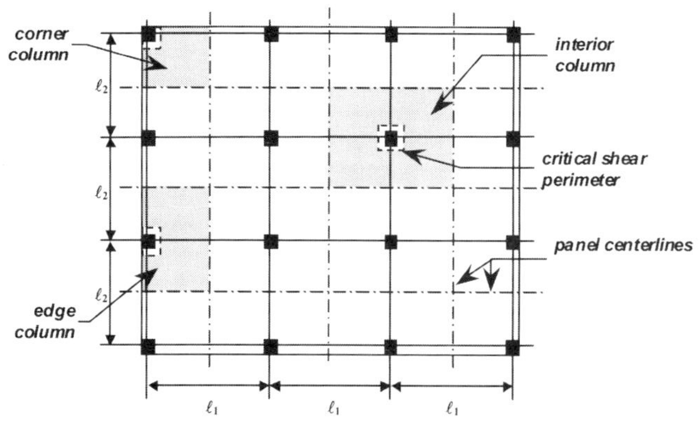

Fig. 9.6.1—Tributary areas for punching shear.

$$d \geq \frac{V_u}{\phi \ell_2 2\sqrt{f_c'}} \quad (9.7.3.1)$$

$$\left(d \geq \frac{6V_u}{\phi \ell_2 \sqrt{f_c'}} \text{ (SI)}\right)$$

where ϕ = 0.75. The largest value of *d* determined for all directions at all columns (interior, edge, and corner) should be used. If the critical section cuts across a zone with a drop panel and a zone without a drop panel, the weighted average value of *d* should be used.

9.7.3.2 *Waffle-slabs*—In waffle-slabs, the transverse dimension of the capital should be used in Eq. (9.7.3.1) instead of ℓ_2. An additional critical section is at the face of the capital, using the provisions for joists in 8.5 and 8.6 with a clear span equal to the clear distance between faces of capitals. Contribution of the joist shear reinforcement may be included in the beam-action shear strength of the joists that engage the column capital.

9.8—Flexure

9.8.1 *Required moment strength*

9.8.1.1 *General*—The factored moments should be computed following (1) through (6):

(1) The slab is divided into design strips in both directions;

(2) Each strip includes a column or support centerline and is bounded by panel centerlines on each side (Fig. 9.8.1.1a);

(3) Where the strip is adjacent to and parallel with a slab edge, it should include the edge column or support centerline and should be bounded by the slab edge on one side and the panel centerline on the other side (Fig. 9.8.1.1a);

(4) Total factored moments should be calculated according to 9.8.1.2 (Fig. 9.8.1.1b and 9.8.1.1c) for all spans of all strips in both directions;

(5) Total factored moments should be divided into positive and negative moments using 9.8.1.3 (Fig. 9.8.1.1b and 9.8.1.1c); and

(6) Positive and negative moments determined from 9.8.1.3 should be distributed to the column strip using 9.8.1.4, and to the two one-half middle strips using 9.8.1.5 (Fig. 9.8.1.1b and 9.8.1.1c).

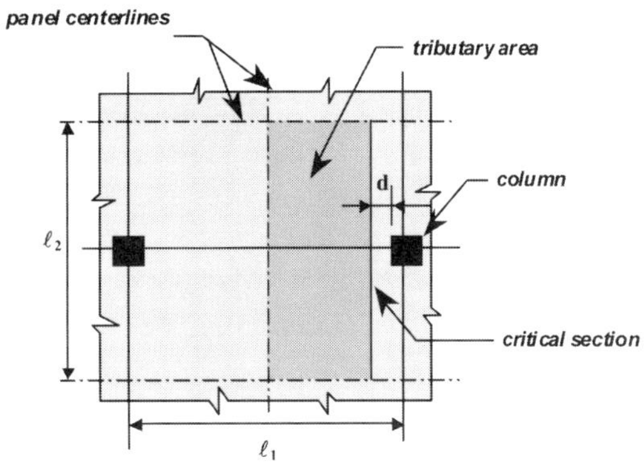

Fig. 9.7.2—Tributary area for beam-action shear.

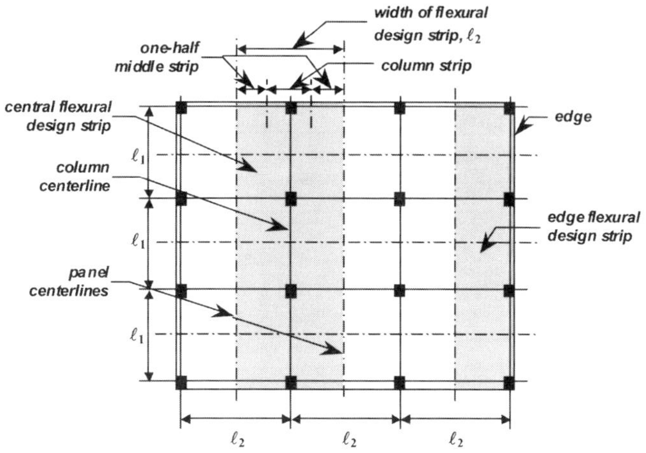

Fig. 9.8.1.1a—Definition of design strips.

9.8.1.2 *Total factored moments*—The absolute sum of positive and average negative factored moments in a span should be calculated from Eq. (9.8.1.2).

$$M_o = \frac{q_u \ell_2 \ell_n^2}{8} \qquad (9.8.1.2)$$

In Eq. (9.8.1.2), (a) through (f) should be followed:

(a) Where transverse span of panels on either side of the centerline of supports varies, ℓ_2 in Eq. (9.8.1.2) should be taken as the average of adjacent transverse spans;

(b) When the span is adjacent and parallel to a slab edge, the distance from the edge to panel centerline should be substituted for ℓ_2 in Eq. (9.8.1.2);

(c) Clear span ℓ_n should extend from face to face of columns, capitals, brackets, or walls;

(d) Value of ℓ_n should not be less than $0.65\ell_1$;

(e) Circular or regular polygon-shaped supports should be treated as square supports with the same area; and

(f) Redistribution of negative and positive factored moments by 10 percent is permitted provided that the sum remains the same.

9.8.1.3 *Negative and positive factored moments*—Negative factored moments should be located at the face of rectangular supports. Circular-shaped supports should be treated as square supports with the same area.

In an interior span, the total factored moment M_o should be distributed as follows:

(a) Negative factored moment .... $0.65M_o$

(b) Positive factored moment .... $0.35M_o$

In an end span, the total factored moment M_o should be distributed as follows:

(c) Interior negative factored moment .... $0.70M_o$

(d) Positive factored moment .... $0.50M_o$

(e) Exterior negative factored moment .... $0.30M_o$

Negative moment reinforcement should be designed to resist the larger of the two negative factored moments determined for spans framing into a common support.

9.8.1.4 *Factored moments in column strips*—Column strips should be proportioned to resist the following portions of the moments determined in 9.8.1.3:

(a) Interior negative factored moments......... 75 percent

(b) Positive factored moments .... 60 percent

(c) Exterior negative factored moments ...... 100 percent

Where column width or wall support extends a distance equal to or greater than three-quarters of the span length ℓ_2 used to compute M_o, negative moments should be considered uniformly distributed across ℓ_2.

9.8.1.5 *Factored moments in middle strips*—The portion of negative and positive factored moments not resisted by column strips should be assigned to the corresponding one-half middle strips. Each middle strip should be proportioned to resist the sum of the two moments assigned to it. A middle strip adjacent to and parallel with an edge and supported by a wall should be proportioned to resist twice the moment assigned to the adjacent one-half middle strip of the first row of interior supports.

9.8.1.6 *Cantilevers*—Cantilevers of solid slabs should be designed in accordance with 7.6 and joists of waffle-systems should be designed in accordance with 8.6.

9.8.1.7 *Unbalanced moments*—Factored unbalanced moment ΔM_u should be taken as $0.3M_o$ at edge supports, and for interior supports it should be taken as the difference of the column strip negative moments of the adjacent interior spans plus an additional value, $\Delta M_{u\text{-}ad}$, determined from Eq. (9.8.1.7)

$$\Delta M_{u-ad} = \frac{q_\ell \ell_2 \ell_n^2}{14} \qquad (9.8.1.7)$$

where ℓ_n is the larger of the adjacent clear spans, and q_ℓ is the service live load.

9.8.1.8 *Transfer of unbalanced moments at supports*—Transfer of the unbalanced moments ΔM_u from 9.8.1.7 at the slab-column joints should be solely through flexure in the slab, provided (a) through (d) are met (9.5.4.4):

(a) Transfer width for resisting unbalanced moments in flexure should be assumed to be one and one-half slab or drop panel thickness, $1.5h$, outside opposite faces of the column or capital;

(b) Negative moment column strip reinforcement within the transfer width should be adequate to resist ΔM_u by either closer bar spacing or additional reinforcement;

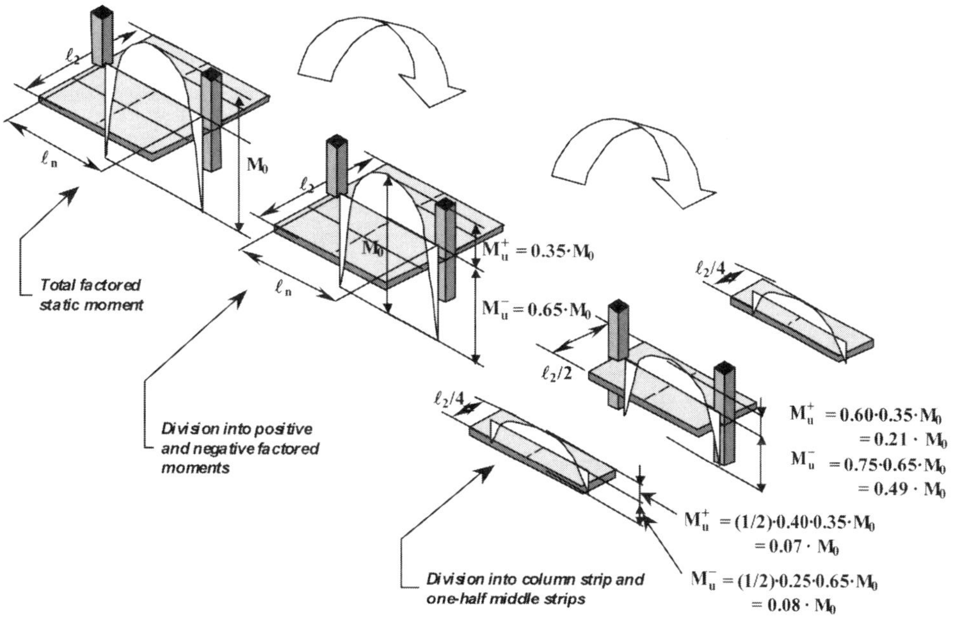

Fig. 9.8.1.1b—Factored moments: interior spans.

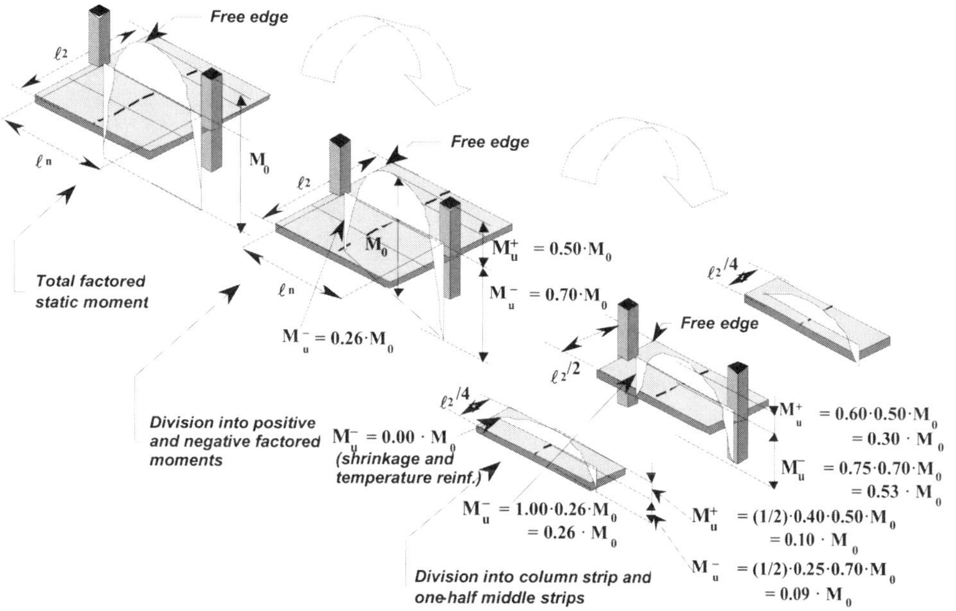

Fig. 9.8.1.1c—Factored moments: exterior spans.

(c) Negative moment reinforcement ratio ρ in the transfer width should not exceed three-quarters of ρ_{max} of 7.3.4.3; and

(d) The increase of the punching shear force acting on the slab is evaluated using the procedure of 9.5.4.4 (9.6).

9.8.2 *Longitudinal flexural reinforcement in solid slabs of slab-column systems*

9.8.2.1 *Positive moment reinforcement*—In both directions, determine the positive moment reinforcement ratio ρ using the values of M_u^+ determined from 9.8.1 (Fig. 9.8.1.1b and 9.8.1.1c) for the strip under consideration. For positive moments in the column strip, the strip width *b* should be as defined in 9.3.2. For positive moments in the one-half middle strips on each side, the strip width should be consistent with the width used to distribute moments as indicated by 9.8.1.5. Positive moment reinforcement should comply with 9.4.2 (Fig. 9.4.2.6).

9.8.2.2 *Negative moment reinforcement*—In both directions, the negative moment reinforcement ratio ρ, in the direction of the span, should be determined using values of M_u^- determined from 9.8.1 (Fig. 9.8.1.1b and 9.8.1.1.c) for the strip under consideration. For negative moments in the column strip, the strip width *b* should be as defined in 9.3.2. For negative moments in the one-half middle strips on each side, the strip width should be consistent with the width used to distribute moments as indicated by 9.8.1.5. Negative moment reinforcement should comply with 9.4.3 (Fig. 9.4.2.6). In the column strip at supports, the unbalanced moment transfer of 9.8.1.7 and 9.8.1.8 should be met. For unequal strip widths meeting at a support, the reinforcement from each side leading to the larger negative moment reinforcement ratio ρ should be used.

9.8.3 *Joist reinforcement of waffle-slab systems*

9.8.3.1 *Positive moment reinforcement in joists of waffle-slabs*—Positive moments used to calculate reinforcement for each joist should be determined using values of M_u^+ from 9.8.1, divided by the number of joists within the strip under consideration. T-beam effect as indicated in 8.4.10 may be used. Positive moment reinforcement should comply with 8.6.5.1, except that the reinforcement cutoff should comply with 9.4.2.

9.8.3.2 *Negative moment reinforcement in joists of waffle-slabs*—Negative moments used to calculate reinforcement for each joist should be determined using values of M_u^- from 9.8.1, divided by the number of joists within the strip under consideration. Negative moment reinforcement should comply with 8.6.5.2, except that the reinforcement cutoff should comply with 9.4.3.

9.8.3.3 *Shear reinforcement in joists of waffle-slabs*—Values of V_u at the supports should be determined using 9.7.3.2. Shear reinforcement should comply with 8.5, 9.5.5, and 9.7.3.2. Stirrup spacing s should be defined for the different regions within the span. Minimum stirrup spacing as indicated by 8.5.2.3 should be used. The first stirrup should not be placed further than $s/2$ from the support face, with s being the stirrup spacing at the support.

9.9—Calculation of support reactions

9.9.1 *Vertical reactions at columns and walls*—Vertical reactions of the supporting members, R_u, should be determined as the total factored design load q_u multiplied by the area bound by panel centerlines around the supporting member (Fig. 9.6.1).

9.9.2 *Column and wall moments from vertical loading*—Columns and walls built integrally with a slab system should resist unbalanced moments due to factored loads on the slab system. Unbalanced moments ΔM_u should be determined from 9.8.1.7, and should be distributed to the columns above and below the slab using the procedure of 8.7.6.3.

CHAPTER 10—COLUMNS

10.1—General

Columns should be designed using Chapter 10. This chapter applies to members reinforced with longitudinal bars and lateral ties, and members reinforced with longitudinal bars and a continuous spiral reinforcement. Rectangular and circular sections are permitted.

10.2—Loads

10.2.1 *Loads to be included*—For columns belonging to frames or slab-column systems, the tributary loads from each floor plus self-weight of the column should be included. Tributary loads should conform to Chapter 4 and the particular loads of each tributary member type (Fig. 10.2.1a and Fig. 10.2.1b).

10.2.2 *Dead load and live load*—Values of P_d for dead load and P_ℓ for live load should be accumulated independently and combined at each floor. P_d should include self-weight of the column. Self-weight should use the dead load factor of the corresponding load combination equations in 4.2.1. Self-weight should be applied at the lower part of the column in that story (Fig. 10.2.1b).

10.2.3 *Required strength*—Values of the factored load P_u and factored moment M_u should be determined at the upper and lower part of each column at each floor. The direction in plan along which the moments M_{ux} and M_{uy} act should be noted (Fig. 10.2.1b).

10.3—Dimensional limits

10.3.1 *General*—In addition to Chapter 10, columns should comply with the general dimensional limits set forth in 1.3. Columns should be vertical and should be continuous down to the foundation. Column section shape should be either rectangular or circular.

10.3.2 *Dimensional limits*

10.3.2.1 *Minimum cross-sectional dimensions for rectangular columns*—Cross-sectional dimensions for rectangular columns should comply with (a) and (b) (Fig. 10.3.2.1):

(a) The least cross-sectional dimension should not be less than 10 in. (250 mm); and

(b) Ratio of the long cross-sectional dimension to the short dimension should not exceed 3, except in slab-column systems, where it should not exceed 2 (9.3.6).

10.3.2.2 *Minimum cross-sectional dimension for circular columns*—Columns with a circular cross section should have a diameter of at least 12 in. (300 mm) (Fig. 10.3.2.2).

10.3.3 *Distance between lateral supports*

10.3.3.1 *General*—It should be assumed that the floor system provides column lateral restraint in both horizontal directions at all supported levels (Fig. 10.3.3.1).

10.3.3.2 *Interior columns*—For interior columns, the column cross section dimension parallel to the direction of the support should not be less than 1/10 of the clear vertical distance between lateral supports, h_n (Fig. 10.3.3.1).

10.3.3.3 *Edge columns*—For edge columns, column cross section dimension perpendicular to the edge should not be

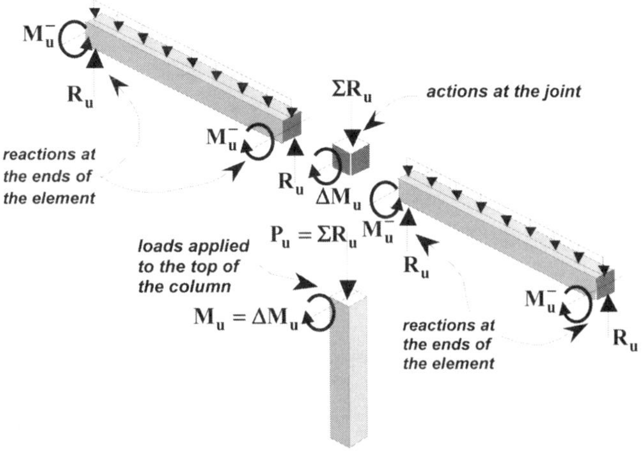

Fig. 10.2.1a—Factored column loads and moments from single floor and in one direction.

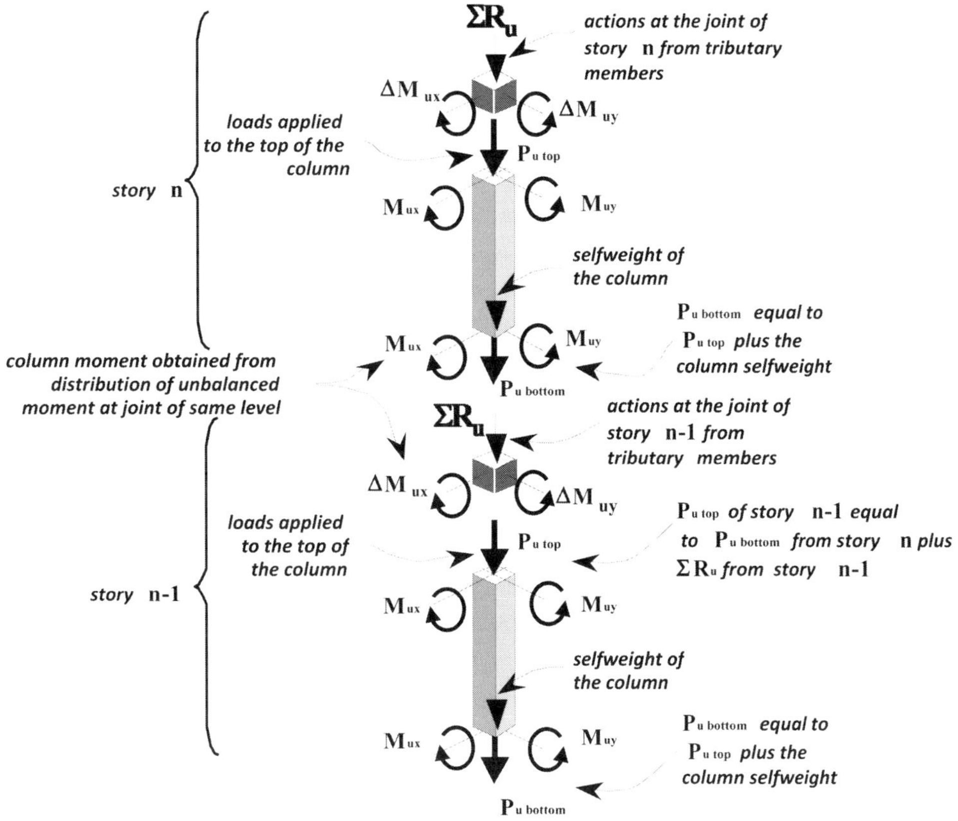

Fig. 10.2.1b—Factored column loads and moments from multiple floors.

less than one-ninth of the clear vertical distance between lateral supports, h_n (Fig. 10.3.3.1).

10.3.3.4 *Corner columns*—For corner columns, minimum column cross section dimension should be one-eighth of the clear vertical distance between lateral supports, h_n (Fig. 10.3.3.1).

10.3.4 *Column built monolithically with wall*—Effective cross section of a tied or spirally reinforced column that is monolithic with a concrete wall should be taken not greater than 1-1/2 in. (40 mm) outside the tie or spiral reinforcement or the lateral wall faces (Fig. 10.3.4).

10.4—Reinforcement details

10.4.1 *General*—Column reinforcement should be of the types described in Chapter 10 and should comply with 10.4.2 to 10.4.4.

10.4.2 *Longitudinal reinforcement*

10.4.2.1 *Description and location*—Longitudinal reinforcement should be provided at the periphery of the column section. Longitudinal reinforcement should be located as close to the column lateral surfaces as practicable following transverse reinforcement and concrete cover (5.4.1 and 10.4.2.13). The area of longitudinal reinforcement should be adequate to resist the simultaneous action of a factored axial load and factored moments acting about the two main column axes (Fig. 10.2.1b).

10.4.2.2 *Minimum and maximum area of longitudinal reinforcement*—Total area of column longitudinal reinforcement, A_{st}, should not be less than 0.01 or more than 0.06 times the gross area A_g of section:

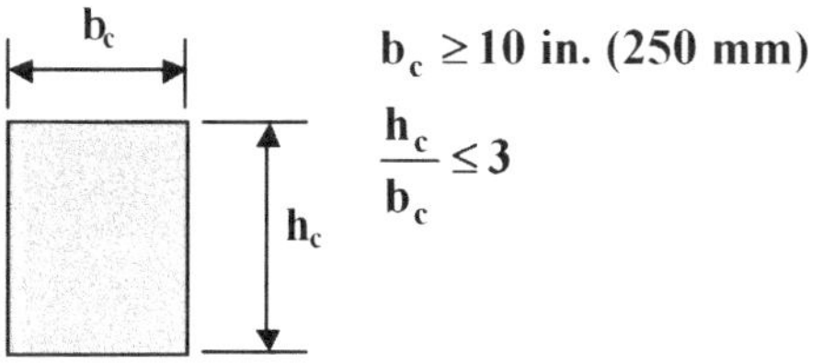

Fig. 10.3.2.1—Minimum cross-sectional dimensions for rectangular columns.

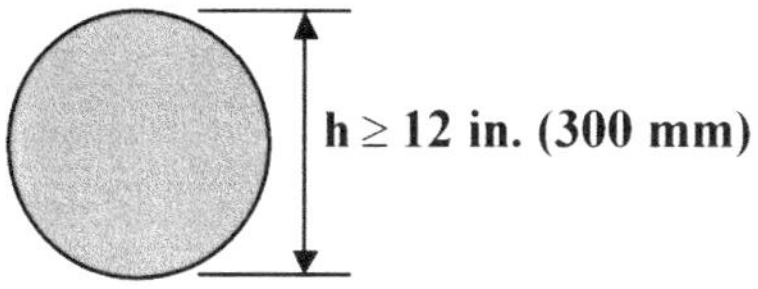

Fig. 10.3.2.2—Minimum cross-sectional dimension for circular columns.

$$0.01 \le \rho_\ell \left(= \frac{A_{st}}{A_g} \right) \le 0.06 \qquad (10.4.2.2)$$

This guide limits the column longitudinal reinforcement ratio to 6 percent because of reinforcement congestion concerns.

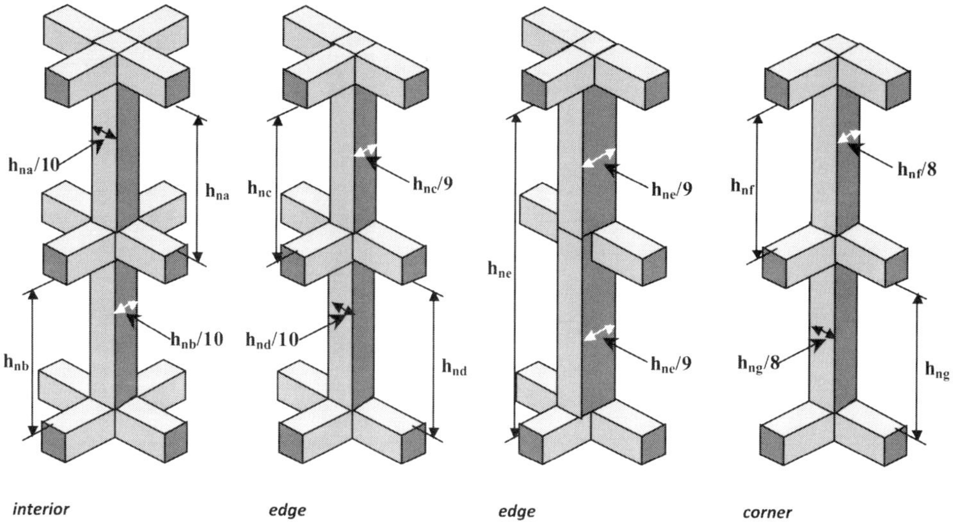

Fig. 10.3.3.1—Lateral restraint for columns.

10.4.2.3 *Minimum diameter of longitudinal bars*—Minimum diameter of longitudinal bars in columns should be 5/8 in. (16 mm).

10.4.2.4 *Minimum number of longitudinal bars*—In square and rectangular columns with ties, there should be a minimum of four bars (at least one longitudinal bar in each corner), and in round columns with spirals, a minimum of six longitudinal bars.

10.4.2.5 *Distribution of longitudinal bars*—Column longitudinal bar spacing along all column faces should be approximately equal.

10.4.2.6 *Minimum clear spacing between longitudinal bars*—Clear spacing between longitudinal bars should not be less than $1.5d_b$ nor 1-1/2 in. (40 mm) (Fig. 10.4.2.6).

10.4.2.7 *Clear spacing between parallel lap splices*—Clear spacing limitation between bars should also apply to the clear distance between a contact lap splice and adjacent splices or bars.

10.4.2.8 *Reinforcing bar splicing*—Up to one-half the longitudinal bars at any given section may be lap spliced, as long as only alternate bars along the cross section perimeter are lap-spliced (Fig. 10.4.2.8a and 10.4.2.8b). This practice reduces congestion and provides the column with good ductility at lap locations. Lap splices of longitudinal reinforcement should comply with 5.8.2.1.

10.4.2.9 *End anchorage of reinforcement*—Longitudinal reinforcement at the upper end of columns and at the foundation should extend as close to the edge as practicable considering concrete cover and end with a standard hook.

10.4.2.10 *Longitudinal bar offset*—Offset bent longitudinal bars should conform to (a) through (g):

(a) Slope of inclined portion of an offset bar should not exceed 1 in 6 (Fig. 10.4.2.10);

(b) Portions of bar above and below an offset should be parallel to column axis;

(c) Lateral ties or spirals should be provided at offset bends;

(d) Offset bend lateral ties or spirals should resist 1.5 times the horizontal force component of an offset bar computed assuming a stress equal to f_y;

(e) Lateral ties or spirals should be placed not more than 6 in. (150 mm) from bend points;

(f) Offset bars should be bent before placement in the forms; and

(g) Where a column face is offset from the column below it by more than one-sixth of the depth of the girder or slab, or 3 in. (75 mm), longitudinal bars should not be offset bent. Separate dowels, lap-spliced with the longitudinal bars adjacent to the offset column faces, should be provided. Lap splices should conform to 5.8.2.1.

10.4.2.11 *Maximum number of longitudinal bars in each rectangular column face*—Calculation of the maximum number of longitudinal bars in a column face should consider the longitudinal and transverse bar diameters, concrete cover (5.4), maximum nominal coarse aggregate size, and minimum clear spacing between bars. When these calculations are not performed, (a) and (b) should be used:

(a) For column dimension b_c greater than 12 in. (300 mm), the maximum number of bars in a column face should be determined using Eq. (10.4.2.11)

$$\text{maximum number of bars per face } \leq \frac{b_c}{3} \left(\frac{b_c}{75}\ \text{(SI)}\right) \qquad (10.4.2.11)$$

where b_c is in inches (mm) (Table 10.4.2.11).

(b) Maximum of three longitudinal bars should be used in the column faces whose dimension b_c is less than or equal to 12 in. (300 mm) (Table 10.4.2.11).

10.4.2.12 *Maximum number of longitudinal bars in circular column*—The calculation for the maximum number of longitudinal bars in circular columns should consider the concrete cover (5.4), maximum nominal coarse aggregate size, minimum clear spacing between bars, and longitudinal

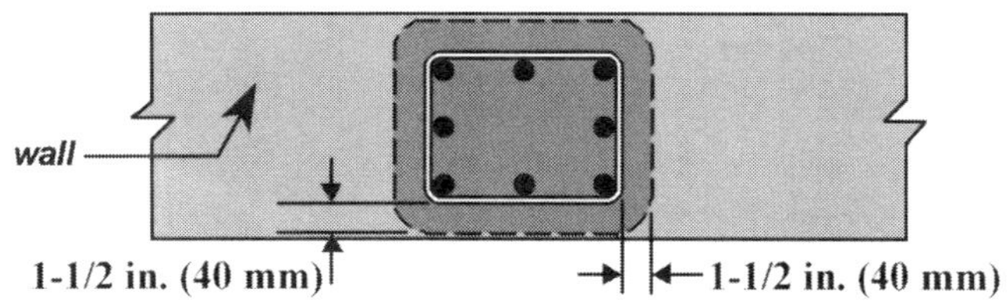

Fig. 10.3.4—Effective cross section of columns built monolithically with a wall.

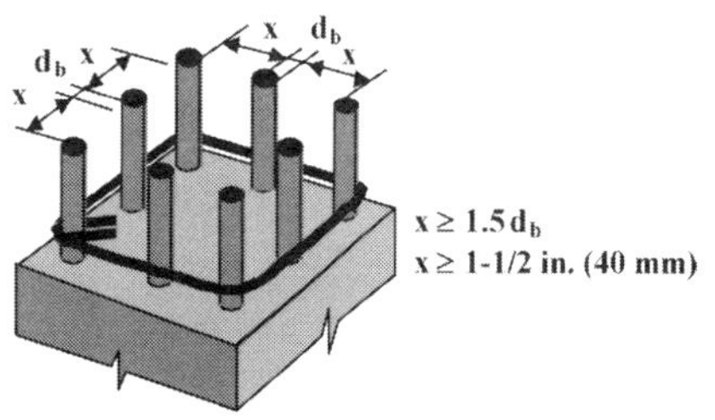

Fig. 10.4.2.6—Clear spacing between longitudinal bars in columns.

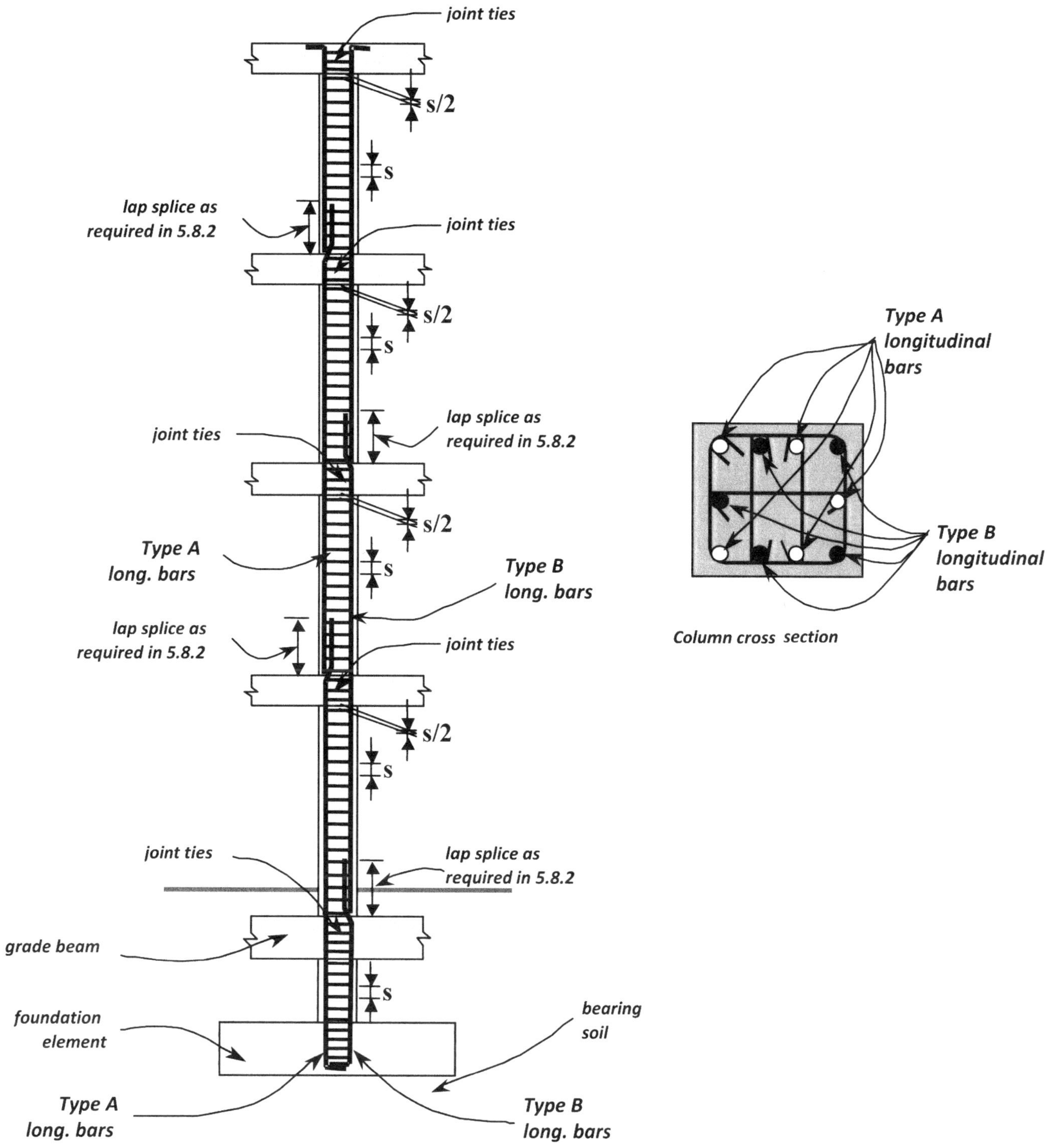

Fig. 10.4.2.8a—Typical column reinforcement layout.

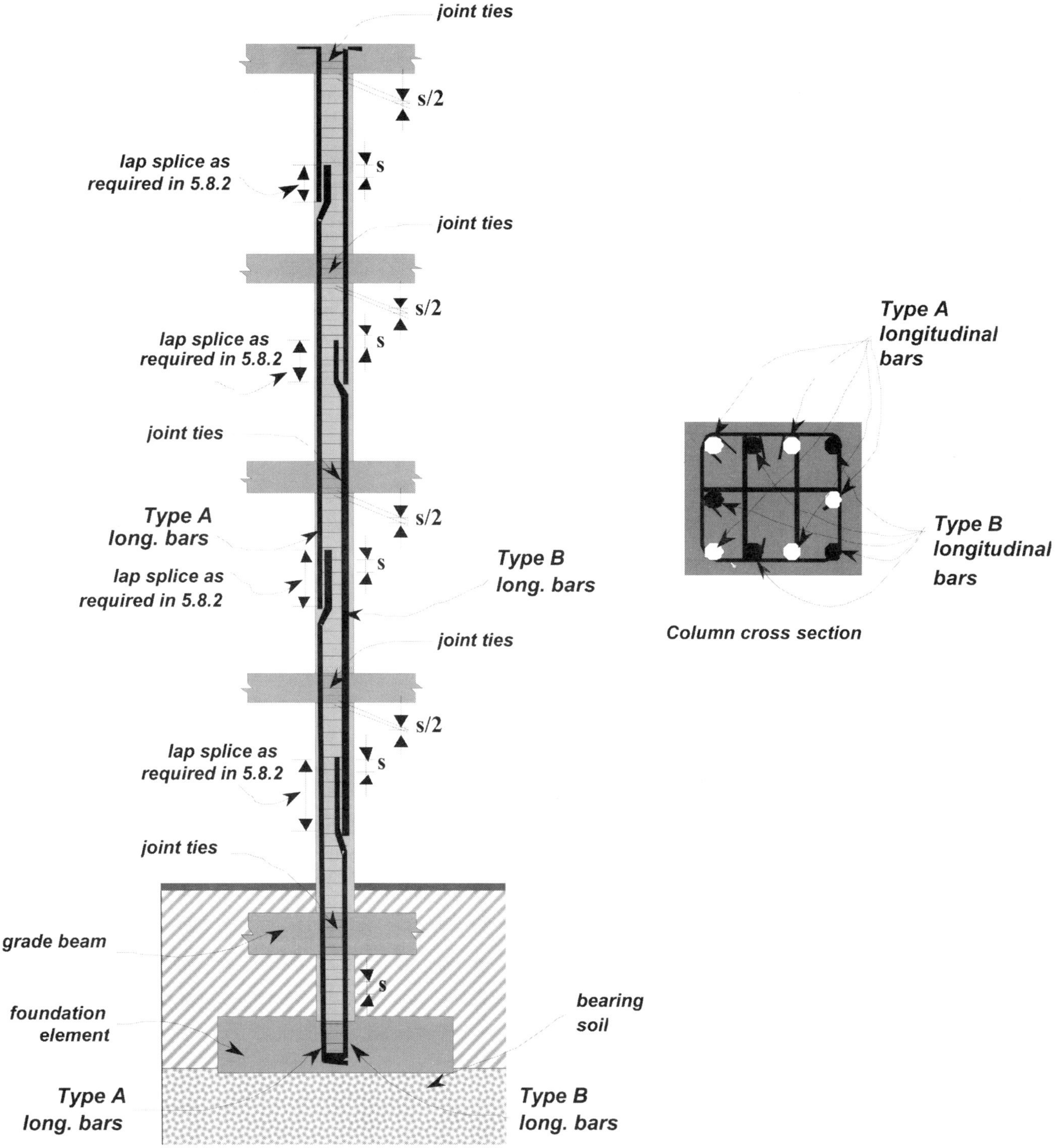

Fig. 10.4.2.8b—Typical column reinforcement layout in high seismic zones.

and transverse bar diameters. When these calculations are not performed, the maximum number of bars should be determined using Eq. (10.4.2.12), where h is the column diameter (Table 10.4.2.12).

$$\text{maximum no. of bars } \leq h - 6 \left(\leq \frac{h}{25} - 6 \text{ (SI)} \right) \quad (10.4.2.12)$$

10.4.2.13 *Value of* d_c *and* d *to use in columns*—Calculation of d_c should consider the concrete cover (5.4) and longitudinal and transverse bar diameters.

The following values can be used for columns: d_c = 2.5 in. (60 mm) for interior exposure, and d_c = 3 in. (75 mm) for exterior exposure.

10.4.3 *Transverse reinforcement*

10.4.3.1 *General*—Columns should have tie reinforcement or spiral reinforcement conforming to 10.4.3.2 or

10.4.3.3, respectively. Columns located in seismic zones should have confining transverse reinforcement as indicated in Chapter 11. At column-beam joints, the minimum number of ties should be as indicated by 10.4.3.4.

10.4.3.2 *Ties in nonseismic zones*—Column ties should comply with (a) through (e):

(a) All longitudinal column bars should be enclosed by lateral ties. Minimum tie bar diameter should be 3/8 in. (10 mm);

(b) Ties should be arranged so that every corner and alternate longitudinal bar should have lateral support provided by the corner of a tie or a crosstie (Fig. 10.4.3.2a);

(c) Along a tie, no longitudinal bar should be farther than 6 in. (150 mm) clear from a laterally supported longitudinal bar (Fig. 10.4.3.2a);

(d) Vertical spacing of ties s should not exceed the least of 16 longitudinal bar diameters, 48 tie bar diameters, and the least dimension of the column section (Fig. 10.4.3.2b); and

(e) The first tie should be located one-half spacing above the top of the slab, beam, or footing, and the uppermost one should be located no more than one-half tie spacing below the lowest horizontal reinforcement of shallowest member supported above.

10.4.3.3 *Spiral*—Columns spirals should comply with (a) through (f):

(a) All longitudinal column bars should be enclosed by a spiral consisting of an evenly spaced continuous bar. Minimum spiral bar diameter should be 3/8 in. (10 mm);

(b) Clear vertical spacing between spirals should not exceed 3 in. (75 mm), nor be less than 1 in. (25 mm), and should comply with 5.7;

(c) Spirals should have 1.5 extra turns at each end of a spiral unit;

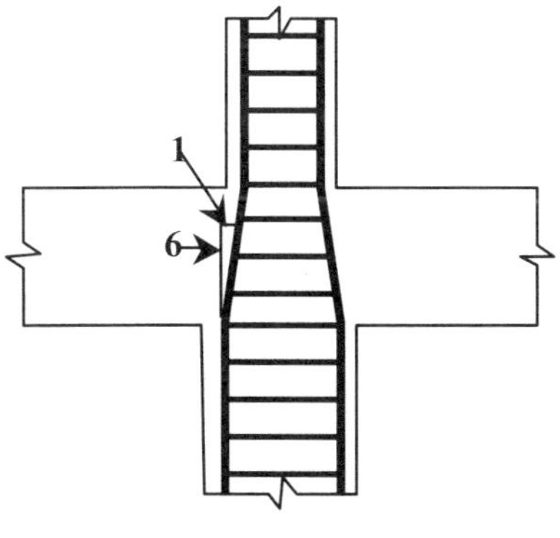

Fig. 10.4.2.10—Longitudinal bar offset.

(d) Splices in spirals should comply with 5.8.2 and end in a hook directed to the column core;

(e) Spirals should extend from the top of footing or slab to the level of lowest horizontal reinforcement of shallowest member supported above. In columns with capitals, spiral should extend to a level at which the diameter or width of capital is two times that of the column; and

(f) Spiral reinforcement ratio ρ_s, defined as ratio of the volume of reinforcement contained in one spiral loop to the volume of concrete in the column confined by the same spiral loop, should be not less than the value given by Eq. (10.4.3.3) (Fig. 10.4.3.3).

$$\rho_s = \frac{A_b \pi (d_{cs} - d_b)}{A_{cs} s} \geq 0.45 \left[\frac{A_g}{A_{cs}} - 1 \right] \frac{f_c'}{f_{yt}} \quad (10.4.3.3)$$

where A_b is the area of the spiral bar or wire; d_b is the spiral bar or wire diameter; d_{cs} is the outside diameter of the spiral; s is the vertical spacing of the spiral; A_{cs} is the area of the confined column core measured from the outside diameter of the spiral $A_{cs} = \pi d_{cs}^2/4$; A_g is the gross column section area; f_c' is the column concrete specified strength; and f_{yt} is the steel yield strength of the spiral.

10.4.3.4 *Column-girder joints*—At column-girder joints of frames, a minimum of three column ties should be provided

Table 10.4.2.11—Maximum number of longitudinal bars in each rectangular column face

Column dimension b_c, in. (mm)	Maximum number of longitudinal bars
b_c < 10 in. (250 mm)	Section not permitted
10 in. (250 mm) ≤ b_c < 12 in. (300 mm)	Three bars
12 in. (300 mm) ≤ b_c	≤ $b_c/3$ [≤ $b_c/75$ (SI)] bars

Table 10.4.2.12—Maximum number of longitudinal bars in circular columns

Column diameter h, in. (mm)	Maximum number of longitudinal bars
h < 12 in. (300 mm)	Section not permitted
12 in. (300 mm) ≤ h	$\leq (h-6)\left(\frac{h}{25} - 6\ (\text{SI})\right)$ bars

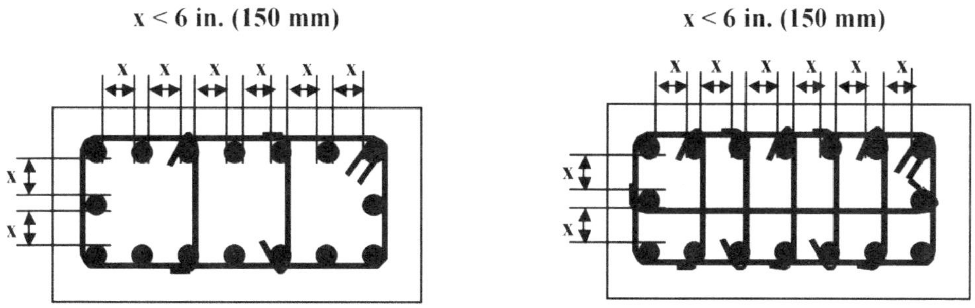

Fig. 10.4.3.2a—Horizontal tie arrangement.

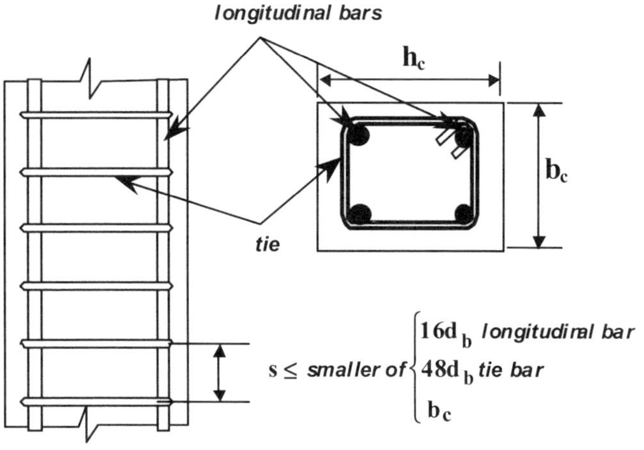

Fig. 10.4.3.2b—Vertical spacing of ties.

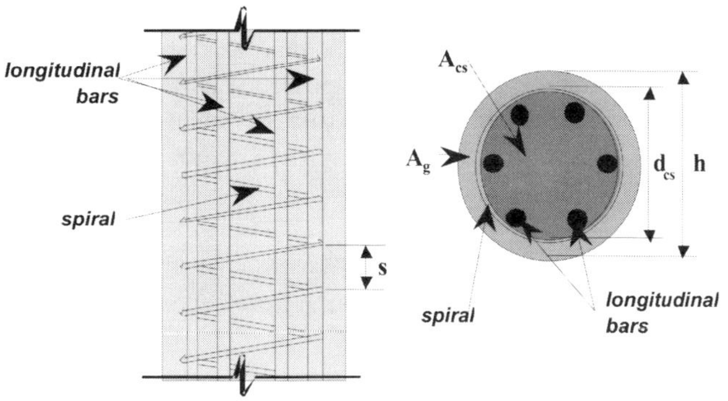

Fig. 10.4.3.3—Spiral reinforcement of column.

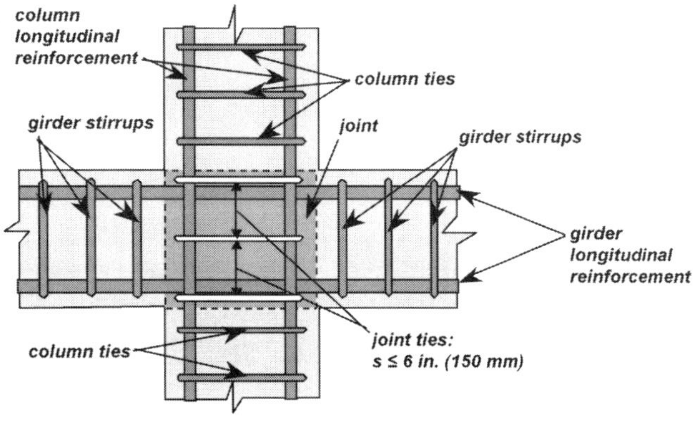

Fig. 10.4.3.4—Column ties in column-girder joints.

within the joint with a maximum vertical spacing between ties of 6 in. (150 mm) (Fig. 10.4.3.4).

10.4.3.5 *Tie hooks*—All column ties should have 135-degree hooks (5.6). Crossties with a 135-degree hook in one end and a 90-degree in the other end may be used. Consecutive crossties engaging the same longitudinal bar should have their 90-degree hooks at opposite sides of the column.

10.4.3.6 *Tie leg splicing*—Column ties should not be lap-spliced.

10.4.4 *Column reinforcement in seismic zones*—In columns that are part of a moment-resisting frame located in seismic zones, reinforcement should also comply with Chapter 11. Columns that are part of slab-column frames in seismic zones should also comply with Chapter 11.

10.5—Flexure

10.5.1 *Factored loads and moments*—Factored axial load P_u and moment M_u, at the section under consideration, should be based on 10.2.

10.5.2 *Trial cross-sectional dimensions and longitudinal reinforcement*

10.5.2.1 *Trial cross-sectional dimensions*—Trial cross-sectional dimensions should be established as follows:

(a) Trial gross cross-sectional area A_g should be determined from Eq. (10.5.2.1a).

$$A_g \approx \frac{2(P_u)_{max}}{f'_c} \quad (10.5.2.1a)$$

(b) For rectangular cross sections, the least dimension b_c should comply with:

$$b_c \begin{cases} \geq 10 \text{ in. (250 mm)} \\ \geq h_c/3 \\ \geq h_n/10 \text{ for central columns} \\ \geq h_n/9 \text{ for edge columns} \\ \geq h_n/8 \text{ for corner columns} \end{cases} \quad (10.5.2.1b)$$

(c) For rectangular cross sections, the larger dimension h_c should comply with:

$$h_c \begin{cases} \geq 10 \text{ in. (250 mm)} \\ \leq 3b_c \\ \geq h_n/10 \text{ for central columns} \\ \geq h_n/9 \text{ for edge columns} \\ \geq h_n/8 \text{ for corner columns} \end{cases} \quad (10.5.2.1c)$$

(d) For circular columns, the diameter h should comply with:

$$h \rightleftharpoons \begin{cases} \geq 12 \text{ in. (300 mm)} \\ \geq h_n/10 \text{ for central columns} \\ \geq h_n/9 \text{ for edge columns} \\ \geq h_n/8 \text{ for corner columns} \end{cases} \quad (10.5.2.1d)$$

10.5.2.2 *Trial longitudinal reinforcement*—Trial area of longitudinal reinforcement, A_{st}, should be established as follows:

(a) For rectangular cross sections, the trial area of longitudinal reinforcement, A_{st}, should comply with:

$$A_{st} \geq \begin{cases} 0.01A_g \\ 4(A_b)_{min.} \end{cases} \text{ (see 10.4.2.3)} \quad (10.5.2.2a)$$

(b) For circular cross sections, the trial area of longitudinal reinforcement, A_{st}, should comply with:

$$A_{st} \geq \begin{cases} 0.01A_g \\ 6(A_b)_{min.} \end{cases} \quad \text{(see 10.4.2.3)} \qquad (10.5.2.2b)$$

10.5.3 *Required moment strength*—Interaction diagrams for column dimensions and reinforcement should be calculated in both directions using 5.12. The design moment strength in both directions should be computed using 5.12.6. Where factored moment M_u at factored axial load P_u exceeds design moment strength, the area of longitudinal reinforcement should be increased without exceeding the maximum reinforcement area permitted by 10.4.2.2 or the maximum number of bars in each column face by 10.4.2.11 or 10.4.2.12. Where limits of 10.4.2.2, 10.4.2.11, or 10.4.2.12 are exceeded, increase the column dimensions. These verifications should be performed at the upper and lower column sections of the same story.

10.5.4 *Biaxial moment strength*—Once column reinforcement is computed for both directions independently, biaxial moment strength should be computed using 5.12.8 at the upper and lower column sections of the same story.

10.6—Shear

10.6.1 *Required shear strength*—Factored shear V_u should be determined from vertical loads and from horizontal loads using load combinations from 4.2.

10.6.1.1 *Factored shear from vertical loads*—Factored shear due to vertical loads should be determined from Eq. (10.6.1.1) for each horizontal direction:

$$V_u = \frac{(M_u)_{top} + (M_u)_{bottom}}{h_n} \qquad (10.6.1.1)$$

where $(M_u)_{top}$ corresponds to the factored moment at the column upper end; $(M_u)_{bottom}$ corresponds to the factored moment at the column lower end; and h_n is the clear distance between column lateral supports.

10.6.1.2 *Factored shear from horizontal loads*—Factored shear V_u due to horizontal loads should be determined from the horizontal loads prescribed in Chapter 4 and applied to the structure as prescribed in Chapter 11 using appropriate load combinations from 4.2.

10.6.2 *Design shear strength*

10.6.2.1 *General*—The design shear strength should equal or exceed factored beam-action shear, which accompanies column moments and occurs in both horizontal directions, having a constant value in the clear distance between floors.

10.6.2.2 *Shear strength*—Column section shear strength ϕV_n should be determined following the procedure in 5.13.4 for beam-action shear using Eq. (10.6.2.2).

$$\phi V_n = \phi(V_c + V_s) \qquad (10.6.2.2)$$

In Eq. (10.6.2.2), ϕV_c is concrete shear strength, ϕV_s is transverse reinforcement shear strength, and $\phi = 0.75$.

10.6.2.3 *Contribution of the concrete to shear strength*—The contribution of concrete to shear strength should be computed using Eq. (10.6.2.3) with $\phi = 0.75$.

$$\phi V_c = \phi 2\sqrt{f_c'}bd$$
$$\left(\phi V_c = \phi\frac{\sqrt{f_c'}}{6}bd \text{ (SI)}\right) \qquad (10.6.2.3)$$

In Eq. (10.6.2.3), d should be taken as the appropriate value, based either on h_c or b_c, in the direction of the shear and b in direction normal to the shear. In circular columns, the product of b times d should be taken as $0.8h^2$, with h being the column diameter.

10.6.2.4 *Contribution of the transverse reinforcement to the shear strength*—Transverse reinforcement shear strength should be computed in each direction using Eq. (10.6.2.4). The design shear strength should be based on Eq. (5.13.3) and (10.6.2.2). Where Eq. (5.13.3) is not satisfied, tie spacing s should be reduced.

$$\phi V_s = \phi\left[\frac{A_v f_{yt} d}{s}\right] \qquad (10.6.2.4)$$

where A_v corresponds to area of the tie legs parallel to the shear direction and s to the largest tie vertical spacing within the clear height of the column; f_{yt} is the yield strength of the tie legs; and $\phi = 0.75$. For circular stirrups or spiral reinforcement, A_v is twice the stirrup or spiral bar or wire area A_b.

10.6.2.5 *Biaxial shear strength verification*—When the column is subjected to shear in the two horizontal directions simultaneously, it should comply with Eq. (10.6.2.5):

$$\sqrt{\left[\frac{(V_u)_x}{(\phi V_n)_x}\right]^2 + \left[\frac{(V_u)_y}{(\phi V_n)_y}\right]^2} \leq 1.0 \qquad (10.6.2.5)$$

where $(V_u)_x$ and $(V_u)_y$ correspond to the factored shear in the direction of the x- and y-axes; and $(\phi V_n)_x$ and $(\phi V_n)_y$ correspond to the shear strength determined from Eq. (10.6.2.2) for the appropriate direction x or y.

10.7—Calculation of foundation reaction

10.7.1 *Vertical load reaction*—Vertical load reaction R_u at the foundation should be equal to the value of P_u at the column lower end directly above the foundation.

10.7.2 *Moment reaction*—Unbalanced moment reaction ΔM_u at the foundation should be equal to the value of M_u at the column lower end directly above the foundation. This unbalanced moment should be distributed to the grade beams and the foundation members as prescribed in Chapter 14.

CHAPTER 11—SEISMIC RESISTANCE

11.1—Special reinforcement details for seismic zones

11.1.1 *General*—The following should be used for the structural members located in seismic zones as defined by 4.11.2.3.

11.1.2 *Frame girders*

11.1.2.1 *Dimensional limits*—Minimum width b_w of girders should be 10 in. (250 mm), and the girder should comply with 8.7.2.

11.1.2.2 *Longitudinal reinforcement*—In addition to 8.7.5, (a) through (f) should be met:

(a) At least two top and bottom longitudinal bars should be provided;

(b) At any section, the ratios of positive and negative moment reinforcement should be equal to or greater than the minimum indicated by 8.4.5;

(c) At any section, the positive and negative moment reinforcement ratios should not exceed 0.025;

(d) Area of positive moment reinforcement at the joint face should not be less than one-half the area of negative moment reinforcement at the same joint face;

(e) Area of positive and negative moment reinforcement at any section should not be less than one-fourth the maximum area of negative moment reinforcement at either joint face; and

(f) Lap splices should not be used in beam-columns joints and the confinement zones defined in 11.1.2.3(a). Full length of the lap splice should be confined with closed stirrups as defined in 11.1.2.3(b), with stirrup spacing not exceeding the smaller of $d/4$ and 4 in. (100 mm).

11.1.2.3 *Transverse reinforcement*—In addition to 8.5, (a) through (e) should be met:

(a) Transverse reinforcement should be hoops (confinement stirrups) at both ends of the girder over a distance equal to twice the member depth h, measured from the face of the supporting member toward midspan (Fig. 11.1.2.3);

(b) Hoops should be closed stirrups at least 3/8 in. (10 mm) diameter, with hooks as defined in 5.6(d) and complying with 10.4.3.2, column ties. Crossties should comply with 5.6(e);

(c) The first hoop should be located no farther than 2 in. (50 mm) from the support face;

(d) Hoop spacing should not exceed the smaller of $d/4$ and 5 in. (125 mm); and

(e) For the central length of the girder span between confinement zones, transverse reinforcement should be closed stirrups with hooks complying with 5.6(d), and the maximum stirrup spacing should be $d/2$.

11.1.2.4 *Shear strength*—In addition to 8.5, (a) through (e) should also be met:

(a) Additional factored shear force ΔV_e, corresponding to the probable moment strength development of the span at the joint face, should be determined as the larger value from Eq. (11.1.2.4a) and Eq. (11.1.2.4b) (Fig. 11.1.2.4a);

$$\Delta V_e = \frac{\left(M_{pr}^{+}\right)_{left} + \left(M_{pr}^{-}\right)_{right}}{\ell_n} \qquad (11.1.2.4a)$$

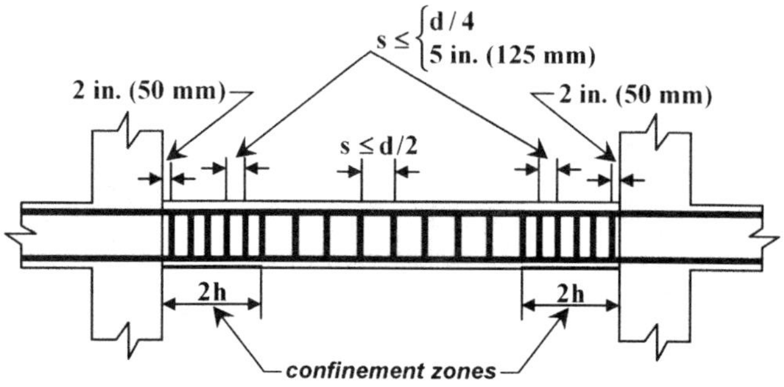

Fig. 11.1.2.3—Confinement stirrup spacing.

$$\Delta V_e = \frac{\left(M_{pr}^{-}\right)_{left} + \left(M_{pr}^{+}\right)_{right}}{\ell_n} \qquad (11.1.2.4b)$$

(b) In Eq. (11.1.2.4a) and (11.1.2.4b), M_{pr}^+ and M_{pr}^- correspond to positive and negative probable moment strength at joint faces, determined from Eq. (8.4.8.1) and using the corresponding longitudinal reinforcement yield strength, f_{ypr}, instead of f_y ($f_{ypr} = 1.25f_y$), and a strength reduction factor $\phi = 1.0$;

(c) The largest value of ΔV_e determined from Eq. (11.1.2.4a) or Eq. (11.1.2.4b) should be added to V_u at the support face, and the shear diagram of 8.5.4.6 should be recalculated (Fig. 11.1.2.4b);

(d) Transverse reinforcement for shear should be determined as prescribed in 8.5.4.5, except where ΔV_e is greater than V_u for gravity loads at the support face in computing shear reinforcement, the contribution of concrete to shear strength should be taken as zero ($\phi V_c = 0$) in the confinement zones indicated in 11.1.2.3(a); and

(e) Hoops described in 11.1.2.3 should be considered effective shear reinforcement.

11.1.3 *Columns*

11.1.3.1 *Dimensional limits*—The restrictions of 10.3 should apply, and 10.3.2.1 should be modified by (a) and (b):

(a) The least cross-sectional dimension should not be less than 12 in. (300 mm); and

(b) Ratio of the longer to the shorter cross-sectional dimension should not exceed 2.5.

11.1.3.2 *Longitudinal reinforcement*—Section 10.4.2 should apply, and 10.4.2.8 should be modified to restrict the splice locations to the center half of the member length.

11.1.3.3 *Minimum moment strength of columns*—Column moment strength should satisfy Eq. (11.1.3.3), unless the full length of a column is provided with transverse reinforcement complying with 11.1.3.4.

$$\Sigma M_c \geq \frac{6}{5}\Sigma M_g \qquad (11.1.3.3)$$

where ΣM_c is the sum of nominal moment strengths M_n of columns framing into a joint; and ΣM_g is the sum of nominal moment strengths M_n of girders framing into the same joint.

Column moment strength should correspond to the minimum moment strength computed using the appropriate equation of Eq. (5.12.6d) and (5.12.6e) for the range of factored axial loads P_u that act on the column. Moment strengths should be added in such a manner that column moments oppose beam moments. Equation (11.1.3.3) should be satisfied for beam moments acting in both directions in the vertical plane of the frame considered (Fig. 11.1.3.3).

11.1.3.4 *Tie transverse reinforcement*—When ties (stirrups) are used as column transverse reinforcement, 10.4.3 and (a) through (h) should be met.

(a) Transverse reinforcement should be hoops (confinement ties) (Fig. 11.1.3.4a) over the length of the confinement zone ℓ_o measured from the joint face at both column ends. The distance ℓ_o should not be less than the largest of the column cross-sectional dimension, one-sixth the clear length of the column, or 20 in. (500 mm);

(b) Hoops should be closed single or overlapping ties with hooks as defined in 5.6(d) and complying with 10.4.3.2, column ties;

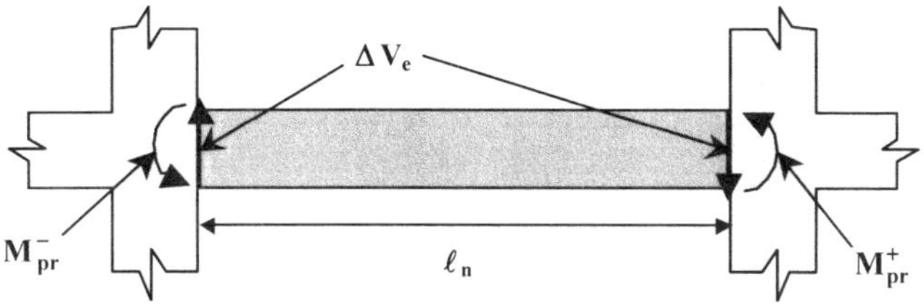

Fig. 11.1.2.4a—Calculation of ΔV_e.

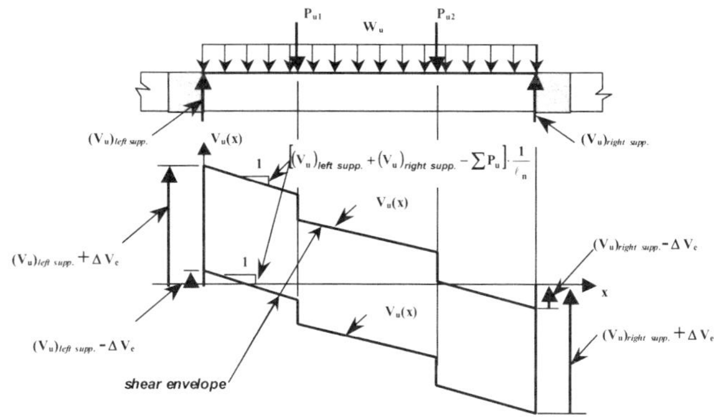

Fig. 11.1.2.4b—Calculation of the shear envelope in the girder.

(c) Crossties that comply with 5.6(e) and have the same bar diameter and spacing as the hoops are permitted. Each crosstie should engage a peripheral longitudinal reinforcing bar. Consecutive crossties should be alternated end-to-end along the longitudinal reinforcement;

(d) Horizontal distance, measured center-to-center, between legs of the peripheral hoops and crossties, and between crossties, should not exceed the larger of 8 in. (200 mm) and one-half the smallest cross-sectional dimension. If the number of hoop legs and crossties determined exceeds the number of longitudinal bars located in the cross section face, additional longitudinal bars should be provided (Fig. 11.1.3.4b);

(e) In confinement zones, maximum hoop spacing measured along the member axis should not exceed the larger of 4 in. (100 mm) and the value determined from Eq. (11.1.3.4).

$$s \le \frac{A_b f_{yt}}{0.6 f_c'} \left(s \le \frac{A_b f_{yt}}{15 f_c'} \text{ (SI)} \right) \qquad (11.1.3.4)$$

where A_b is the hoop and crosstie bar area, and f_{yt} is the nominal yield strength of the hoop and crosstie (Fig. 11.1.3.4a);

(f) The first hoop should be located no farther than 2 in. (50 mm) from the joint face;

(g) When reinforcement as indicated above is not placed throughout the column clear length, transverse reinforcement in the central part of the column clear length between confinement zones should be hoops of the same diameter, yield strength f_{yt} and number of crossties used in the confinement zones, and maximum center-to-center spacing should not exceed the smaller of six times the longitudinal column bar diameter d_b and 6 in. (150 mm) (Fig. 11.1.3.4a); and

(h) Longitudinal reinforcement splices should be located in the central zone of the column with hoop confinement provided over the entire splice length, and not more than one-half of total longitudinal bars should be spliced (Fig. 11.1.3.4c).

11.1.3.5 *Spiral reinforcement*—When spirals are used as column transverse reinforcement, 10.4.3.3 and (a) through (c) should be met:

(a) Transverse reinforcement should be a spiral complying with 11.1.3.5 over the confinement zone length ℓ_o not less than the largest of the column cross-sectional dimensions, one-sixth the clear length of the member, and 20 in. (500 mm), measured from the joint face at both column ends;

(b) The spiral volumetric ratio should not be less than computed by Eq. (10.4.3.3) and Eq. (11.1.3.5);

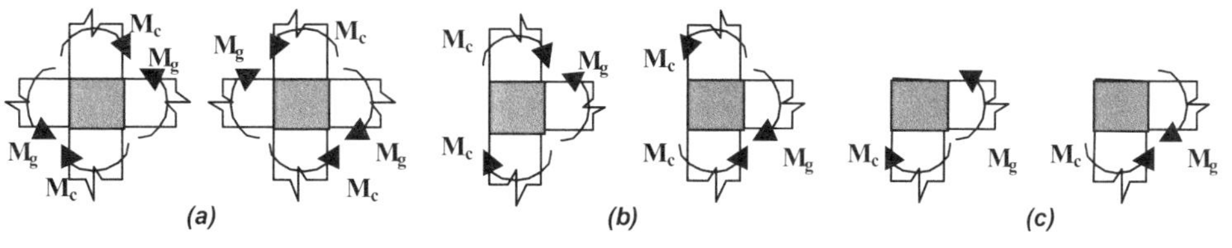

Fig. 11.1.3.3—Minimum moment strength of columns.

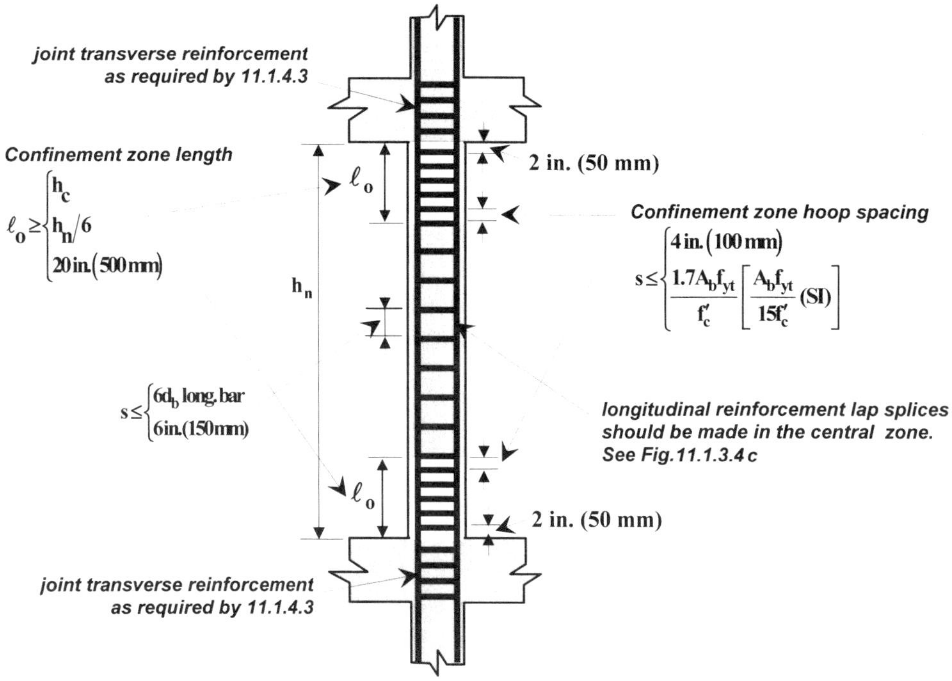

Fig. 11.1.3.4a—Confinement hoop spacing in columns.

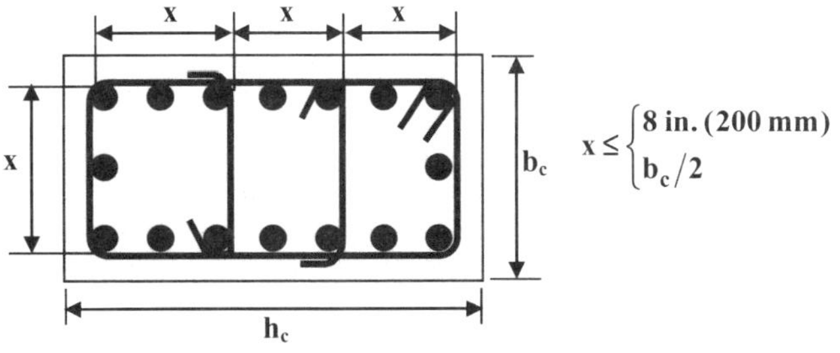

Fig. 11.1.3.4b—Arrangement of hoop legs (confinement ties) and crossties.

$$\rho_s = \frac{A_b \pi (d_{cs} - d_b)}{A_{cs} s} \geq 0.12 \frac{f'_c}{f_{yt}} \qquad (11.1.3.5)$$

(c) Outside confinement zones, the maximum center-to-center spiral spacing should not exceed the smaller of six times the longitudinal column bar diameter d_b and 6 in. (150 mm).

11.1.3.6 *Shear strength*—Section 10.6 and (a) through (d) should be met:

(a) Factored shear ΔV_e, corresponding to probable column moment strength at the joint faces, should be determined using Eq. (11.1.3.6) for both principal directions in plan (Fig. 11.1.3.6a);

$$\Delta V_e = \frac{(M_{pr})_{top} + (M_{pr})_{bottom}}{h_n} \qquad (11.1.3.6)$$

(b) In Eq. (11.1.3.6), M_{pr} is the probable moment strength at the joint faces, determined using f_{ypr} instead of f_y ($f_{ypr} = 1.25f_y$) and a strength reduction factor $\phi = 1$. Column moment strength should correspond to the maximum probable moment strength computed using the appropriate equation of Eq. (5.12.6d) and (5.12.6e), for the range of factored axial loads P_u that act on the column. Factored shear for the column, ΔV_e, need not exceed the value determined from the joint shear based on probable girder moment strength M_{pr} at the joint as determined in 11.1.2.4 (Fig. 11.1.3.6b);

(c) Transverse reinforcement for shear should be determined as prescribed in 10.6, except that the concrete contribution to the shear strength should be taken as ($\phi V_c = 0$) in the confinement zones indicated in 11.1.3.4(a) and 11.1.3.5(a); and

(d) Hoops (confinement ties) or spirals indicated by 11.1.3.4 and 11.1.3.5 should be considered effective as shear reinforcement.

11.1.4 *Frame joints*

11.1.4.1 *General*—For frame joints located in seismic zones, 11.1.4 should apply, and 10.4.3.4 should not apply.

11.1.4.2 *Column dimensions at a joint*—Column dimension parallel to the beam should not be less than 20 times the largest longitudinal girder bar diameter d_b, where longitudinal girder bars extend through the column-girder joint.

11.1.4.3 *Transverse reinforcement*—Transverse reinforcement conforming to (a) and (b) should be provided within the column-girder joint:

(a) Horizontal transverse hoops (confinement ties) with the same area and spacing indicated by 11.1.3.4 should be provided within the column-girder joint. This area within the joint is the confined core. Where girders, having a width equal to or greater than three-fourths of the column width, frame into the joint at all four sides, the hoop spacing can be twice that indicated by 11.1.3.4, but not exceed 6 in. (150 mm); and

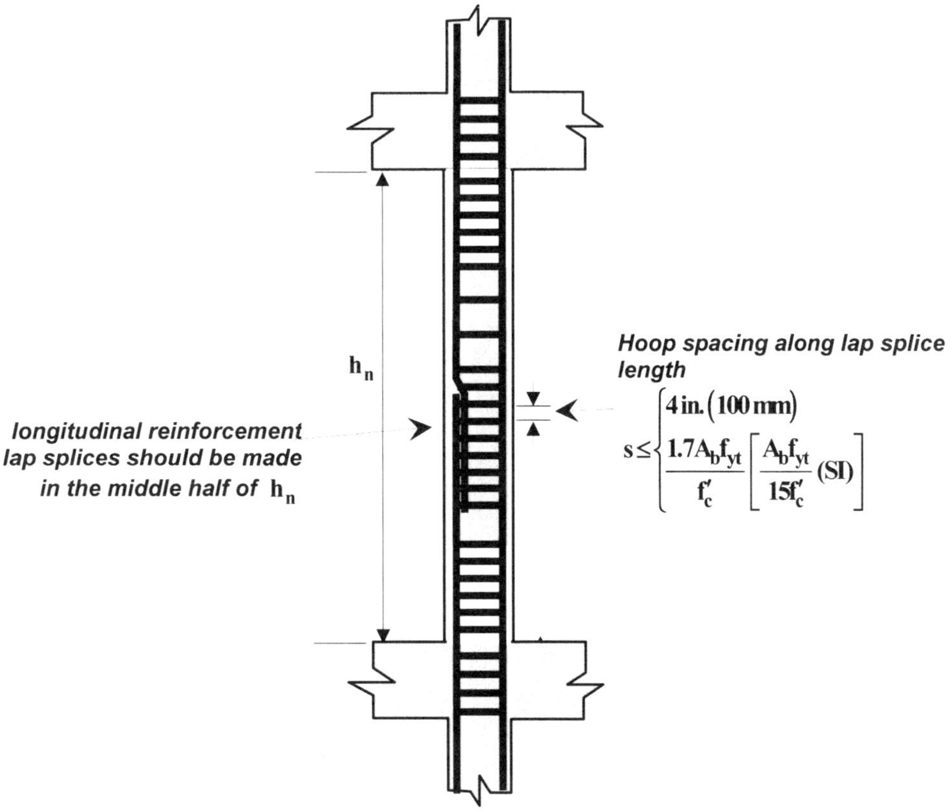

Fig. 11.1.3.4c—Confinement hoop spacing in columns for splices.

(b) Where longitudinal girder reinforcement is located outside the confined core defined in (a), vertical hoops for girders as indicated by 11.1.2.3 should be provided to confine it.

11.1.4.4 *Joint shear strength*—Horizontal joint shear strength should equal or exceed the factored shear that develops due to the probable moment strength of columns and girders at the joint (Fig. 11.1.4.4a). Items (a) through (d) apply.

(a) Factored shear at the joint, V_u, should be determined for both principal directions using Eq. (11.1.4.4a) for joints where girders frame into both sides, and using Eq. (11.1.4.4b) where girders frame into one side only.

$$V_u = f_{ypr}\left(A_s + A'_s\right)_{girder} - \left(\Delta V_e\right)_{column} \quad (11.1.4.4a)$$

$$V_u \ge \begin{cases} f_{ypr}\left(A_s\right)_{girder} - \left(\Delta V_e\right)_{column} \\ f_{ypr}\left(A'_s\right)_{girder} - \left(\Delta V_e\right)_{column} \end{cases} \quad (11.1.4.4b)$$

In Eq. (11.1.4.4a) and (11.1.4.4b), $(A_s)_{girder}$ corresponds to girder longitudinal reinforcement area. The shear developed

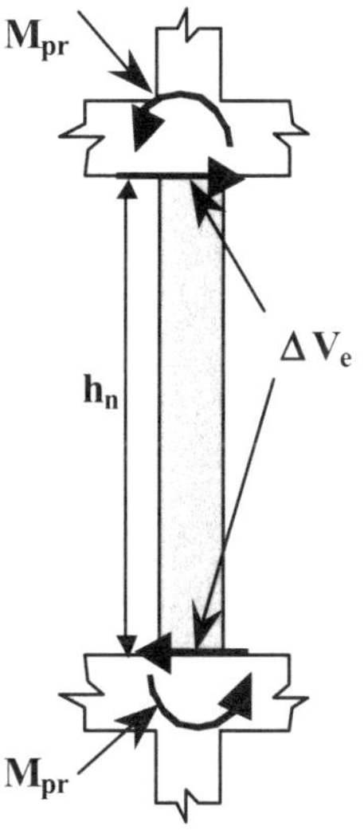

Fig. 11.1.3.6a—Calculation of ΔV_e for columns.

from the column moment strength should be determined from 11.1.3.6;

(b) Shear strength at the critical plane in the joint should be (for A_j, refer to Fig. 11.1.4.4b and 11.1.4.4c):

For joints confined on all four faces $\phi V_n = \phi 20\sqrt{f'_c}A_j$

$$\left[\phi V_n = \phi 1.7\sqrt{f'_c}A_j \text{ (SI)}\right]$$

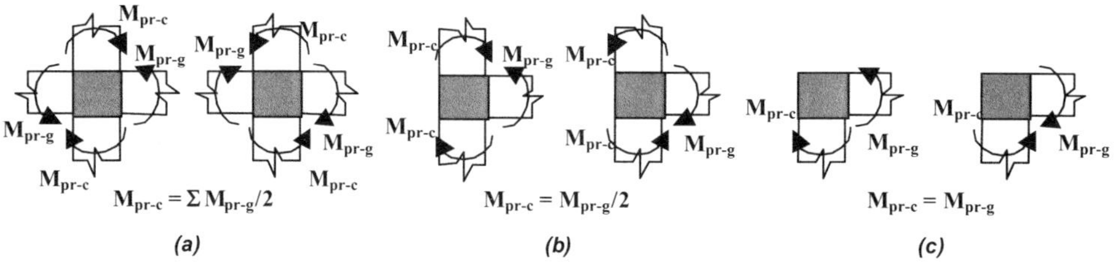

Fig. 11.1.3.6b—Maximum M_{pr} *for columns required for column shear* ΔV_e.

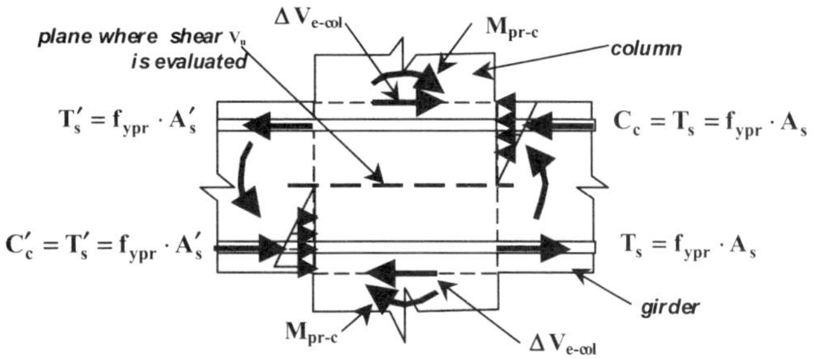

Fig. 11.1.4.4a—Joint shear determination.

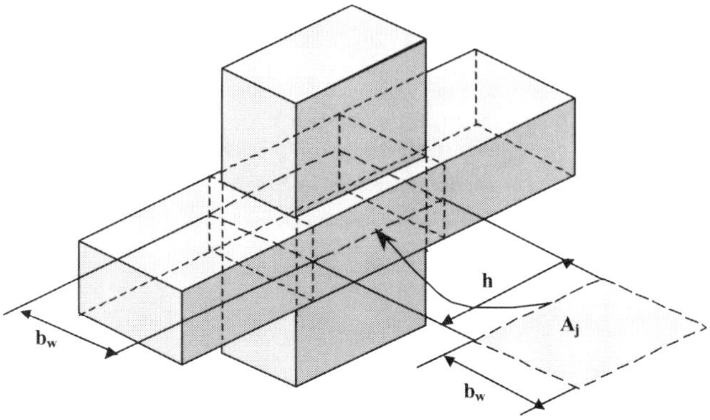

Fig. 11.1.4.4b—Definition of A_j *for girder wider than columns.*

For joints confined on three faces or on opposite faces

$$\phi V_n = \phi 15\sqrt{f_c'}A_j$$

$$\left[\phi V_n = \phi 1.25\sqrt{f_c'}A_j \text{ (SI)}\right]$$

For all other joints

$$\phi V_n = \phi 12\sqrt{f_c'}A_j$$

$$\left[\phi V_n = \phi 1.0\sqrt{f_c'}A_j \text{ (SI)}\right]$$

(c) A girder that frames into a joint face is considered to provide confinement to the joint if at least three-fourths the joint face is covered by the girder; and

(d) A_j is the effective cross-sectional area within the joint, in a plane parallel to the plane of the reinforcement generating the shear and is equal to the product of joint depth and the effective joint width. Joint depth is equal to the column dimension parallel to the girder direction. Joint effective width is equal to the girder width for girders equal to or larger than column width (Fig. 11.1.4.4b). For girders narrower than column width, joint effective width is equal to the smaller of the girder width plus joint depth, or girder width plus twice the smaller perpendicular distance from the girder longitudinal axis to the column side, without exceeding column width (Fig. 11.1.4.4c).

11.1.4.5 *Girder reinforcement anchorage*—Longitudinal reinforcement terminating at the joint should end in a 90-degree standard hook located within the column confined core (11.1.4.3(a)). Anchorage distance should comply with 5.8.3.

11.1.5 *Walls*

11.1.5.1 *General*—Reinforced concrete walls located in seismic zones should comply with Chapter 12 and of 11.1.5.

11.1.5.2 *Boundary elements*—Boundary elements in structural walls should comply with (a) through (h).

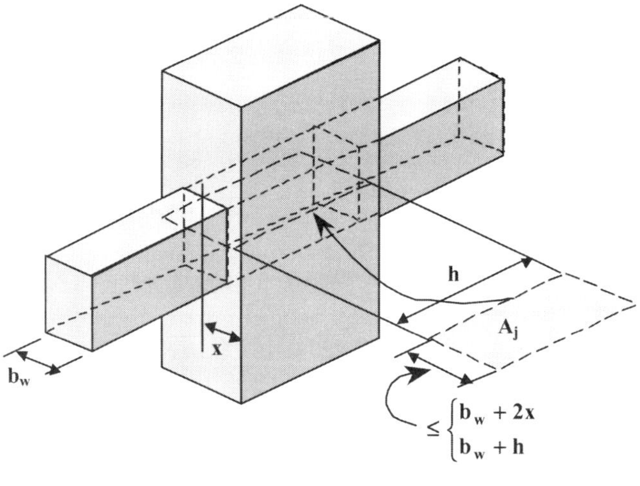

Fig. 11.1.4.4c—Definition of A_j *for girder narrower than column.*

(a) Where the maximum compressive extreme fiber stress f_{cu}, calculated from Eq. (11.1.5.2a) from load combinations that include seismic effects, exceeds $0.2f_c'$, one of the following should be provided:

- Boundary elements at both ends of structural walls; or
- Confinement transverse reinforcement complying with 11.1.3.4 for the entire wall

$$f_{cu} = \frac{P_u}{A_g} + \frac{6M_u}{\ell_w^2 b_w} \qquad (11.1.5.2a)$$

(b) Boundary elements may be discontinued where f_{cu} is less than $0.15f_c'$;

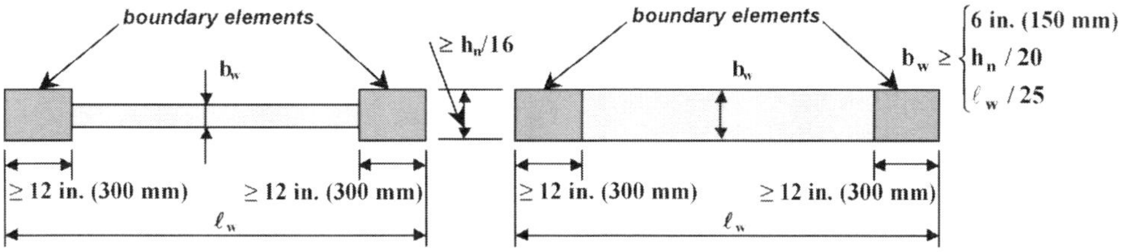

Fig. 11.1.5.2—Boundary element dimensions.

(c) Boundary element thickness should not be less than the larger of $h_n/16$ or b_w, and should have a horizontal length in the wall direction not less than 12 in. (300 mm) at each end (12.3.2.1 and Fig. 11.1.5.2).

(d) Boundary elements should have transverse reinforcement as specified for columns in 11.1.3.4, with the spacing in 11.1.3.4(e) not exceeded at any point along the height of the boundary element;

(e) Boundary elements should be proportioned to resist all factored gravity loads on the wall, including tributary loads and self-weight, as well as the vertical force resulting from resisting seismic overturning moments, following 12.2.3. Compressive factored axial load on boundary elements, P_{cu}, should be determined from Eq. (11.1.5.2b), and the tensile factored axial load P_{tu} on the boundary elements from Eq. (11.1.5.2c):

$$P_{cu} = \frac{P_u}{2} + \frac{M_u}{(\ell_w - 12\text{ in.})} \quad \left[P_{cu} = \frac{P_u}{2} + \frac{M_u}{(\ell_w - 300\text{ mm})}\text{ (SI)}\right] \tag{11.1.5.2b}$$

$$P_{tu} = \frac{P_u}{2} - \frac{M_u}{(\ell_w - 12\text{ in.})} \quad \left[P_{tu} = \frac{P_u}{2} - \frac{M_u}{(\ell_w - 300\text{ mm})}\text{ (SI)}\right] \tag{11.1.5.2c}$$

(f) Longitudinal reinforcement should be proportioned for factored compression axial load P_{cu} using Eq. (5.12.3.1) and (5.12.3.2a). In Eq. (5.12.3.1), A_g should be replaced by the boundary element area. The reinforcement ratio should not exceed the limits of 12.4.4.3. When longitudinal reinforcement determined from Eq. (11.1.5.2b) and Eq. (11.1.5.2c) exceeds these limits, boundary element size should be increased. When the boundary element size is increased, P_{cu} and P_{tu} should be adjusted in Eq. (11.1.5.2b) and Eq. (11.1.5.2c);

(g) Longitudinal reinforcement should be proportioned for the absolute value of the factored tensile axial load P_{tu}—a value greater than zero in Eq. (11.1.5.2c) means no tensile loads—using Eq. (5.12.5); and

(h) Where boundary elements also serve as columns within a frame, they should be proportioned as columns using Chapter 10.

11.1.5.3 *Shear strength*—Structural wall shear strength should comply with 12.6.

11.1.6 *Slab-column systems*

11.1.6.1 *General*—Slab-column systems located in seismic zones should comply with 11.1.6 and Chapter 9. This structural system should not be used in seismic zones without using reinforced concrete walls to resist seismic lateral loads and limit lateral deformation imposed by seismic acceleration.

11.1.6.2 *Lateral force moments*—Factored lateral force moments, as determined from 4.15.4.2, should be added to the column strip moments, using appropriate load combinations. Reinforcement corresponding to factored lateral force moments should be placed within the effective width zone defined in 9.8.1.8(a).

11.1.6.3 *Column strip reinforcement*—Items (a) through (e) and Chapter 9 should be observed:

(a) At least one-half the top and bottom column strip reinforcement at the support should be placed within the effective slab width defined in 9.8.1.8(a);

(b) At least one-quarter the top column strip reinforcement at the support should be continuous throughout the span;

(c) Continuous bottom column strip reinforcement should be at least one-third the top reinforcement at the support;

(d) At least one-half of all bottom reinforcement at midspan should be continuous and should have the development length prescribed in 5.8.1 at the support face as defined in 9.4.2.8; and

(e) At discontinuous slab edges, all top and bottom reinforcement at the support should comply with the development length prescribed in 5.8.1 at the support face.

11.1.6.4 *Waffle-slabs*—In waffle-slabs, joists located in the column strip and all joists that connect to capitals, should have a minimum width b_w of 6 in. (150 mm), and should be provided with hoops spaced at not more than $d/4$ throughout the capital and extending from the capital face a minimum distance $2d$ into the span.

11.2—Interaction with nonstructural elements

11.2.1 *General*—Supporting standards of 1.4 should be used for design of nonstructural elements. Notwithstanding, in certain cases nonstructural elements interact with structural members, affecting their behavior and introducing loads and moments not covered by approximate procedures of this guide. This situation is especially important in structures subjected to seismic or other lateral loads. The potential of life-threatening situations caused by nonstructural element failures during strong ground motion should be addressed by the structural or architectural designer.

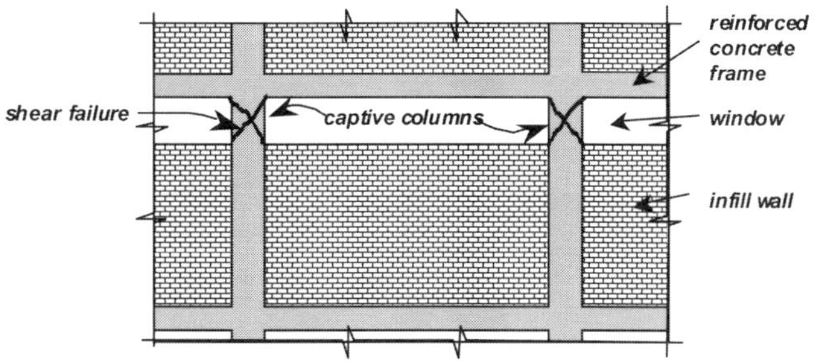

Fig. 11.2.3.1—Captive or short column effect.

11.2.2 *Masonry infills*—The influence of masonry infills on the structure's lateral force behavior should be addressed by the designer, either by considering such influence in the design process or by isolating the masonry from the structure. In the second alternative, appropriate measures should be taken to maintain the out-of-plane stability of the masonry when subjected to seismic or wind lateral loads.

11.2.3 *Captive or short columns*

11.2.3.1 *Description*—Historically, the most damaging effect of reinforced concrete frames interacting with nonstructural elements has been caused by captive or short columns. A captive column exists where a structural or nonstructural wall confines the column bottom, but leaves a gap at the top, below the horizontal structural member (Fig. 11.2.3.1). This type of wall arrangement is very common in buildings where the gap is provided for lighting purposes, such as educational buildings. This situation introduces a large, unanticipated shear force in the column when the structure is subjected to lateral loads.

11.2.3.2 *Corrective measures*—Two alternative corrective measures should be considered:

(a) Separate the infill wall from the column by providing a gap. The gap between wall and column should be approximately 1.5 percent of the story height h_{pi}. The masonry wall should be anchored to prevent its overturning when subjected to out-of-plane lateral loads; or

(b) Locate a much shorter window in the central part of the span, thus having the infill masonry wall against the column along its full height. In this alternative, the distance between the column face and the window should be at least twice the vertical dimension of the gap left by the window (Fig. 11.2.3.2).

If neither (a) nor (b) is satisfied, hoops should be provided over the full column length as indicated by 11.1.3.4, and column shear strength should be determined from 11.1.3.6 using the vertical window dimension instead of h_n in Eq. (11.1.3.6).

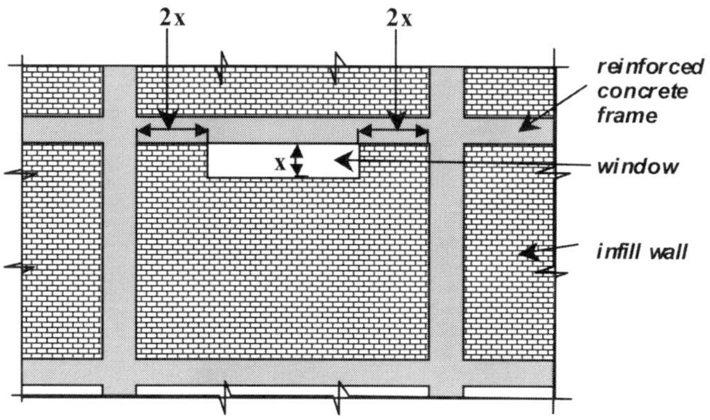

Fig. 11.2.3.2—Alternative to account for the short or captive column effect.

CHAPTER 12—REINFORCED CONCRETE WALLS

12.1—General

In-plane and out-of-plane design of reinforced concrete walls should be performed using Chapter 12. Soil-retaining walls should be designed following Chapter 14.

12.2—Loads

12.2.1 *Loads to be included*—Loads for reinforced concrete walls should be established from Chapter 4. Loads that should be included in the design are (Fig. 12.2.1):

(a) Live and dead loads from tributary structural members from each floor located above. Tributary loads should be established from Chapter 4 and the loads of each tributary member type;

(b) Wall self-weight; and

(c) Lateral loads from wind, seismic, or soil lateral pressures.

12.2.2 *Dead load and live load*—P_d should include wall self-weight. The wall self-weight corresponding to each floor may be applied at the lower part of the wall in that floor. The unbalanced moment due to vertical loads should be determined from the supported member (8.7.6 and 9.9.2).

12.2.3 *Lateral loads*—Applied factored horizontal shear V_u at each story and in both principal directions should be determined from Chapter 4. The factored lateral force moments M_u should be established at the upper and lower part of the wall in both principal plan directions at each story by following (a), (b), and (c):

(a) Factored horizontal shear at each story x, V_{xu}, should be determined for the wall from Chapter 4;

(b) Factored lateral force applied at each story x, F_{xu}, should be determined as the difference in factored shear between the two successive stories, V_{xu} and $V_{(x+1)u}$; and

(c) Factored lateral force moments M_{xu} at each story x should be determined using Eq. (12.2.3) (Fig. 12.2.3).

$$M_{xu} = \sum_{i=x}^{n}\left[F_{iu}\cdot\left(h_i - h_x\right)\right] \quad (12.2.3)$$

12.2.4 *Factored loads*—Factored loads P_u and V_u and moments M_u should be established at the upper and lower part of the wall in each story. Horizontal loads should be divided into in-plane and out-of-plane loads (Fig. 12.2.4).

12.3—Dimensional limits

12.3.1 *General*—In addition to Chapter 12, walls should comply with general dimensional limits set forth in 1.3 and 11.1.5, should be rectangular, except as permitted by 12.3.2.2,

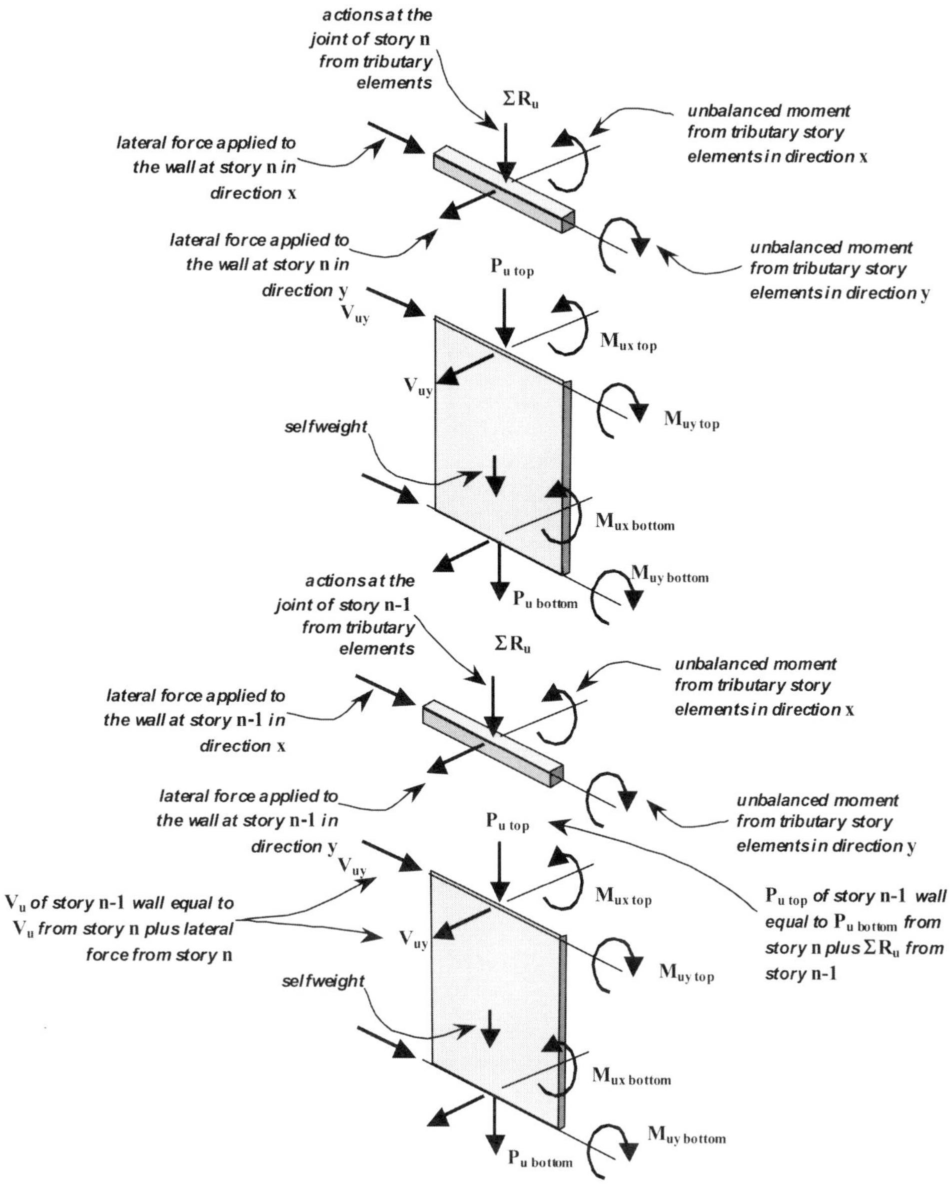

Fig. 12.2.1—Reinforced concrete wall with applied factored loads and moments.

should be aligned vertically, and should be continuous to the foundation. Walls resisting lateral loads are restricted in the maximum axial load they can carry by Section 4.11.3.1, limiting the axial load to that corresponding to the balanced point as indicated by Eq. (5.12.4.1a).

12.3.2 *Dimensional limits*

12.3.2.1 *Minimum thickness*—Minimum wall thickness (Fig. 12.3.2.1) should be the larger of 6 in. (150 mm), or 1/25 of the wall length ℓ_w. At changes of wall thickness in contiguous stories, 11.1.5.2(c) should apply.

12.3.2.2 *Columns embedded in walls*—A column may be embedded in a wall instead of increasing wall thickness over the entire wall length. The increase of thickness may be placed at one side of the cross section. Transverse column dimension should meet 10.3.3.

12.3.3 *Distance between lateral supports*—It should be assumed that lateral support is provided by the floor system at all levels connected to the wall (Fig. 12.3.2.1) in the two horizontal directions. Clear vertical distance between lateral supports, h_n, should not exceed 20 times the wall thickness.

12.3.4 *Beams on top of walls*—Beams or girders should be provided for the full horizontal wall length at every floor and roof supported by the wall. These beams or girders should conform to 8.7.2.3 and conform to 4.15.5, collector elements.

12.4—Reinforcement details

12.4.1 *General*—Reinforcement types should be as described in Chapter 12 and should comply with 12.4.2 to 12.4.6.

12.4.2 *Maximum reinforcement spacing*—Maximum vertical and horizontal reinforcement should be the lesser of three times the reinforced concrete wall thickness and 12 in. (300 mm) (Fig. 12.4.2).

12.4.3 *Number of curtains of reinforcement*—Two curtains, or layers, of reinforcement, having both vertical

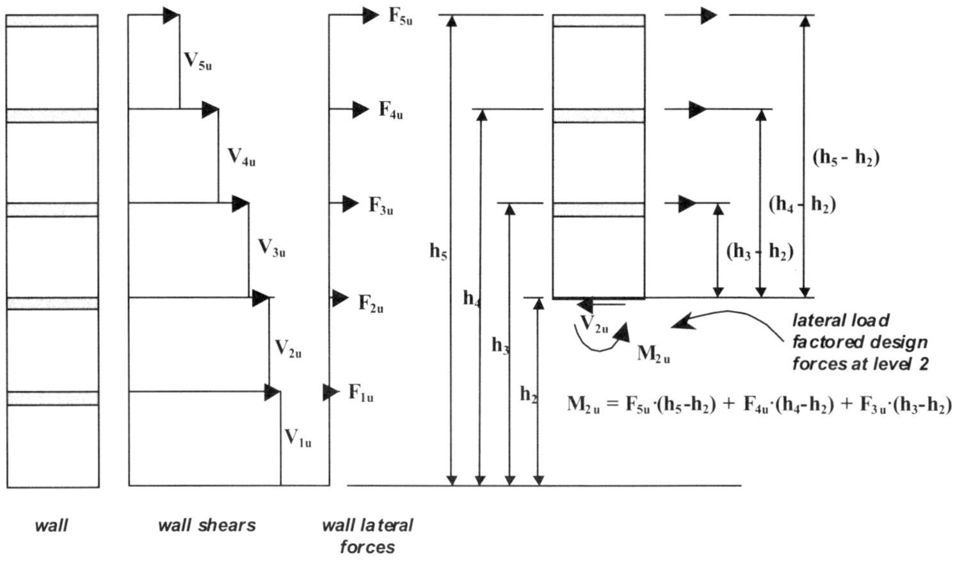

Fig. 12.2.3—Calculation of factored lateral force moments.

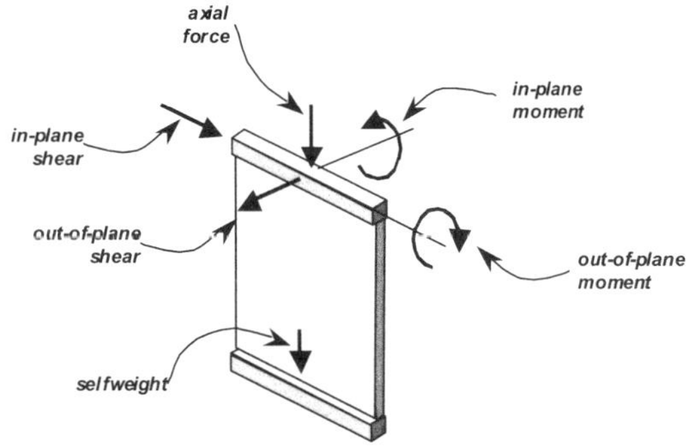

Fig. 12.2.4—In-plane and out-of-plane loads.

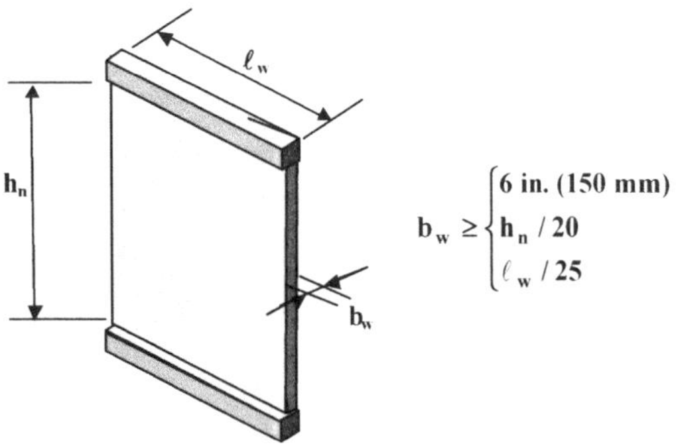

Fig. 12.3.2.1—Minimum cross-sectional dimensions for rectangular reinforced concrete walls.

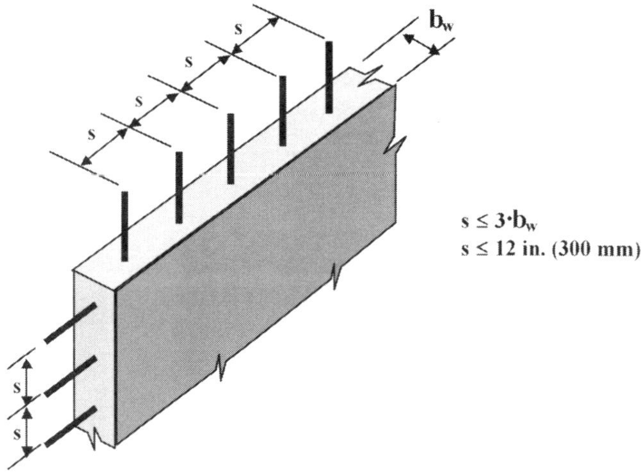

Fig. 12.4.2—Reinforcement spacing in reinforced concrete walls.

and horizontal reinforcement, should be placed in cases (a), (b), and (c):

(a) Wall thickness exceeds 10 in. (250 mm);

(b) Vertical reinforcement ratio ρ_{vw} exceeds 0.01 (12.4.5.2); and

(c) In-plane factored shear force V_u in the wall exceeds ϕV_c as given by Eq. (12.6.2.2).

When two reinforcement curtains are placed, each should have approximately half the reinforcement. Minimum cover to a curtain should be 1-1/4 in. (30 mm), and maximum cover to a curtain should be one-third the wall thickness. For exterior exposure, minimum cover should be 2 in. (50 mm).

In all other cases, only one reinforcement curtain, located in the wall center, may be used.

12.4.4 *Vertical reinforcement*

12.4.4.1 *General*—Vertical reinforcement should consist of one or two curtains of bars or welded wire reinforcement placed parallel to the wall face. The area of vertical reinforcement should be sufficient to resist the combination of factored axial loads and factored moments acting about the two main wall axes.

12.4.4.2 *Minimum area of vertical reinforcement*—Minimum vertical reinforcement ratio ρ_{vw} of steel area to gross concrete horizontal section area should be 0.0025.

12.4.4.3 *Maximum area of vertical reinforcement*—Maximum vertical reinforcement ratio ρ_{vw} of steel area to

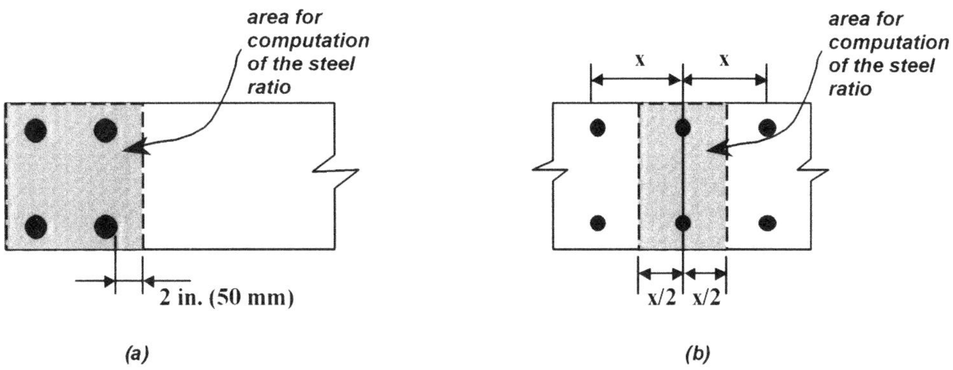

Fig. 12.4.4.4—Calculation of vertical reinforcement ratio.

gross reinforced concrete wall horizontal section area should be 0.06. Where the ratio ρ_{vw} exceeds 0.01, vertical reinforcement should be enclosed with ties as prescribed for columns in 10.4.3.1.

$$0.0025 \leq \rho_{vw} \left(= \frac{A_{st}}{b_w \ell_w} \right) \leq 0.06 \qquad (12.4.4.3)$$

12.4.4.4 *Calculation of vertical reinforcement ratio*—Items (a) and (b) should apply to walls where area and vertical reinforcement spacing varies or columns are embedded:

(a) Vertical reinforcement ratio ρ_{vw} should not exceed the maximum vertical reinforcement ratio defined by 12.4.4.3, and should be evaluated in an area where the vertical reinforcement is concentrated, as shown in Fig. 12.4.4.4(a). The vertical reinforcement ratio ρ_{vw} should be evaluated over an area bound by wall faces and 2 in. (50 mm) beyond the reinforcing bars as measured along the wall length where the vertical reinforcement is concentrated (10.3.4); and

(b) Vertical reinforcement ratio ρ_{vw} should not be less than the minimum vertical reinforcement ratio set forth in 12.4.4.3 where the vertical reinforcement concentration is reduced by increasing the bar spacing or by decreasing the vertical bar diameter, as shown in Fig. 12.4.4.4(b).

12.4.4.5 *Reinforcement splicing*—Lap splices of vertical wall reinforcement should comply with the lap splice length of 5.8.2. It is permitted to lap splice 100 percent of the vertical reinforcement at any given section, except at the supported member of the floor system.

12.4.4.6 *End anchorage of reinforcement*—At the wall top and at the bottom (in the foundation), vertical reinforcement should extend as close to the edge as practicable considering concrete cover and end in a standard hook.

12.4.5 *Horizontal reinforcement*

12.4.5.1 *General*—Horizontal reinforcement should consist of one or two layers of bars or welded wire reinforcement placed parallel with the wall face. The amount of horizontal reinforcement should be that needed to resist factored in-plane shear at the wall section. For heavily reinforced walls, as described in 12.4.4.3, transverse reinforcement should be provided as in columns.

12.4.5.2 *Walls with transverse reinforcement*—Where vertical reinforcement ratio ρ_{vw} exceeds 0.01, vertical reinforcement should be enclosed by ties complying with 10.4.3.1, including vertical spacing limits (12.4.4.3).

12.4.5.3 *Minimum horizontal reinforcement area*—Minimum horizontal reinforcement ratio ρ_t, of steel area to gross concrete vertical section area, should be 0.0025.

12.4.5.4 *Reinforcement splicing*—Lap splices of horizontal reinforcement should comply with the lap splice length defined by 5.8.2.

12.4.5.5 *End anchorage of reinforcement*—Horizontal reinforcement terminating at wall edges should have a standard hook that engages the edge vertical reinforcement or U-shaped stirrups of same size and spacing as the horizontal reinforcement should be provided. If U-shaped stirrups are provided, they should be spliced to the horizontal reinforcement.

12.4.6 *Reinforced concrete wall reinforcement in seismic zones*—In walls that are part of the lateral-force-resisting system in seismic zones, reinforcement should comply with Chapter 11 (11.1.5).

12.5—Flexure

12.5.1 *Factored loads*—Factored axial loads P_u and moments M_u at section under consideration should be determined following 12.2.

12.5.2 *Trial vertical reinforcement*—A trial area of vertical reinforcement, A_{st}, should be the minimum vertical reinforcement ratio of 12.4.4.2.

12.5.3 *Required moment strength*—Interaction diagrams for trial wall dimensions and reinforcement should be calculated in both directions using 5.12. Total vertical reinforcement area A_{st} should be divided into total extreme face steel area A_{se} and total side face steel area A_{ss}, for the direction considered, as indicated by 5.12.4.1. The design moment strength in both directions should be computed using 5.12.6. When the factored moment M_u at factored axial load P_u exceeds the design moment strength at factored axial load level P_u, the vertical reinforcement area should be increased. When the reinforcement area exceeds the limit of 12.4.4.3, increase the wall dimensions. After adjusting the dimensions, the wall self-weight should be corrected and the wall design strength should be recalculated. These calculations should be performed at the upper and lower sections of the same story.

12.6—Shear

12.6.1 *Required shear strength*—Factored in-plane shear V_u should be determined from vertical loads and from horizontal loads as indicated by 12.2.

12.6.2 *Shear strength of reinforced concrete walls*

12.6.2.1 *General*—The information in 5.13 should apply to the design of reinforced concrete walls for shear. Items (a) through (d) should be used:

(a) The design for shear perpendicular to the wall face (out-of-plane shear) should be in accordance to provisions for solid slabs in 7.4. The design for in-plane shear should be performed following 12.6.2;

(b) The wall should be vertically continuous from roof to foundation and have no openings for windows or doors;

(c) The wall should have vertical and horizontal distributed reinforcement, conforming to the minimum area and maximum spacing values of 12.4; and

(d) The design shear strength ϕV_n should be computed using Eq. (12.6.2.1).

$$\phi V_n = \phi(V_c + V_s) \qquad (12.6.2.1)$$

where ϕV_c is the concrete contribution to the design shear strength, and ϕV_s is the reinforcement contribution to the design shear strength, and $\phi = 0.75$.

12.6.2.2 *Contribution of concrete to shear strength*—At each critical location to be investigated, only the wall web should be considered to compute ϕV_c, using Eq. (12.6.2.2).

$$\phi V_c = \phi 2\sqrt{f'_c} b_w \ell_w$$
$$\left(\phi V_c = \phi \frac{\sqrt{f'_c}}{6} b_w \ell_w \text{ (SI)}\right) \qquad (12.6.2.2)$$

where b_w is the wall web width, ℓ_w its horizontal length, and $\phi = 0.75$.

12.6.2.3 *Contribution of reinforcement to shear strength*—Contribution of horizontal web reinforcement to the wall's design shear strength should be:

$$\phi V_s = \phi(\rho_t f_y b_w \ell_n) \qquad (12.6.2.3)$$

where ρ_t is the ratio of horizontal reinforcement, f_y is its yield strength, and $\phi = 0.75$.

12.6.2.4 *Design of shear reinforcement*—Where factored shear V_u exceeds ϕV_c, the ratio of horizontal reinforcement should not be less than the amount determined from Eq. (12.6.2.4a), with $\phi = 0.75$.

$$\rho_t \geq \frac{V_u - \phi V_c}{\phi f_y b_w \ell_w} \qquad (12.6.2.4a)$$

In addition, (a), (b), and (c) should be met:

(a) Two reinforcement curtains of both vertical and horizontal reinforcement should be placed;

(b) Vertical steel ratio ρ_{vw} should not be less than the horizontal steel ratio ρ_t where h_w/ℓ_w is less than 2; and

(c) Value of ϕV_n should not exceed the value given by Eq. (12.6.2.4b), with $\phi = 0.75$.

$$\phi V_n = \phi(V_c + V_s) \leq \phi 10\sqrt{f'_c} b_w \ell_w$$
$$\left(\phi V_n = \phi(V_c + V_s) \leq \phi \frac{5}{6}\sqrt{f'_c} b_w \ell_w \text{ (SI)}\right) \qquad (12.6.2.4b)$$

12.6.3 *Shear strength*—Wall shear strength should be computed as follows:

(a) Out-of-plane wall shear strength should be computed in accordance with the provisions for solid slabs in 7.4. When factored shear V_u exceeds ϕV_c as given by Eq. (7.4.2b), using the wall horizontal length ℓ_w instead of b, wall thickness should be increased and the wall self-weight corrected; and

(b) In-plane shear strength should be computed in accordance with 12.6.2. Where factored shear V_u exceeds ϕV_n as given by Eq. (12.6.2.1), the amount of horizontal reinforcement should be increased, complying with 12.6.2.4. Where factored shear V_u exceeds ϕV_n as given by Eq. (12.6.2.4b), wall thickness should be increased and the wall self-weight corrected.

12.7—Calculation of reactions at foundation

12.7.1 *Vertical load reaction*—Vertical load reaction R_u at the foundation should be equal to the value of P_u at the lower end of the wall directly above the foundation.

12.7.2 *Moment reaction*—Unbalanced moment reaction ΔM_u in each principal direction at the foundation should be equal to the value of M_u in that direction at the lower end of the wall directly above the foundation. Unbalanced moments should be distributed to the foundation member as prescribed in Chapter 14.

12.8—Core walls

12.8.1 *General*—Core walls under this guide are walls composed of planar elements whose horizontal section has the shape of an I, U, T, or C. All of Chapter 12 and Section 11.1.5 for structures located in seismic zones are applicable to core walls with the exceptions noted in current 12.8.

12.8.2 *Embedded columns at corners*—All corners of intersecting planar wall segments and all edges of a core wall should be provided with an embedded column that complies with the dimensional and reinforcement limits for columns. If applicable, these columns should comply with 11.1.5.2, boundary elements.

12.8.3 *Loads*—Loads on the core wall should be computed as indicated in 12.2 with the exception that in 12.2.4 instead of in-plane and out-of-plane distinction for the horizontal loads, these loads should be called the principal axis in plan *x* and *y*, as shown in Fig. 12.8.3.

12.8.4 *Dimensional limits*—The dimensional limits as set forth in 12.3.2 apply to core walls as indicated for each of the planar wall components of the core wall.

12.8.5 *Reinforcement details*—Reinforcement details in 12.4 apply to core walls.

12.8.6 *Flexure*—Factored axial loads P_u and moments M_u at section under consideration should be determined following 12.2.

For core walls, the factored moment M_u should be computed in both principal directions with respect to an axis located in the center of the section of the wall in the direction of interest.

Moment strength for core walls should be computed as follows:

(a) When the web or webs of the core walls are in compression and the flange in tension due to the flexural moment (Fig. 12.8.6(a)), the same procedure prescribed for planar walls should be used as presented in 5.12. Dimension h corresponds to the length of the wall in the direction of the horizontal forces causing the moment ℓ_w. Dimension b corresponds to the sum of the widths of the webs located in the direction of the horizontal forces causing the moment. Vertical reinforcement should be distributed as A_{se} and A_{ss}, as indicated in 5.12.4.1, including wall and embedded column vertical reinforcement as appropriate, depending on their location in the wall section.

(b) When the web of the core walls is in tension and the flange in compression due to the flexural moment (Fig. 12.8.6(b)), it is permitted due to the restriction imposed in axial load by Section 4.11.3.1 to disregard the axial load P_u when computing the flexural strength ϕM_n and use only the longitudinal reinforcement of the embedded columns in tension as the flexural reinforcement to compute the required moment strength using the same procedure as presented for T-beams in 8.4.10.4. The limits for effective flange of 8.4.10 should be complied with interpreting the span length as the height of the wall measured from the base of the wall at the foundation to the section of interest. The effective flange width presented in 8.4.10.3 for isolated T-beams should be used when the horizontal section of the core wall is T-shaped.

12.8.7 *Shear*—Factored, V_u should be determined from vertical loads and from horizontal loads as indicated by 12.2. The shear strength ϕV_n of the core wall should be computed following 12.6.2. Only the planar wall components that resist in-plane shear in the direction of the horizontal forces contribute to the core wall shear strength. With reference to Fig. 12.8.3, for shear in the direction of the x-axis direction, only planar wall element C contributes to the core wall shear strength, and for shear in the y-axis direction, both planar wall elements A and B contribute to the core wall shear strength.

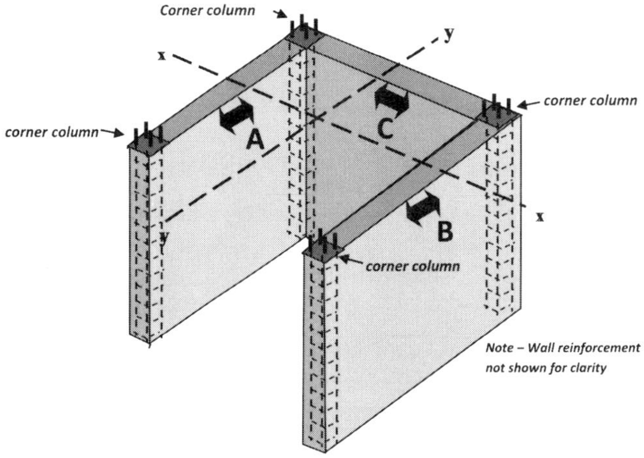

Fig. 12.8.3—Typical core wall showing embedded corner columns.

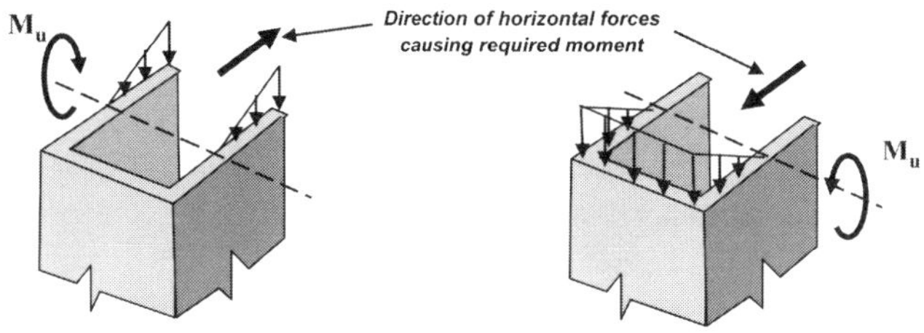

(a) Flange in tension *(b) Flange in compression*

Fig. 12.8.6—Procedure for computing the moment strength of core wall.

CHAPTER 13—OTHER STRUCTURAL MEMBERS

13.1—Stairways and ramps

13.1.1 *General*—The design of stairways and ramps should be governed by 13.1, in addition to Chapters 1 and 3.

13.1.2 *Types of stairways and ramps*—Reinforced concrete stairways and ramps consist of an inclined slab supported at its ends, with steps formed upon its upper surface for stairways.

The unsupported span of a stairway slab should be kept reasonably short. If no break occurs in the flight between floors, intermediate beams, supported by the building's structural framework, should be used. If the stairway between floors is divided into two or more flights, beams should support the intermediate landing, or the intermediate slab may be suspended from a beam at the upper floor level by means of rod hangers (small columns acting in tension). Where conditions permit, the intermediate slab could be supported directly by the building's exterior reinforced concrete walls.

13.1.3 *Dimensional limits*

13.1.3.1 *General*—The building code having jurisdiction, or the corresponding referenced standard from 1.4, governs the needed number of stairways and many other stair-related details. In the absence of prevailing codes or standards, 13.1.3.2 through 13.1.3.7 can be used.

This section provides the maximum distance from the most remote point in the floor area to the stairway, minimum stairway width, maximum height of any straight flight, maximum height (or rise) of a single step, minimum distance (the run) between vertical faces of two consecutive steps, the relation between rise and run for safety and convenience in climbing, and accessibility recommendations for people with disabilities.

13.1.3.2 *Minimum stair width*—Minimum stair width and minimum landing dimension should be 4 ft (1.2 m).

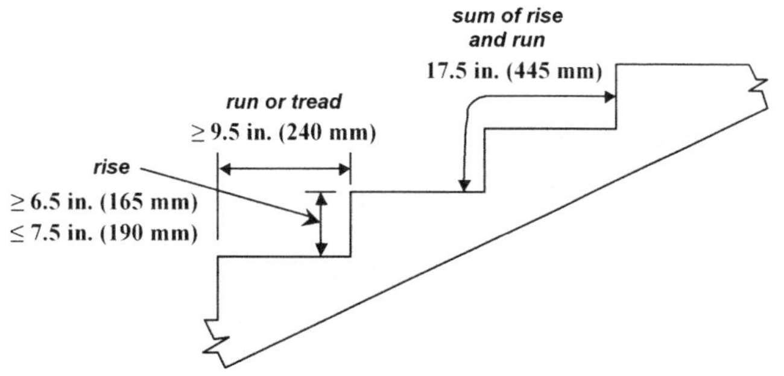

Fig. 13.1.3.3—Maximum step rise and minimum step rise and thread in stairways.

13.1.3.3 *Maximum step rise and minimum step rise and tread*—Maximum stair step rise should be 7-1/2 in. (190 mm). A stair step rise less than 6-1/2 in. (165 mm) should not be used in building internal stairways. Minimum run or tread width, exclusive of nosing, should be 9-1/2 in. (240 mm) (Fig. 13.1.3.3).

The following recommendation may be used: for steps without nosing, the sum of rise and run is 17-1/2 in. (445 mm), but rise is not less than 6-1/2 in. (165 mm) nor more than 7-1/2 in. (190 mm).

13.1.3.4 *Maximum height of a straight flight between landings*—For stairways serving as exits from places of assembly, a maximum height of a straight flight between landings should be 8 ft (2.5 m). For other cases, a maximum height of a straight flight between landings should be 12 ft (3.5 m).

13.1.3.5 *Number of stairways*—The number of stairways is governed by the stair slab width, the number of probable occupants on each floor, and the floor area dimensions. The number of stairways should be as indicated by (a), (b), and (c):

(a) Distance from any point in an open floor area to nearest stairway or exit should not exceed 100 ft (30 m);

(b) Corresponding distance along corridors in a particular area should not exceed 130 ft (40 m); and

(c) Combined width of all stairways in any story should accommodate at one time the total number of persons occupying the largest floor area served by the stairs above the floor area under consideration on the basis of one person for each 2 ft (0.6 m) of stair width and 1.5 treads on the stairs, and one person for each 3 ft^2 (0.3 m^2) of floor area on the landings and halls within the stairway enclosure.

13.1.3.6 *Fireproofing enclosure*—In buildings over 40 ft (12 m) in height, and in commercial buildings regardless of height, the stairways should be completely enclosed by fire-resistant partitions with a minimum 1-hour fire rating with at least one stairway continuing to the roof. An open ornamental stairway may be used from the main entrance floor to the floor next above, provided it is not the only stairway.

13.1.3.7 *Access for people with disabilities*—Disabled people should have full access to all public buildings (1.3.1). Each public building should provide continuous, unobstructed routes by which physically disabled persons can park their vehicles or leave public transportation, move to the building, enter, reach virtually any point in the building, and have access to such interior amenities as workplaces, retail counters, ticket windows, drinking fountains, toilet and washroom fixtures, and public telephones. The unobstructed route should not include stairs, and ramps should be provided. Maximum ramp slope should not exceed a slope of 1 vertical to 12 horizontal.

13.1.4 *Structural design*

13.1.4.1 *General*—A reinforced concrete stairway consists of an inclined slab supported at its ends by members of the floor system, or auxiliary members between floors, with stairway steps formed on the upper surface. Stairways and ramps should meet 13.1.4.2 to 13.1.4.8.

13.1.4.2 *Design load definition*—The design load for stairway and ramp slabs should be established from Chapter 4. Gravity loads should include dead loads and appropriate live loads for stairways. Dead load q_d should include the slab and steps self-weight, and the flat and standing nonstructural elements as defined in 4.5.3. Live load q_ℓ should be determined by 4.6. The factored total load q_u should be the largest value determined from load combinations for q_d and q_ℓ using Eq. (4.2.1a) and Eq. (4.2.1b).

13.1.4.3 *Reinforcement details*—Reinforcement should comply with 7.3, solid slabs, and 7.7.3, single-span one-way solid slabs. Where the stairway or ramp is continuous with a horizontal slab, longitudinal reinforcement should be continued straight within the slab until it reaches the opposite external slab surface where it should be bent and either continued to the opposite external surface or developed as indicated in 5.8.1 (Fig. 13.1.4.3).

13.1.4.4 *Dimensional requirements*—Dimensional limits for single-span one-way slabs given in 7.7.1 should be met.

13.1.4.5 *Required moment strength*—Required moment strength should be determined following 7.7.2. The span should be equal to the horizontal distance between supports. The influence of the slab slope and the slab bend on the factored moment calculation should be disregarded.

13.1.4.6 *Shear strength*—Factored shear should be computed following 7.7.4.

13.1.4.7 *Support reaction*—Support reactions should be computed following 7.7.5.

13.1.4.8 *Design of the supports*—Supporting members should be designed for reactions determined in 13.1.4.7.

13.1.4.9 *Construction details*—Usual practice is to construct stairways after the main structural framework of the building has been completed. Recesses are left in the beams to support the stairway slab and dowels are provided to furnish the necessary anchorage. Occasionally, however, stairways are constructed at the same time as the floors, and so negative moment reinforcement is furnished at the stairway slab support, as in any continuous-beam construction. Steps are usually monolithic with the slab, but they can be molded after the main slab is in place. In the latter instance, provisions should be made for securing the step to the slab through small-diameter dowel bars.

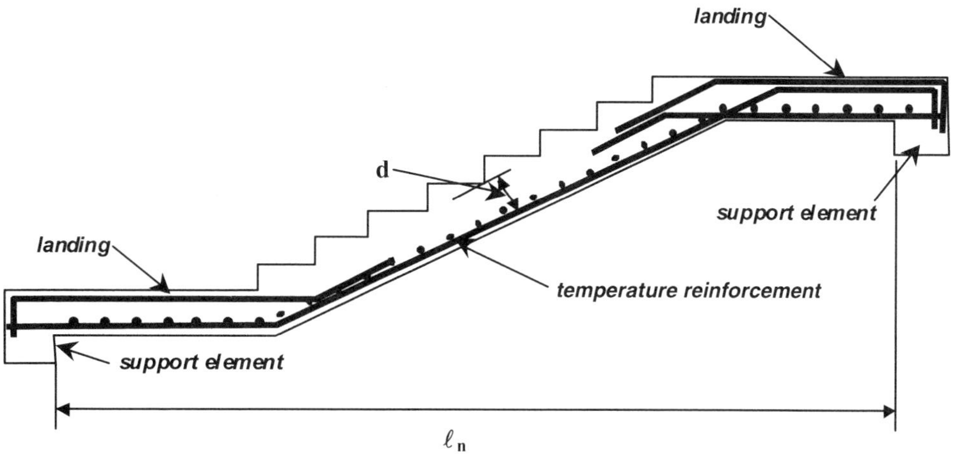

Fig. 13.1.4.3—Typical reinforcement details.

13.2—Small water tanks (for potable water storage)

13.2.1 *General*—Section 13.2.1 provides simplified guidelines for the design and construction of small water storage tanks intended for the sole use of the building occupants. These tanks should be either supported by the structure or, where supported by the soil, should as part of a building be on the surface, or partially or totally buried. Section 4.13.2.3 should apply. Section 13.2 should not be used in the design and construction of general environmental and sanitary structures. This guide assumes the engineer has limited experience with water tanks, and is not intended to replace ACI 350-06 for the design of stand-alone water storage tanks. ACI 350-06 is recommended for the design of larger water storage tanks and related environmental structures.

13.2.2 *Load definition*

13.2.2.1 *Loads to be included*—Loads for tanks should be established from Chapter 4. Loads (a) through (f) should be included:

(a) Dead loads: self-weight of tank components, and all dead loads supported by the tank's top slab, such as structural and nonstructural elements, earth-fill over buried tanks, and fixed equipment loads;

(b) Live loads from tributary structural members supported by the tank, live loads supported by the tank's top slab, and live loads on the surface of the earth-fill over buried tanks;

(c) Lateral earth pressure in buried tanks;

(d) Buoyancy force on buried or partially buried, empty tanks;

(e) Internal fluid pressure from liquid contained by the tank; and

(f) Seismic loads from tanks supported by the building, such as a tank located on the roof or at any other story of the building. Note that a full water tank can retain a large mass of water and can therefore create a significant seismic load.

13.2.2.2 *Factored loads*—Factored loads should be determined using loads included in 13.2.2.1 and combined using load combination equations of 4.2. Resulting factored loads should be increased by 30 percent (multiplied by 1.3). For tanks within the scope of this guide, this calculation is a satisfactory simplification of the ACI 350-06 detailed calculation for environmental durability. Due care should be taken in determining which loads can act simultaneously with other loads.

13.2.3 *Dimensional limits*

13.2.3.1 *Concrete cover*—Minimum concrete cover for reinforcement in the tank foundation should be 3 in. (75 mm), and minimum cover for all other reinforcement in tanks should be 2 in. (50 mm).

13.2.3.2 *Joints*—Joints should be used to limit volumetric concrete changes due to drying shrinkage, creep, and temperature variation. All contraction or construction joints should have waterstops. Maximum spacing between joints should not exceed 12 ft (4 m).

13.2.3.3 *Minimum thickness of walls and slabs*—Tank walls and slabs should have a minimum thickness of 8 in. (200 mm). Walls and slabs with clear spans greater than 12 ft (3.5 m) should have a minimum thickness of 12 in. (300 mm).

13.2.4 *Flexure*

13.2.4.1 *Minimum reinforcement ratio*—Minimum reinforcement ratio ρ_ℓ in any direction should be 0.0030.

13.2.4.2 *Minimum reinforcing bar diameter*—Minimum reinforcing bar diameter d_b should be 3/8 in. (10 mm) for tank walls and slabs up to 9 in. (225 mm) in thickness, and 1/2 in. (13 mm) for tank wall and slabs thicker than 9 in. (225 mm).

13.2.4.3 *Maximum bar spacing*—Maximum bar spacing in regions of maximum moment should be 6 in. (150 mm).

13.2.4.4 *Flexural reinforcement area*—Flexural reinforcement area should be determined as for solid slabs, using 7.7.

13.2.5 *Shear*—Shear strength should be computed using 7.7.4, solid slabs.

13.2.6 *Concrete in tanks*—Concrete used in tank construction should meet (a) through (e):

(a) Minimum *w/c* should be 0.40;

(b) Outside finish should be as smooth as practicable and inside surface should be a broom finish to reduce slip hazard;

(c) Maximum nominal aggregate size should be as permitted by 5.7. It is recommended to use the largest aggregate possible to reduce shrinkage;

(d) Concrete should be compacted as soon as it is deposited in the forms; and

(e) Minimum moist curing time should be 7 days, keeping concrete damp during that time. This can be achieved by maintaining wall forms in place.

CHAPTER 14—FOUNDATIONS

14.1—Soil investigation

14.1.1 *Borings*—A properly planned field exploration and laboratory investigation should be conducted by a qualified geotechnical engineer who has experience in the location of the proposed building. It should conform to 14.1.1.1, 14.1.1.2, or both:

14.1.1.1 *Hand auger*—Several types of cutting points are advanced manually or by power equipment. The depth of these holes is generally limited to 50 ft (15 m), with 20 ft (6 m) being the most common.

14.1.1.2 *Percussion drilling*—A standard penetration test (SPT) can be performed by means of a percussive tool attached to the rig. The SPT is widely used and extensively referenced.

14.1.2 *In-place tests*—Methods of investigation should provide information referring to items in 14.1.2.1, 14.1.2.2, and 14.1.3.

14.1.2.1 *Penetration tests*—Two types of penetration tests should be considered:

(a) SPT—The SPT consists of counting the number of blows N necessary to produce a 12 in. (300 mm) penetration of a standard split spoon into the soil. Standard weight used in the test is 140 lb (64 kg), and it is dropped from a standard height of 30 in. (750 mm); and

(b) Cone penetration test (CPT)—The CPT consists of pushing a 60-degree cone of 1.55 in.^{2} (1000 mm^2) base area into the soil at a constant rate of 0.79 in./s (20 mm/s). The force needed to advance the cone divided by the base area is the cone resistance q_c.

14.1.2.2 *Load tests*—Where it is useful, investigate the bearing capacity of a terrain by means of an elementary load test, complying with (a) through (e):

(a) Test should be conducted in a pit having a width at least three times that of the load plate;

(b) After each load increment, the corresponding deformation is observed and recorded;

(c) Load-settlement relationship should be determined either to the point of failure or to a load value three times that anticipated for service;

(d) Unfactored allowable soil capacity should be determined as a value having a factor of safety of 3 for the least favorable situation or that producing a settlement not larger than 1 in. (25 mm), whichever is less; and

(e) Explore the soil beneath footings to a depth of three times its largest size to overcome test misinterpretations.

14.1.3 *Soils report*—The soils report should contain, as a minimum, information related to:

(a) Location;

(b) Site topography;

(c) Stratigraphy;

(d) Groundwater elevation;

(e) Ground surface elevation; and

(f) Local conditions needing special consideration such as soil strength, compressibility, expansion potential, comments on collapsibility, liquefaction potential, and local history of the performance of typical construction methods.

14.2—Allowable soil-bearing capacity

14.2.1 *Allowable bearing capacity for granular soils*

14.2.1.1 *Standard penetration test*—The allowable bearing capacity q_a, in lb/ft^2 (kPa [kN/m^2]) for granular soils, should be estimated as a product of the blow count N, measured by means of the SPT, using the relationship described by Eq. (14.2.1.1).

$$q_a = 220N \quad [q_a = 11N \text{ (SI)}] \qquad (14.2.1.1)$$

14.2.1.2 *Cone penetration test*—For medium-size granular soil, when cone penetration data are available, allowable bearing capacity q_a should be evaluated by means of Eq. (14.2.1.2).

$$q_a = 0.027q_c \qquad (14.2.1.2)$$

14.2.2 *Allowable bearing capacity for cohesive soils*—Allowable bearing capacity q_a for cohesive soils should be taken as:

$$q_a \approx q_u = 2s_u \qquad (14.2.2a)$$

When SPT data are available, the relation expressed in Eq. (14.2.2b) should be used for obtaining allowable bearing capacity q_a, in lb/ft^2 (kPa [kN/m^2]), for soils not very soft (values determined are not reliable for very soft soils).

$$q_a = 250N \quad [q_a = 12.5N \text{ (SI)}] \qquad (14.2.2b)$$

When CPT data are available, the relation expressed in Eq. (14.2.2c) should be used only when collateral means of bearing capacity confirmation exist.

$$q_a \approx \frac{q_c}{12} \qquad (14.2.2c)$$

14.2.3 *Procedures to obtain allowable bearing capacity*—Procedures to evaluate allowable bearing capacity q_a, described in 14.2.1 and 14.2.2, should be used in cases that conform to (a) through (g):

(a) A formal soil-bearing capacity recommendation from a geotechnical engineer does not exist or site exploration has not been performed;

(b) Strength parameters correspond to the mean properties of the bearing stratum or layer supporting loads applied through foundation to the soil;

(c) Ground surface is nearly flat;

(d) Base of structure is horizontal;

Table 14.2.3—Maximum permitted allowable bearing capacities

Soil	Bearing capacity q_a, lb/ft^2	Bearing capacity q_a, kPa (kN/m^2)
Alluvial soil	≤ 1000	≤ 50
Soft clay	1500	75
Firm clay	2000	100
Wet sand	2500	125
Sand and clay mixed	3000	150
Fine dry sand (compact)	4000	200
Hard clay	5000	250
Coarse dry sand (compact)	6000	300
Sand and gravel mixed (compact)	7000	350
Gravel (compact)	8000	400
Soft rock	12,000	600
Hard pan or hard shale	16,000	800
Medium rock	20,000	1000
Hard rock	30,000	1500

(e) Footing depth is the same magnitude as its width;

(f) Water level is below a depth of twice the width of the largest footing; and

(g) In stratified or multi-layered soils, the smallest value of the corresponding soil strengths is used.

Values given in Table 14.2.3 correspond to average bearing capacities for various types of soils at a depth of 3 ft (1 m). These values may be used as verification for values determined through exploration.

14.2.4 *Allowable bearing capacity increase for wind and seismic*—Allowable bearing capacity q_a, determined through procedures in Chapter 14, or from geotechnical investigation, should be increased by one-third when being computed for loads that include effects of wind or seismic.

14.3—Settlement criteria

To define an allowable bearing capacity q_a for a certain soil in a certain location, a reasonable settlement should be determined.

A reasonable settlement varies between 1 and 2 in. (25 and 50 mm), and the methods stated should produce values in that range. Local precedent can be used where no additional information is available; however, should site conditions be such that precedent is no longer applicable because particular soils are very soft or because other difficult conditions are foreseen, then a complete geotechnical analysis should be conducted.

14.4—Dimensioning foundation members

Minimum horizontal foundation area contacting bearing soil should be computed by dividing the sum of all unfactored vertical loads tributary to the foundation member by the allowable soil-bearing capacity q_a.

14.5—Spread footings

14.5.1 *General*—Isolated footings should be designed using 14.5. Wall footings should comply with 14.5 and 14.6. This section applies to footings that can be subjected to flexural moments from the supported column or wall. The analysis and design of solid mats and deep foundations, such as piles, are beyond the scope of this guide.

14.5.2 *Load definition and area of the footing*

14.5.2.1 *Loads to be included*—The design of spread footings should include the loads in (a) to (c), as shown in Fig. 14.5.2.1.

(a) The soil fill weight on top of the footing, and all dead and live loads applied to the fill, plus the footing self-weight. It may be assumed that the footing has the same density as the soil for a first trial, to be corrected as the footing size is defined. The sum of the pressure on the bearing soil is called the overburden pressure q_o.

(b) The unfactored load effects transferred by the column or wall to the footing from all sources, including dead, live, wind, seismic, and other loads as prescribed by Chapter 4, expressed as unfactored axial loads, moments, and shears.

(c) As an alternate to (b), unfactored axial load may be obtained from unfactored unit loads, including self-weight, multiplied by column or wall tributary area from all floors supported by it. Factored wind effects should be converted to unfactored values by dividing factored values by 1.6, and seismic effects converted to unfactored values by dividing factored values by 1.43.

14.5.2.2 *Maximum unfactored vertical load*—Unfactored axial loads applied by the column or wall through the footing to the underlying soil, overburden loads transmitted by the fill, the fill weight, and the footing self-weight should be combined according to load cases of 4.2 without applying load factors. Cases (a) and (b) should be investigated:

(a) P_v, maximum vertical unfactored load excluding wind or seismic; and

(b) P_{ov}, maximum vertical unfactored load including lateral load (wind or seismic) overturning effects.

14.5.2.3 *Minimum footing area*—Minimum footing area A_f should be determined as the greater area obtained from Eq. (14.5.2.3a) or Eq. (14.5.2.3b).

(a) For maximum vertical unfactored load P_v, excluding lateral loads (wind or seismic) overturning effects:

$$A_f = \frac{P_v}{(q_a - q_o)} \quad (14.5.2.3a)$$

(b) For maximum vertical unfactored load P_{ov}, including lateral loads (wind or seismic) overturning effects:

$$A_f = \frac{P_{ov}}{(1.33q_a - q_o)} \quad (14.5.2.3b)$$

14.5.2.4 *Factored soil reaction*—Factored net soil reaction q_{un} should be computed by dividing the largest factored axial load at the column or wall base, P_{ub}, by the footing area. The footing area should conform to 14.5.2.3 and 14.5.3.

14.5.3 *Dimensional limits*—Footing dimensions should comply with 14.5.3.1 through 14.5.3.7.

14.5.3.1 *Plan shape*—Spread footings should be square or rectangular in plan.

14.5.3.2 *Symmetry*—Spread footings should have the column or wall located at the footing area centroid. This

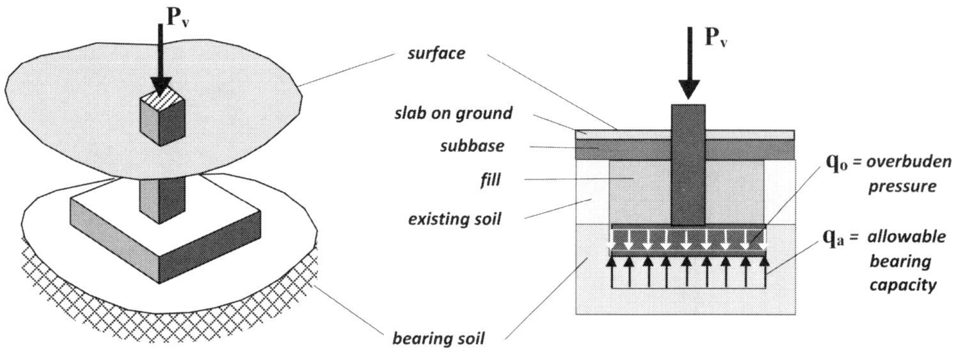

Fig. 14.5.2.1—Loads acting on footings.

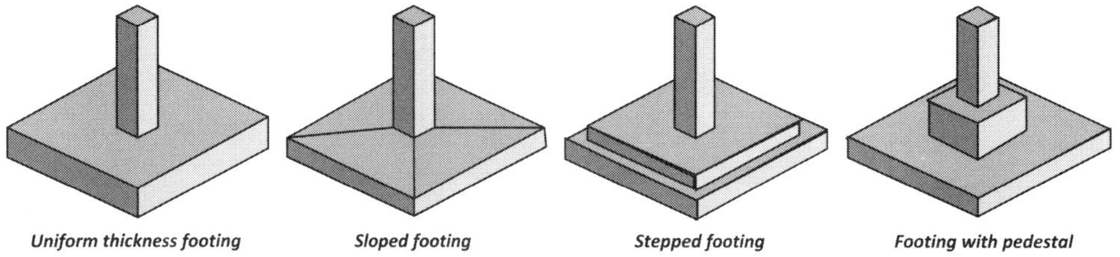

Fig. 14.5.3.6—Types of spread footings.

is waived if induced moments and additional soil pressure meet the limits in 14.5.7.

14.5.3.3 *Minimum depth to bearing soil*—Minimum vertical distance from the soil surface to footing bottom (footing-soil interface) should be 3 ft (1 m).

14.5.3.4 *Minimum footing area*—Minimum footing dimension in plan should be 3 ft (1 m).

14.5.3.5 *Minimum footing thickness*—Footing depth above bottom reinforcement should not be less than 6 in. (150 mm).

14.5.3.6 *Sloped or stepped footings*—In sloped or stepped footings, angle of slope or depth and location of steps should satisfy design and dimensional limits at every section. Sloped or stepped footings designed as a unit should be constructed to ensure action as a unit. Figure 14.5.3.6 shows the types of spread footings within the scope of this guide.

14.5.3.7 *Footings supporting circular-shaped columns*—For locations of critical section for moments, shear, and development of reinforcement in footings, circular-shaped concrete columns should be treated as square columns with the same area.

14.5.4 *Reinforcement details*

14.5.4.1 *General*—Reinforcement type and characteristics in spread footings should be as described in 14.5.4.

14.5.4.2 *Location of reinforcement*—Reinforcement should be provided in the lower part of spread footings in both directions. Reinforcement should be located as close to footing bottom as practicable following the concrete cover of 5.4.1. In rectangular column footings, reinforcement parallel to the shorter side should be located above reinforcement parallel to the longer side. In wall footings, reinforcement perpendicular to the plane of the wall should be located under the reinforcement parallel to the wall.

14.5.4.3 *Minimum clear spacing between parallel bars*—Minimum clear spacing between parallel bars in a layer should be the largest bar nominal bar diameter d_b, but not less than 1 in. (25 mm).

14.5.4.4 *Maximum spacing*—In spread footings, reinforcement should be spaced no farther apart than three times the footing thickness, nor 12 in. (300 mm). In square column footings and in wall footings, reinforcement should be distributed uniformly across the entire footing width. In rectangular column footings, in the long direction, reinforcement should be uniformly distributed across the width, and in the short direction, reinforcement should be factored by the value given in Eq. (14.5.4.4) and uniformly distributed across the entire footing width.

$$\text{factor} = \frac{2\beta_f}{(\beta_f + 1)} \tag{14.5.4.4}$$

14.5.4.5 *Minimum reinforcement area*—Minimum reinforcement ratio ρ_{min} in any direction should be 0.0018.

14.5.4.6 *Maximum reinforcement area*—Maximum reinforcement ratio ρ_{max} should be that given for slabs in 7.3.4.3.

14.5.4.7 *Development of reinforcement*—Reinforcement at critical sections should be developed on each side of that section by embedment length or hook (Fig. 14.5.4.7). Critical sections should be assumed at the same locations as defined in 14.5.6.1 for maximum factored moment and at all other vertical planes where section or reinforcement changes occur.

14.5.4.8 *Reinforcement cutoff*—In spread footings, reinforcement at the moment critical section in either direction should be maintained over the full length of the footing in that direction.

14.5.4.9 *End anchorage of reinforcement*—Reinforcing bars should end in a standard hook at the footing edge if needed by calculation of the development length at critical section.

14.5.4.10 *Value of* d_c *and* d *to use in footing*—The calculation of d_c, distance from extreme tension fiber to the centroid of the tension reinforcement, should consider the concrete cover for members cast and permanently exposed to earth, the bar diameters used, and perpendicular reinforcement placed under the considered reinforcement.

The following values of d_c may be used to compute d as $d = h - d_c$. For reinforcement parallel to the long direction of footings and for wall footings, d_c = 3.5 in. (90 mm). For reinforcement parallel to the short direction of column footings, d_c = 4.5 in. (115 mm).

14.5.4.11 *Details for transfer of column or wall loads to footing*—Loads and moments at the base of column, wall, or pedestal should be transferred to the supporting pedestal or footing by bearing on concrete and by reinforcement. The following should apply:

(a) Bearing stress due to factored loads at the contact surface between supported and supporting member should not exceed concrete bearing strength for either surface as given in 5.14;

(b) The column compressive force that exceeds the footing concrete bearing strength should be transferred to the footing by reinforcement;

(c) Footing depth should be adequate to allow the column longitudinal reinforcement to end in a standard hook and to develop the anchorage distance of 5.8.3. Wall vertical reinforcement should have the embedment length given in 5.8.1;

(d) If uplift is expected, anticipated tension between the column and footing should be transferred by vertical reinforcement only; and

(e) If a pedestal is used, vertical reinforcement (including column reinforcement) in the pedestal should be not less than 0.005 times gross concrete area and should have ties in accordance with 10.4.3.2.

14.5.5 *Shear*

14.5.5.1 *General*—Footing shear strength should be computed using 9.5 for the two types of shear that occur near the supported column, pedestal, or wall, and at changes of footing thickness in stepped footings:

(a) Punching-shear or two-way action shear in column footings; and

(b) Beam-action shear that accompanies flexural moments.

14.5.5.2 *Location of critical sections for punching shear*—Punching-shear critical section should be located at a distance $d/2$ from the face of column, pedestal, or stepped footing thickness change (9.5.4).

14.5.5.3 *Punching shear strength*—Required punching shear strength is equal to the largest factored axial load P_{ub} at the column base. P_{ub} might be reduced by the product of the area bound by the critical section times the factored net soil reaction q_{un}, from 14.5.2.4.

14.5.5.4 *Punching shear design strength*—Punching shear strength is determined using the same methods as slab-column systems of 9.5.4 for interior columns.

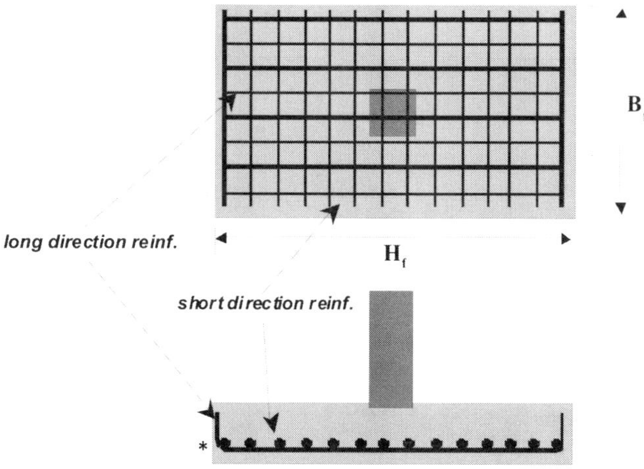

Fig. 14.5.4.7—Distribution of reinforcement in rectangular spread footings.

14.5.5.5 *Minimum footing depth as required by punching shear*—A trial footing depth should be determined from Eq. (14.5.5.5).

$$d \geq \sqrt{\left(\frac{h_c + b_c}{4}\right)^2 + \frac{P_u}{\phi 12\sqrt{f'_c}}} - \left(\frac{h_c + b_c}{4}\right) \tag{14.5.5.5}$$

$$\left[d \geq \sqrt{\left(\frac{h_c + b_c}{4}\right)^2 + \frac{P_u}{\phi\sqrt{f'_c}}} - \left(\frac{h_c + b_c}{4}\right) \text{(SI)}\right]$$

Final dimensions should meet 9.5.4.

14.5.5.6 *Critical section locations for beam-action shear*—Critical section for beam-action shear should be located at a distance d from the face of column, wall, pedestal, or stepped footing thickness change (9.5.5).

14.5.5.7 *Beam-action shear strength*—Beam-action required shear strength V_u should be the product of the factored net soil reaction q_{un}, from 14.5.2.4 and the tributary area of the footing beyond the critical section as defined in 14.5.5.6 (Fig. 14.5.5.7).

14.5.5.8 *Beam-action shear design strength*—Beam-action shear strength is determined using the slab-column system approach of 9.5.5.

14.5.5.9 *Beam-action shear*—The effective depth d for punching shear should be used for initially establishing location of the beam-action shear critical section, the tributary area, and the design strength ϕV_c from Eq. (9.5.5). When design strength is less than required strength, the value of d should be increased. For wall footings, 14.6.2 should be used.

14.5.6 *Flexure*

14.5.6.1 *Critical sections for moments*—Maximum required moment strength for an isolated footing should be computed at face of column, wall, or pedestal (Fig. 14.5.6.1).

14.5.6.2 *Required moment strength*—Required moment strength M_u at any footing section should be the cantilever

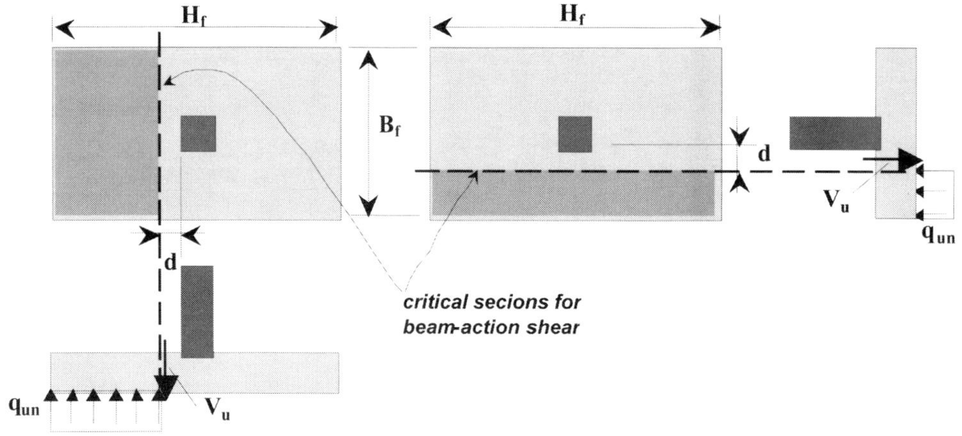

Fig. 14.5.5.7—Tributary areas for beam-action shear.

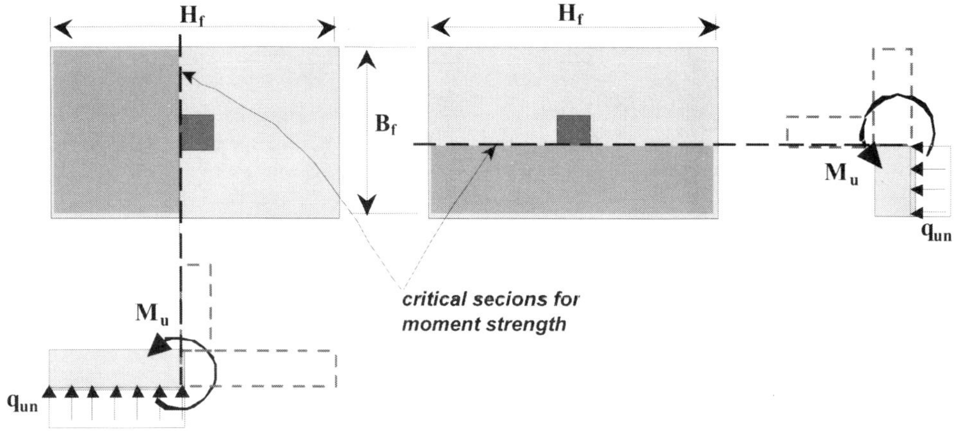

Fig. 14.5.6.1—Critical sections for moments.

moment of the factored net soil reaction q_{un}, acting over entire footing area on one side of that vertical plane using Eq. (14.5.6.2) (Fig. 14.5.6.2).

$$M_u = \frac{q_{un}\ell_n^2 b}{2} \qquad (14.5.6.2)$$

14.5.6.3 *Approximate design moment strength*—For footings, the design moment strength of 5.11.4 should be approximated using Eq. (14.5.6.3), with $\phi = 0.90$.

$$\phi M_n = \phi 0.85 A_s f_y d \qquad (14.5.6.3)$$

14.5.6.4 *Calculating the flexural reinforcement ratio*—Flexural reinforcement ratio, $A_s/(bd)$, should be determined from Eq. (14.5.6.4).

$$\rho = \frac{A_s}{bd} \geq \frac{M_u}{\phi 0.85 f_y bd^2} \qquad (14.5.6.4)$$

If ρ is less than ρ_{min}, from 14.5.4.5, A_s should be increased. If ρ is greater than ρ_{max}, as established in 7.3.4.3, the member's effective depth d should be increased.

14.5.6.5 *Longitudinal reinforcement*—Longitudinal reinforcement in both plan principal directions having the corresponding area determined from 14.5.6.4 and meeting 14.5.4 should be provided in the lower part of the footing (Fig. 14.5.4.7).

14.5.7 *Footings subjected to external moments*

14.5.7.1 *General*—Footings subject to moments, transmitted by column or wall, or due to eccentricity of the column or wall with respect to the footing centroid (14.5.3.2), should be designed according to 14.5. Initial sizing of the footing should be performed using procedures of 14.5.2.3. Initial size should be adjusted to meet 14.5.7.

14.5.7.2 *Eccentricity*—Eccentricity in both directions, e_H and e_B, parallel to H_f and parallel to B_f, respectively, should be determined from (a), (b), and (c):

(a) Divide moments (14.5.2.1(b)) transmitted by the column or wall by the axial load that accompanies them in the same load case. This calculation may be performed for either unfactored or factored moments and axial loads;

(b) For eccentric column or wall locations, as the distance, measured parallel to each of the principal directions, from the column or wall centroid to footing centroid; and

(c) For moments transmitted by a column or wall eccentrically located, by adding the eccentricities from (a) and (b) with consideration of the moment sign.

14.5.7.3 *Uplift check*—Where the resultant of the vertical force transmitted by the column or wall is located inside the footing kern (Fig. 14.5.7.3), there is no uplift, and soil stresses should be computed according to 14.5.7.4. If the resultant is outside the kern, increase footing size or 14.5.7.5 should be met.

14.5.7.4 *Compliance with the soil allowable bearing capacity: No uplift*—Allowable bearing capacity should not be exceeded under the applied loads and moments. Compliance should be verified using Eq. (14.5.7.4a) or Eq. (14.5.7.4b).

(a) For maximum vertical unfactored load P_v, excluding lateral loads' (wind or seismic) overturning effects:

$$\frac{P_v}{B_f H_f}\left(1+6\left[\frac{e_H}{H_f}+\frac{e_B}{B_f}\right]\right) \le (q_a - q_o) \qquad (14.5.7.4a)$$

(b) For maximum vertical unfactored load P_{ov}, including lateral loads' (wind or seismic) overturning effects:

$$\frac{P_{ov}}{B_f H_f}\left(1+6\left[\frac{e_H}{H_f}+\frac{e_B}{B_f}\right]\right) \le (1.33\, q_a - q_o) \qquad (14.5.7.4b)$$

Where allowable bearing capacity is not exceeded, the value of q_{un} should be computed using Eq. (14.5.7.4c):

$$q_{un} = \frac{P_u}{B_f \cdot H_f}\left(1+6\left[\frac{e_H}{H_f}+\frac{e_B}{B_f}\right]\right) \qquad (14.5.7.4c)$$

14.5.7.5 *Compliance with the soil allowable bearing capacity: Uplift*—Allowable bearing capacity should not be exceeded under applied loads and moments. When the resultant of the vertical loads transmitted by the column or wall is located outside the footing kern, this resultant should coincide with the resultant force of the soil pressure diagram (Fig. 14.5.7.5). Suitable values of and H_f and B_f should be selected to comply with Eq. (14.5.7.5a) or Eq. (14.5.7.5b).

(a) For maximum vertical unfactored load P_v, excluding lateral loads' (wind or seismic) overturning effects:

$$\frac{2P_v}{3B_f a_w} \le (q_a - q_o) \qquad (14.5.7.5a)$$

(b) For maximum vertical unfactored load P_{ov}, including lateral loads' (wind or seismic) overturning effects:

$$\frac{2P_{ov}}{3B_f a_w} \le (1.33 q_a - q_o) \qquad (14.5.7.5b)$$

If allowable bearing capacity is not exceeded, the value of q_{un} should be computed using Eq. (14.5.7.5c):

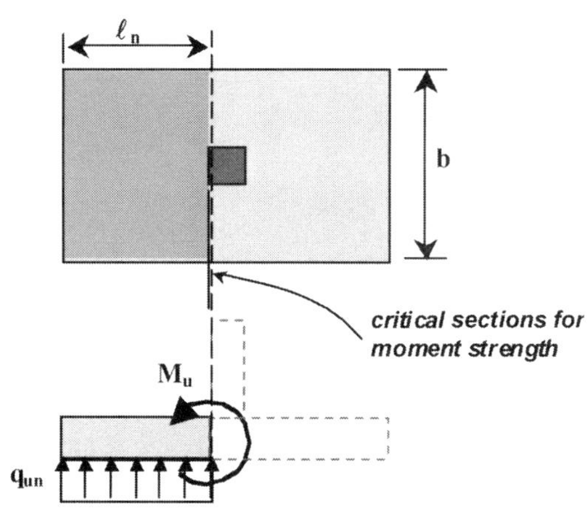

Fig. 14.5.6.2—Calculation of required moment strength.

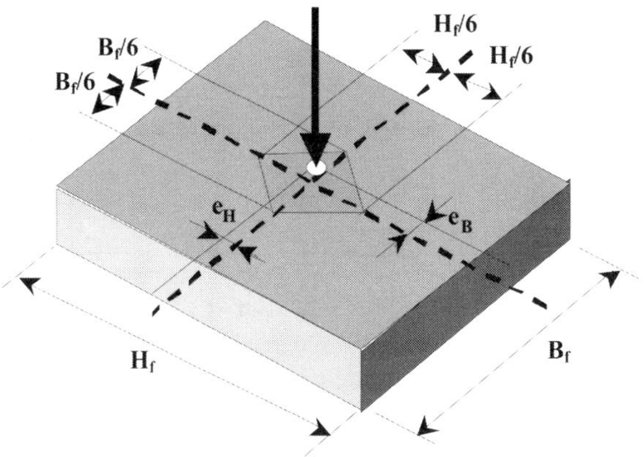

Fig. 14.5.7.3—Location of the resultant to avoid footing uplift.

$$q_{un} = \frac{2P_u}{3B_f a_w} \qquad (14.5.7.5c)$$

If allowable bearing capacity is exceeded by the eccentric applied load, combined footings or the distribution of moments to grade beams should be investigated. For wall footings, 14.6.4 should apply.

14.6—Wall footings

14.6.1 *General*—Continuous footings under walls should be designed in accordance with 14.5 and 14.6.

14.6.2 *Shear*—Wall footing shear should comply with 14.5.5 and with the exceptions listed in 14.6.2.1 through 14.6.2.2.

14.6.2.1 *Punching shear*—Punching shear in wall footings should be disregarded.

14.6.2.2 *Beam-action shear*—Procedures of 14.5.5.6 and 14.5.5.7 should be followed except that the beam-action shear required strength should be computed per unit length of footing at the wall face, and for simplicity, using Eq. (14.6.2.2a) (Fig. 14.6.2.2). Minimum footing effective depth d, as controlled by beam-action shear, should be determined from Eq. (14.6.2.2b).

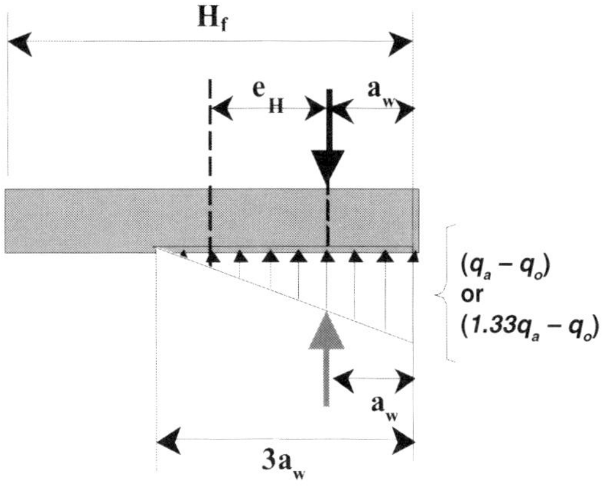

Fig. 14.5.7.5—Soil pressure with uplift.

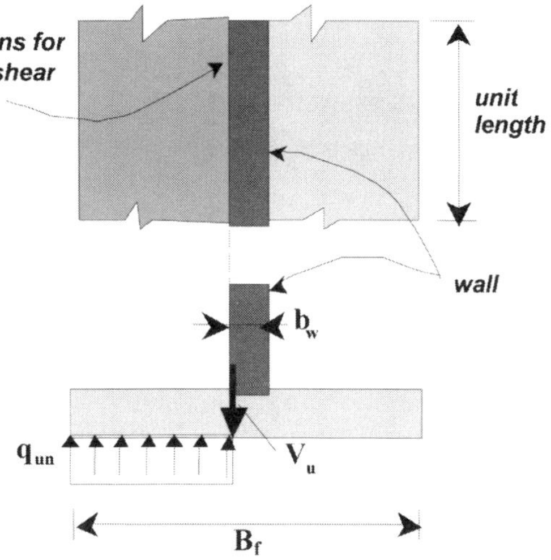

Fig. 14.6.2.2—Beam-action shear in wall footings.

$$V_u = q_{un}\left(\frac{B_f - b_w}{2}\right)\ell \quad (14.6.2.2a)$$

$$d \geq \frac{q_{un}(B_f - b_w)}{\phi 4\sqrt{f_c'}} \quad \left[d \geq \frac{q_{un}12(B_f - b_w)}{\phi\sqrt{f_c'}} \text{ (SI)}\right] \quad (14.6.2.2b)$$

14.6.3 *Flexure*—Wall footings should comply with 14.5.6 and (a) and (b):

(a) Seismic zone wall boundary elements that have an edge located within one-half the footing depth from a footing edge should have transverse reinforcement in accordance with 11.1.3.4 below the footing top. This reinforcement should extend into the footing no less than the footing depth; and

(b) Where seismic effects create uplift loads in wall boundary elements or columns, top footing reinforcement should be provided in the same amount and spacing as the bottom reinforcement.

14.6.4 *Wall moments at the foundation*—Lateral loads are resisted primarily by walls; therefore, it is possible that wall-overturning moments at the foundation produce eccentricities outside the footings dimensioned for vertical load effects only. For large eccentricities, extending the footing under a grade beam should be investigated (Fig. 14.6.4a). The grade beam should extend symmetrically at both wall ends. If the symmetrical extension is not possible, this solution should not be used.

Soil bearing should be computed using 14.5.7.5 with extended footing dimensions. The footing should be designed using procedures of 14.5 and 14.6. The grade beam should be designed for factored moment and factored shear determined from Eq. (14.6.4a) and (14.6.4b), using 8.7 and 14.2.

$$M_u = \frac{P_u H_f}{3} \quad (14.6.4a)$$

$$V_u = P_u \quad (14.6.4b)$$

The factored moment from Eq. (14.6.4a) divided by the strength reduction factor should be used to compute top and bottom beam reinforcement. Transverse reinforcement determined for the factored shear given in Eq. (14.6.4b) should be used throughout the beam length. Splices of longitudinal beam reinforcement should be located under the wall near its center. Reinforcement layout should be as shown in Fig. 14.6.4b.

14.7—Combined footings

14.7.1 *General*—Combined footings are used when a column is located at the property line, thus making it impossible to build a concentric footing. Combining two columns in the same footing allows for a uniform soil reaction under the footings. Figure 14.7.1 shows several types of combined footings. Types (a) to (d) are not covered in this guide. Type (e) links two spread footings through a grade beam, and is covered in 14.7.2.

14.7.2 *Combined footings linked by grade beams*

14.7.2.1 *Reactions and general dimensions*—A combined footing should consist of two spread footings linked by a grade beam. Figure 14.7.2.1 shows column loads and soil reactions under the footings and the relevant dimensions. The reactions should be determined from equilibrium:

$$R_1 = \frac{2P_1\ell_s}{2\ell_s + h_c - H_f} \quad (14.7.2.1a)$$

$$R_2 = P_1 + P_2 - R_1 \quad (14.7.2.1b)$$

14.7.2.2 *Footing design*—Each footing should be designed as a spread footing for the reactions determined in 14.7.2.1, using 14.5. Punching shear calculations for eccentrically loaded footings should use edge column methods.

14.7.2.3 *Beam design*—The grade beam should be designed using 8.7 and 14.2 for the factored shear and moment determined from Eq. (14.7.2.3a) and Eq. (14.7.2.3b).

$$V_u = P_u \frac{H_f - h_c}{2\ell_s + h_c - H_f} \quad (14.7.2.3a)$$

where P_u corresponds to the maximum factored axial force of the left side column. The required moment strength of the beam should be the negative moment (producing tension in the upper part of the grade beam) that varies, as shown in Fig. 14.7.2.3. The factored moment that should be used is:

$$M_u^- = \frac{P_u}{2}\left(H_f - h_c\right) \quad (14.7.2.3b)$$

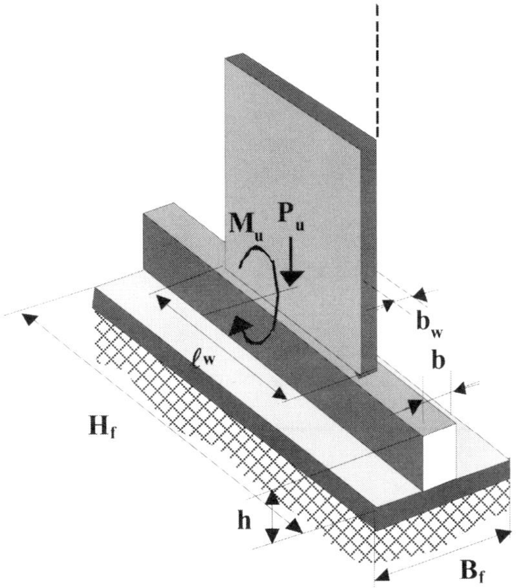

Fig. 14.6.4a—Wall footing extended by use of a grade beam.

14.8—Piles and caissons

Piles and caissons should be dimensioned and designed in accordance with the codes and standards listed in 1.4 in addition to geotechnical report recommendations.

14.9—Footings on piles

Footings on piles and pile caps should be dimensioned and designed in accordance with the codes and standards listed in 1.4 in addition to geotechnical report recommendations.

14.10—Foundation mats

14.10.1 *General*—When spread footing area exceeds one-half the projection area of the building foundation, a foundation mat should be investigated. Only foundation mats with beams are covered in this guide.

Figure 14.10.1a shows a foundation mat where grade beams are located on top of the slab in contact with the bearing soil and needing an additional slab for the first floor. In this case, joists the same depth as the grade beams may be used. Figure 14.10.1b shows a foundation mat with grade beams located under the slab and in contact with the bearing soil. Each alternative has its merits from construction and cost perspectives, although the design procedure is essentially the same for both cases. For the case shown in Fig. 14.10.1a, the formwork used to cast the floor slab will be lost, but there is the advantage of having space under it for utilities and ducts. For the case shown in Fig. 14.10.1b, no formwork is lost and the soil may be used to form the grade beams if it is properly covered by a lean concrete.

14.10.2 *Load definition and foundation mat area*

14.10.2.1 *Loads to be included*—The design of foundation mats should include loads in (a) through (c).

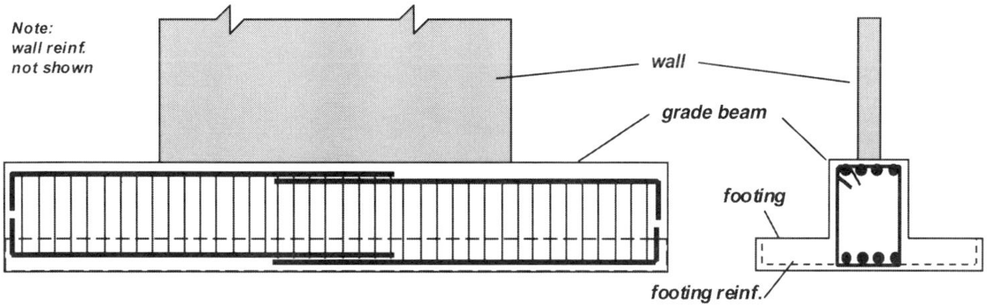

Fig. 14.6.4b—Wall footing grade beam reinforcement layout.

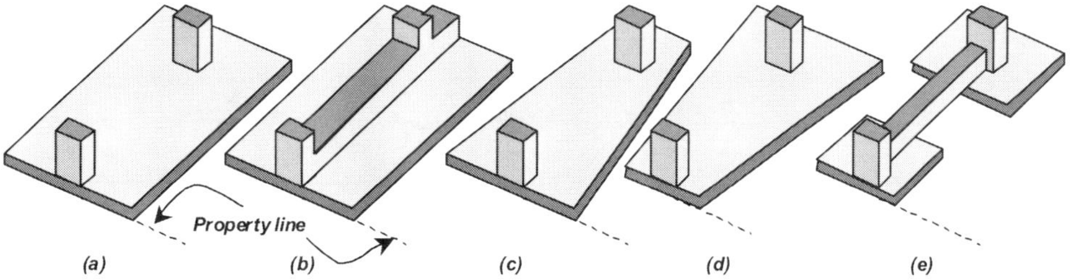

Fig. 14.7.1—Types of combined footings.

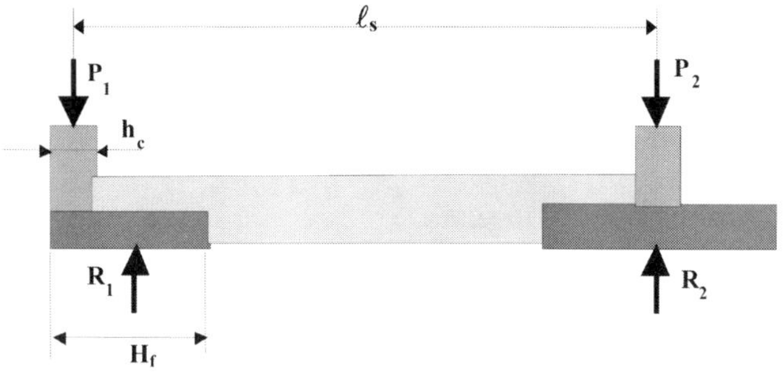

Fig. 14.7.2.1—Combined footing loads, reactions, and dimensions.

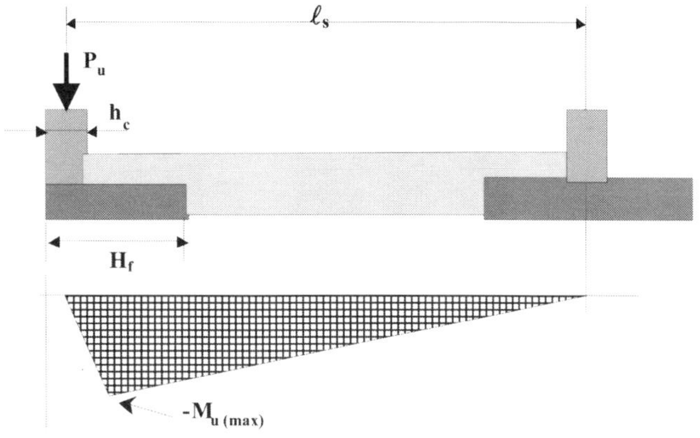

Fig. 14.7.2.3—Moment variation in grade beam of combined footing.

(a) All unfactored dead and live loads applied directly to the mat slab plus the self-weight of the mat. The sum of these loads should be described using q_o;

(b) Unfactored axial loads, moments, and shear loads carried by columns and walls and transferred to the mat foundation from all sources, including dead, live, wind, seismic, and other effects as prescribed in Chapter 4; or

(c) As an alternative to (b), unfactored axial load may be obtained from unfactored unit loads used to compute factored loads, including self-weight, multiplied by column or wall tributary area from all floors supported by it. Factored wind effects should be converted to unfactored values by dividing factored values by 1.6, and seismic effect unfactored values by dividing factored values by 1.43.

14.10.2.2 *Maximum unfactored vertical load*—To obtain total unfactored vertical load P_v, applied to the soil by the mat foundation, unfactored axial loads carried by columns and walls, floor loads, and mat foundation self-weight should be combined according to load cases of 4.2 without applying load factors. The location of the resultant unfactored vertical load should be determined.

14.10.2.3 *Verification of allowable bearing capacity*—Equation (14.10.2.3) should be used to compute soil-bearing pressure. Soil-bearing capacity should not be exceeded.

$$\frac{P_v}{B_f H_f} \leq (q_a - q_o) \quad (14.10.2.3)$$

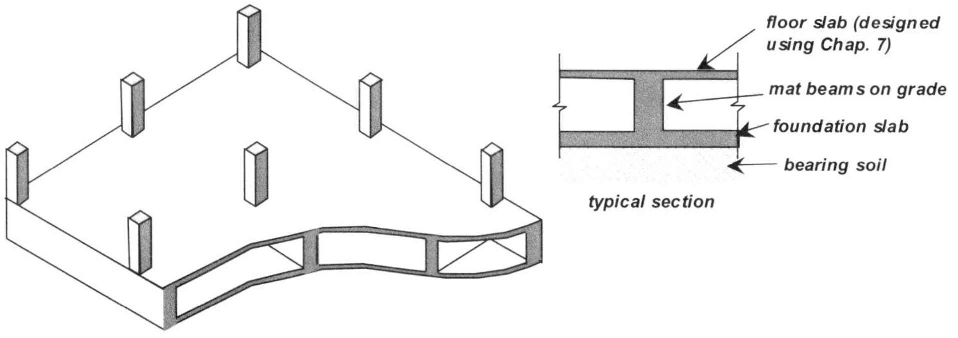

Fig. 14.10.1a—Foundation mat with grade beam on top of the slab.

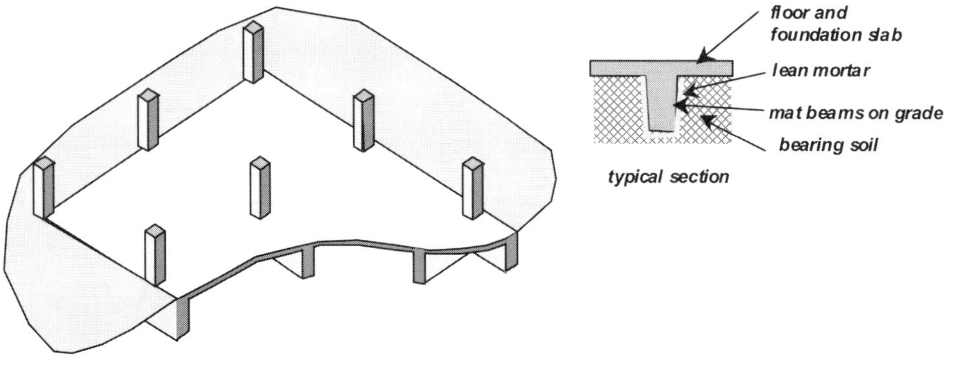

Fig. 14.10.1b—Foundation mat with grade beams under the slab.

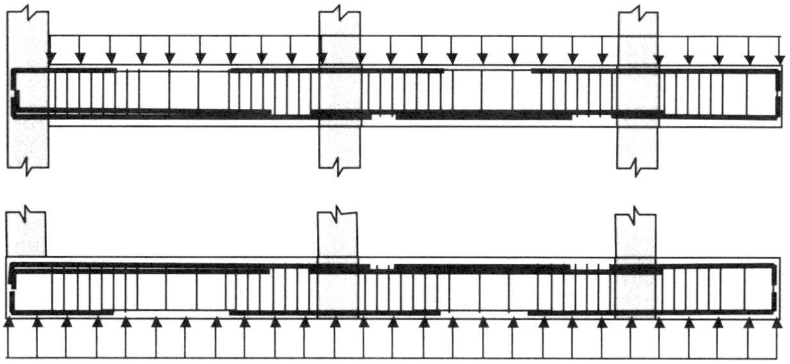

Fig. 14.10.3.1—Reinforcement layout for beams.

14.10.2.4 *Eccentricity of vertical loads*—The distance between the resultant of the vertical loads and the mat area centroid should not exceed the value indicated in the geotechnical report, or in its absence, should not be greater than 5 percent of the largest plan dimension of the mat, H_f. The existence of eccentricity increases the possibility of differential settlement. If eccentricity exceeds the maximum value, it should be reduced by adding or removing appropriate mat sections.

14.10.2.5 *Factored soil reaction*—Factored net soil reaction q_{un} should be determined by dividing the sum of factored axial loads from all columns and walls at the base, P_{un}, by the mat area. This value should be used for design of the foundation slab and the grade beams.

14.10.3 *Design procedures*

14.10.3.1 *General*—The following procedures should apply to the design of the mat foundation. There are several important aspects related to the design of the mat foundation.

(a) Although mat slab layout is similar to a regular floor slab, the order of magnitude of the loads, as compared with a typical floor slab, increases by a factor equal to the number of building stories;

(b) The mat slab, as any foundation element, will be loaded gradually as building construction progresses. These loads will cause deformations and settlements of the mat foundation; and

(c) The loads act in the opposite direction than in floor slabs; therefore, longitudinal reinforcement in the mat members should have their layout reversed. Negative moment reinforcement, usually located in the upper part of floor slabs, should be placed as bottom reinforcement in foundation mats and mat grade beams. For the same reason, positive moment reinforcement, usually located in the bottom part of floor members, should be placed as top reinforcement in the mat foundation (Fig. 14.10.3.1).

14.10.3.2 *Foundation slab*—The foundation slab (member contacting the soil) should be designed as a slab on girder using Chapter 7, loaded by the factored net soil reaction q_{un}. The self-weight should not be included because it is supported directly by the soil. Top and bottom reinforcement layout schemes of Chapter 7 should be reversed.

14.10.3.3 *Mat grade beams*—Mat grade beams should be designed as girders that are part of a moment-resisting frame using Section 8.7. Factored net soil reaction q_{un} should be used as the total load applied directly to the girder and the foundation slab reactions as tributary loads to the girder. Self-weight should not be included because it is supported directly by the soil. Top and bottom reinforcement layout schemes of 8.7 should be used with reinforcement reversed top and bottom. In seismic zones, Section 11.1.2 should apply.

14.10.3.4 *Floor slab, when independent from foundation slab*—When the floor slab is independent from the foundation slab (Fig. 14.10.1a), the floor slab should be designed as a slab on girders using the floor load of Chapter 7. In this case, the self-weight should be considered, and the reinforcement layouts of Chapter 7 are not reversed and should be used as shown.

14.10.3.5 *Values* d_c *and* d *in foundation mats*—The calculation of d_c, the distance from extreme tension fiber to centroid of tension reinforcement, should consider concrete cover, bar diameters, and perpendicular reinforcement under the reinforcement under consideration.

The following values of d_c may be used to compute d as $d = h - d_c$. For one-way solid slabs that are part of the foundation mat and in contact with the soil, and for reinforcement in the short direction in two-way slabs that are part of the foundation mat in contact with the soil, d_c = 3.5 in. (90 mm). For reinforcement in the long direction of two-way slabs in contact with the soil, d_c = 4.5 in. (100 mm). For girders, beams, and joist that are part of the foundation mat, use 14.12.

14.11—Retaining walls

14.11.1 *Types of retaining walls*—Figure 14.11.1 shows several types of retaining walls. Types (a) to (d) are free-standing and should be designed using active earth pressure. Type (e) is restrained at top and bottom; therefore, it should be designed using at-rest earth pressure. Only type (e), basement walls, is considered in this guide. For other types, the referenced documents in 1.4 should be used.

14.11.2 *Lateral earth pressure*

14.11.2.1 *General*—Values from a geotechnical report should be used where backfill has appropriate drainage and where hydrostatic pressure from water accumulation in the backfill is not possible. In remote areas, where a geotechnical report or local geotechnical information is unavailable, values given in 14.11.2 can be used.

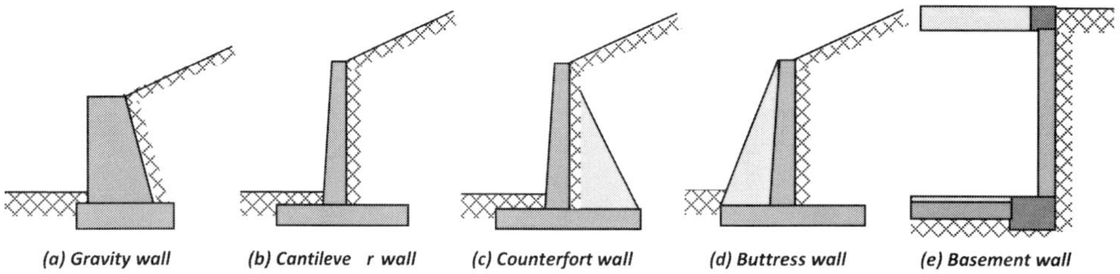

Fig. 14.11.1—Types of retaining walls.

Table 14.11.2.2—Typical values of ϕ_s for dry sands composed primarily of quartz

Density	Blow count (SPT)	Angle of internal friction
Very loose	$N \leq 4$	$\phi_s \leq 28.5°$
Loose	$5 < N \leq 10$	$28.5° < \phi_s \leq 32°$
Medium	$10 < N \leq 30$	$32° < \phi_s \leq 36°$
Dense	$30 < N \leq 50$	$36° < \phi_s \leq 41°$
Very dense	$50 < N$	$41° < \phi_s \leq 46°$

When a portion or the whole of the adjacent soil is below the water table, calculation should be based on the weight of the soil diminished by buoyancy plus full hydrostatic pressure. When designing basement walls and similar approximately vertical structures below grade, provisions should be made for lateral pressure of adjacent soil. Allowance should be made for possible surcharge from fixed or moving loads.

Earth pressure is stated generally in simple linear equations. Although this treatment overlooks some actual behavior characteristics, it is preferred for simplicity. The designer, however, should remember that often, lateral earth pressure distribution is not linear and earth loads tend to migrate from the more flexible to the stiffer portions of the system. Construction stages and procedures have a great influence in this load migration.

14.11.2.2 *Angle of internal friction,* ϕ_s—For soils, the angle of internal friction, ϕ_s, corresponds to the relevant parameter for lateral earth pressure determination. The following guidelines should be used to determine the angle of internal friction of the soil:

(a) For sands, the relationship given by Eq. (14.11.2.2) or the values from Table 14.11.2.2 should be used, where N is the blow count from the SPT:

$$\phi_s = 28.5° + \frac{N}{4} \tag{14.11.2.2}$$

(b) For wet sands, the values of ϕ_s from Eq. (14.11.2.2) or Table 14.11.2.2 should be decreased by one or two degrees;

(c) For gravels and crushed rock with states of density similar to those given in Table 14.11.2.2, the values of ϕ_s should be increased from 2 to 6 degrees;

(d) For dry silts and very silty sands, the values of ϕ_s should be taken between 2 to 6 degrees less than those of Table 14.11.2.2; and

(e) For clays, ϕ_s depends on the drainage condition and loading rate. Typical values should be taken between 20 to 30 degrees. For low drainage situations, values range from 10 to 20 degrees.

14.11.3 *At-rest earth pressure*—At-rest pressure exists in level ground under long-term conditions. That pressure should be evaluated as:

$$p_o = K_o \gamma z \tag{14.11.3a}$$

where

$$K_o = 1 - \sin\phi_s \tag{14.11.3b}$$

14.11.4 *Active earth pressure*—Active earth pressure develops when the wall moves slightly from the earth bank. In this case:

$$p_a = K_a \gamma z \tag{14.11.4a}$$

where K_a is the active earth pressure coefficient and should be calculated using Eq. (14.11.4b).

$$K_a = \frac{1 - \sin\phi_s}{1 + \sin\phi_s} \tag{14.11.4b}$$

14.11.5 *Passive earth pressure*—Passive earth pressure develops when the wall, or any of its parts, is forced against the earth bank. In this case:

$$p_p = K_p \gamma z \tag{14.11.5a}$$

where K_p is the passive earth pressure coefficient and should be calculated using Eq. (14.11.5b).

$$K_p = \frac{1 + \sin\phi_s}{1 - \sin\phi_s} \tag{14.11.5b}$$

14.11.6 *Braced excavations*—Most retaining walls for small buildings are braced excavations where the wall is laterally restrained at the base, the top, and possibly at other levels. In this situation, pressure distribution resembles a rectangle or a trapezoid with the upper triangle missing ($0.25h_s$). For practical cases, design pressure should be evaluated using 14.11.6.1 and 14.11.6.2:

14.11.6.1 *Granular soils*

$$p = 0.65 K_a \gamma h_s \tag{14.11.6.1}$$

Table 14.11.7.1—Minimum design lateral active pressure γK_a

Description of backfill material	Unified soil classification	Design active soil pressure (lb/ft² per ft of depth)	Design active soil pressure (kN/m² per m of depth)
Sand and gravel soil type			
Well-graded, clean gravels—gravel-sand mixtures	GW	120	6.0
Poorly-graded clean gravels—gravel-sand mixtures	GP	120	6.0
Silty gravels—poorly-graded, gravel-sand mixtures	GM	120	6.0
Clayey gravels—poorly-graded, gravel-and-clay mixtures	GC	150	7.5
Well-graded, clean sands—gravelly sand mixtures	SW	120	6.0
Poorly-graded, clean sands—sand-gravel mixtures	SP	120	6.0
Silty sands—poorly-graded sand-silt mixtures	SM	150	7.5
Silt and clay soil type			
Sand-silt clay mix with plastic fines	SM-SC	270	13.5
Clayey sands—poorly-graded sand-clay mixtures	SC	270	13.5
Inorganic silts and clayey silts	ML	270	13.5
Mixture of inorganic silt and clay	ML-CL	270	13.5
Other soil types			
Inorganic clays of low to medium plasticity	CL	320	16.0
Organic silts and silt clays, low plasticity	OL	Unsuitable as backfill	Unsuitable as backfill
Inorganic clayey silts, plastic silts	MH		
Inorganic clays of high plasticity	CH		
Organic clays and silty clays	OH		

14.11.6.2 *Cohesive soils*

$$p = 0.2\gamma h_s \text{ for } s_u \geq 2000 \text{ lb/ft}^2 \text{ (100 kPa)} \quad (14.11.6.2a)$$

$$p = 0.3\gamma h_s \text{ for } 2000 \text{ lb/ft}^2 \text{ (100 kPa)} > s_u \geq 500 \text{ lb/ft}^2 \text{ (25 kPa)} \quad (14.11.6.2b)$$

$$p = 0.4\gamma h_s \text{ for } 500 \text{ lb/ft}^2 \text{ (25 kPa)} \geq s_u \quad (14.11.6.2c)$$

14.11.7 *Minimum soil lateral pressure*

14.11.7.1 *Minimum active soil pressure*—Soil loads in Table 14.11.7.1 should be used as lateral soil pressure unless specified otherwise in a soil investigation report. This pressure corresponds to the active pressure γK_a. Lateral pressure from surcharge loads should be added to the lateral earth pressure load. Lateral pressure should be increased where soils with expansion potential are present at the site as determined by a geotechnical investigation.

In Table 14.11.7.1, lateral soil pressure is given for moist conditions for the specified soils at optimum densities. Actual field conditions should govern. Submerged or saturated soil pressure should include the buoyant soil weight plus the hydrostatic loads.

14.11.7.2 *Minimum lateral pressure at rest*—For rigid walls laterally restrained top and bottom, lateral soil pressure in Table 14.11.7.1 should be increased as follows:

(a) For sand and gravel soil type to 60 lb/ft² (9.5 kN/m²) per ft (m) of depth; and

(b) For silt and clay soil type to 100 lb/ft² (16 kN/m²) per ft (m) of depth.

These cases correspond to the lateral soil pressure at rest, γK_o.

14.11.8 *Lateral pressure on retaining wall*—Lateral pressure caused by the adjacent soil on the retaining wall should be calculated using procedures described in 14.11.8.1 and 14.11.8.2.

14.11.8.1 *Retaining walls not laterally restrained at the top*—For retaining walls not restrained at the top by the building slabs, the lateral pressure at any depth z should be calculated using Eq. (14.11.8.1a).

$$p_z = \gamma K_a z \quad (14.11.8.1a)$$

where p_z corresponds to the lateral pressure, in lb/ft² (kN/m²) at a depth z, in ft (m), measured from the surface; and γK_a should be given by the geotechnical report, or in its absence, minimum values given in Table 14.11.7.1 may be used. Total lateral force applied by the soil to the wall should be determined using Eq. (14.11.8.1b).

$$F_{ac} = \frac{1}{2}\gamma K_a h_s^2 \quad (14.11.8.1.b)$$

where F_{ac} is the total lateral force, and h_s is soil height against the wall measured from the base of the wall footing to soil surface. F_{ac} should be assumed to act at a height $h_s/3$ measured from the base of the wall footing.

14.11.8.2 *Retaining walls laterally restrained at the top*—For retaining walls restrained laterally by building slabs, lateral pressure at any depth, z, should be calculated using Eq. (14.11.8.2a).

$$p_z = \gamma K_o z \quad (14.11.8.2a)$$

where p_z corresponds to lateral pressure, in lb/ft² (kN/m²) at a depth z, in ft (m), measured from the surface; and γK_o should be given by the geotechnical report or, in its absence, minimum values given in 14.11.2 may be used. Total lateral force applied by the soil to the wall should be determined using Eq. (14.11.8.2b).

$$F_o = \frac{1}{2}\gamma K_o h_s^2 \quad (14.11.8.2b)$$

where F_o is the total lateral force, and h_s is the soil height against the wall measured from the base of the wall footing to soil surface. F_o should be assumed to act at a height $h_s/3$ measured from the base of the wall footing.

14.11.9 *Loads for basement walls*

14.11.9.1 *Loads to be included*—Basement retaining wall design should include the following loads:

(a) Out-of-plane earth pressure p_z, as stated in 14.11.6 and 14.11.8.2;

(b) Out-of-plane vertical surcharge pressure located on top of the backfill, p_t, increasing lateral earth pressure in the surcharge value multiplied by K_o;

(c) In-plane loads and moments due to wind and seismic loads, transmitted by supporting elements;

(d) In-plane noncompensated lateral loads induced by soil lateral loads (4.13.2.3); and

(e) Any additional lateral, out-of-plane pressure exerted by the soil caused by the compaction of the backfill of the wall. In many cases, the backfill compaction is more efficiently performed where the worker comfort is highest, at the upper part of the fill; which is why this guide proposes (14.11.6) a conservative, uniform, out-of-plane soil pressure distribution on the wall unless the geotechnical investigation report indicates a different distribution of pressures.

14.11.9.2 *Seismic effects*—Seismic effects, including increased earth pressure, wall in-plane moments and out-of-plane moments, and out-of-plane shear, should be disregarded for a one-story wall, restrained at top and bottom. The in-plane shear transmitted by the slab diaphragm in the upper part of the wall should be considered.

14.11.9.3 *Lateral pressure*—Factored lateral pressures, p_{uw} and p_{utw}, on the wall should be determined by multiplying pressures p_z and p_t, from 14.11.6 and 14.11.8.2 by the appropriate load factor from 4.2.

14.11.10 *General conditions for basement walls*

14.11.10.1 *Support top and bottom*—Basement retaining walls under the scope of this guide should have lateral restraint provided at the top of the wall by the first aerial floor slab and at the bottom of the wall by either (a) the grade beams in high and moderate seismic zones in 14.12 or (b) by the wall footing in low and no seismic risk zones. In both cases, a wall footing should be provided to transmit gravity loads from the following components to the soil: the wall self-weight, vertical reactions of the slab supported by the top of the wall, and any vertical loads transmitted by friction that are produced by the compaction of the backfill and the portion of the backfill that is supported directly by the wall footing. Window openings for lighting and ventilation purposes can cause a captive column problem in seismic zones (11.2.3). In those cases, alternative lighting and ventilation should be used.

14.11.10.2 *Drainage*—The design pressure as stated does not include hydrostatic pressure; therefore, drainage such as through-filters and weep-holes for dissipation of hydrostatic pressure should be used.

14.11.10.3 *Backfill material*—Materials marked as unsuitable for backfill material in Table 14.11.7.1 should not be used.

14.11.10.4 *Surcharge*—Any expected surcharge during construction or use of the structure should be included in the wall design.

14.11.10.5 *Minimum thickness*—Minimum thickness of basement walls should be 7-1/2 in. (190 mm).

14.11.11 *Reinforcement details*—Basement retaining walls should comply with 12.4.

14.11.12 *Flexure*—Required moment strength of the basement retaining wall depends on the out-of-plane soil pressure distribution and the rotational restraint at the top and bottom of the wall. Rotational restraint is provided by the slab at the top of the wall, the beam along the wall length as indicated by 12.3.4, and the wall footing. Based on these circumstances, the following support conditions result in the reinforcement shown in Fig. 14.11.12 with the required moment strength per unit of wall width indicated in Table 14.11.12:

(a) *Minimum rotational restraint top and bottom*—The wall has no rotational restraint provided by elements that could provide it.

(b) *Rotational restraint provided at top and minimum rotational restraint at bottom*—The slab elements of the first aerial floor have sufficient flexural stiffness to inhibit any flexural rotation of the wall at the top, whereas the elements that provide the lateral restraint at the bottom do not have flexural stiffness to inhibit flexural rotation of the retaining wall.

(c) *Minimum rotational restraint at top and rotational restraint provided at bottom*—The slab elements of the first aerial floor are simply supported at the wall and do not provide any flexural stiffness to inhibit flexural rotation of the wall at the top, whereas the wall footing or the elements that provide lateral restraint at the bottom of the wall have sufficient flexural stiffness to inhibit flexural rotation of the wall bottom.

(d) *Rotational restraint provided both top and bottom*—The slab elements of the first aerial floor have sufficient flexural stiffness to inhibit any flexural rotation of the wall at the top and the wall footing or the elements that provide lateral restraint at the bottom of the wall have sufficient flexural stiffness to inhibit flexural rotation of the wall bottom.

Reinforcement area and spacing should be determined using procedures of 7.3 for slabs. Calculation of d_c, the distance from extreme tension fiber to tension reinforcement centroid, should consider concrete cover, bar diameters, and perpendicular reinforcement placed between reinforcement considered and the concrete surface. A value of d_c = 3-1/2 in. (90 mm) should be used to compute d as $d = h - d_c$.

Table 14.11.12—Basement retaining wall required moment strength per horizontal unit length of wall

Wall flexural restraint			Required moment strength for reinforcement type (Fig. 14.11.12)		
Case	Top	Bottom	A	B	C
a)	Free	Free	$M_u = \frac{p_{uw}\ell_n^2}{17} + \frac{p_{utw}\ell_n^2}{8}$	$M_u = \frac{p_{uw}\ell_n^2}{15} + \frac{p_{utw}\ell_n^2}{8}$	
b)	Restrained	Free	$M_u = \frac{p_{uw}\ell_n^2}{17} + \frac{p_{utw}\ell_n^2}{8}$	$M_u = \frac{p_{uw}\ell_n^2}{24} + \frac{p_{utw}\ell_n^2}{14}$	
c)	Free	Restrained		$M_u = \frac{p_{uw}\ell_n^2}{33} + \frac{p_{utw}\ell_n^2}{14}$	$M_u = \frac{p_{uw}\ell_n^2}{15} + \frac{p_{utw}\ell_n^2}{8}$
d)	Restrained	Restrained	$M_u = \frac{p_{uw}\ell_n^2}{30} + \frac{p_{utw}\ell_n^2}{12}$	$M_u = \frac{p_{uw}\ell_n^2}{47} + \frac{p_{utw}\ell_n^2}{24}$	$M_u = \frac{p_{uw}\ell_n^2}{20} + \frac{p_{utw}\ell_n^2}{12}$

14.11.13 *Shear*—Required shear strength per unit horizontal wall length for the basement wall should be determined from:

$$V_u = \frac{p_{uw}\ell_n}{2.5} + \frac{p_{utw}\ell_n}{1.67} \quad (14.11.13)$$

The shear strength of solid slabs in 7.4 should be met.

14.12—Grade beams (foundation beams)

14.12.1 *General*

14.12.1.1 *Description*—Grade beams located under the surface and above the footings (Fig. 14.12.1.1a) should form a grid linking all columns and walls at the building base (Fig. 14.12.1.1b). In mat foundations, mat beams serve the same function.

14.12.1.2 *Function*—Grade beams functions include (a) through (d):

(a) Make the overall building settlement more uniform because they permit redistribution of footing loads and reduce the potential for differential settlements;

(b) Link combined footings (14.7), distributing moments due to eccentric footings or moments transmitted from columns to footings;

(c) Help transmit loads due to wall moments to the soil; and

(d) In seismic zones, form a diaphragm at foundation level that inhibits differential horizontal movements between columns and walls.

14.12.1.3 *Mandatory usage*—The use of a grade-beam grid is mandatory for moderate or high seismic risk zones and zones of no or low seismic risk.

14.12.2 *Loads*

14.12.2.1 *General*—Grade beams dimensions and reinforcement should be based on the main function (14.12.1.2), with allowances made for the other functions. Recommendations of the geotechnical report should be observed.

14.12.2.2 *Differential settlements*—To minimize differential settlements, grade beams should have sufficient stiffness and strength to transfer the loads developed. Stiffness controls for short spans and strength for long spans. The grade beam

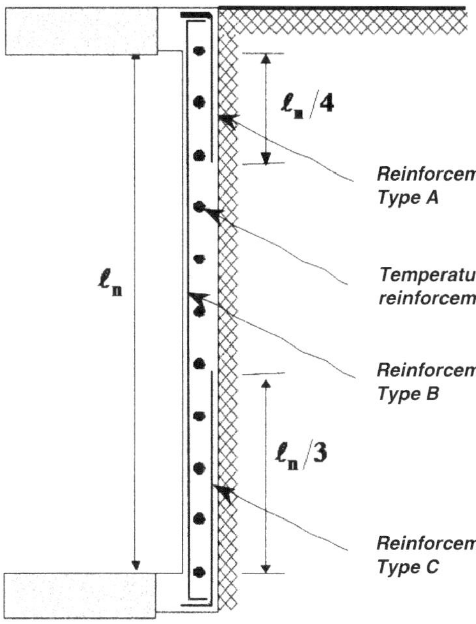

Fig. 14.11.12—Design moments and reinforcement for basement walls.

should be dimensioned for the factored positive and negative moment given by Eq. (14.12.2.2a).

$$M_u = \frac{P_u\ell_s}{160} \quad (14.12.2.2a)$$

where P_u is the largest axial force carried by columns or walls linked by the grade beam, and ℓ_s is the center-to-center distance to that column or wall. Factored shear for the grade beam should be computed from Eq. (14.12.2.2b).

$$V_u = \frac{P_u}{80} \quad (14.12.2.2b)$$

14.12.2.3 *Seismic*—The grade beam should act as a tie between neighboring columns or walls. The grade beam should be capable of resisting, in tension and compression, a

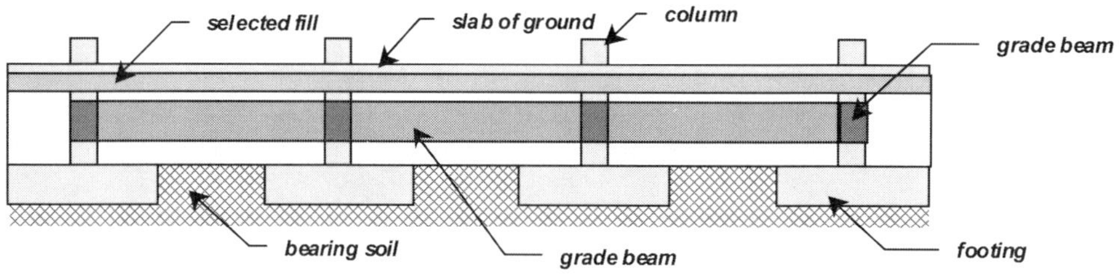

Fig. 14.12.1.1a—Location of grade beams.

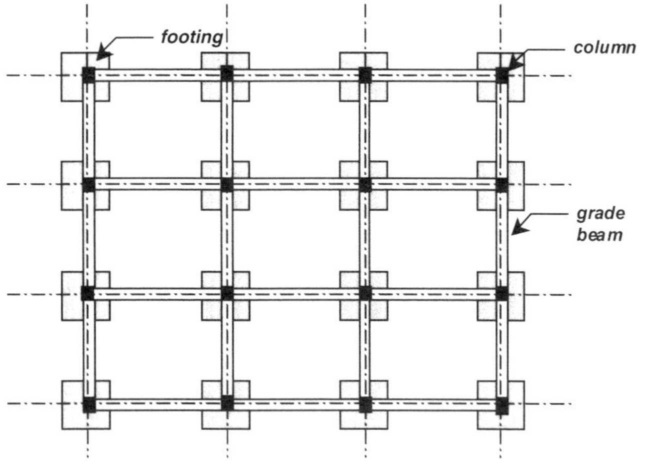

Fig. 14.12.1.1b—Grid of grade beams.

factored axial force equivalent to $0.25A_aP_u$, where P_u is the largest axial force carried by columns or walls linked by the grade beam and A_a is the effective earthquake peak ground horizontal acceleration given in 4.11. For calculation of compression and tension strength, refer to 5.12.3 and 5.12.5.

14.12.3 *Dimensional limits*—The grade beam should be proportioned to resist the loads of 14.12.2, and minimum cross-sectional dimension should be the clear spacing between connected columns divided by 40 in. (1000 mm), or by 20 in. (500 mm) seismic zones, but need not exceed 20 in. (500 mm).

14.12.4 *Reinforcement*—Grade beams should have continuous longitudinal reinforcement developed within or beyond the supported column or anchored at all discontinuities. Closed ties should be provided at a maximum spacing of the lesser of one-half the smallest orthogonal cross-sectional dimension or 12 in. (300 mm). In seismic zones, grade beams should comply with 11.1.2. For grade beams and beams-on-grade, d_c = 4 in. (100 mm) may be used to compute d as $d = h - d_c$.

14.13—Slabs-on-ground

14.13.1 *General*—Section 14.13 covers only interior slabs continuously supported on selected material fill, which in turn is supported by soil. When the foundation slabs-on-ground is used to transfer loads from other parts of the structure to the supporting soil, their design should conform to appropriate sections of Chapter 14. Design of pavements or slabs-on-ground subject to appreciable concentrated loads is not covered in this guide.

14.13.2 *Subbase*—The prepared subbase should consist of a compacted selected material fill made from crushed stone or gravel with an appropriate mixture of fines. Minimum subbase thickness should be 8 in. (200 mm).

14.13.3 *Minimum thickness*—The minimum slab-on-ground thickness should be 4 in. (100 mm).

14.13.4 *Joints*—Contraction joints should be used to provide uniform cracking as the slab-on-ground shortens due to shrinkage and temperature variations. They should be either sawed or hand-formed and extend at least one-fourth the slab thickness. Joint spacing in both directions should be from 6 to 20 ft (2 to 6 m). Reinforcement, if needed, should be continuous across the joint. No panel should be larger than 250 ft^2 (30 m^2) or have the long side exceed the short side by more than 25 percent.

14.13.5 *Reinforcement*—Reinforcing bars or welded-wire reinforcement should be used when the distance between contraction joints exceeds 8 ft (2.5 m). Reinforcement should be placed approximately at one-third the slab thickness measured from the upper surface, complying with 5.4, concrete cover. Minimum reinforcement ratio should be one-half the minimum value given by 7.3.3 for slabs. This reinforcement should be lap spliced complying with 5.8.2. All reinforcement for slabs-on-ground should be supported by bolsters at close-enough centers to prevent deformation by workers or equipment. Lifting, pulling up, or puddling in are not acceptable methods for placing reinforcement.

CHAPTER 15—DRAWINGS AND SPECIFICATIONS

15.1—General

Building construction employs several types of drawings, each depicting an important part of the project. These drawings, which cover all phases of the work necessary for the finished structure, should have the scope given in 15.1.1 through 15.1.5.

15.1.1 *Site drawings*—Site drawings or plot plans depict:

(a) Location of the building on the property;

(b) Utility lines;

(c) Drainage and drainage structures;

(d) Outside walks, steps, driveways, and curbs;

(e) Preconstruction ground elevations and finished grades; and

(f) Location of all site structures, including signs, retaining walls, and paving.

15.1.2 *Architectural drawings*—In many cases, architectural drawings are the basis for the drawings of other disciplines, such as structural, mechanical, and electrical. Architectural drawings depict:

(a) Design occupancy and usage of areas;

(b) Finished appearance of the building in elevations;

(c) Plans and sections completely dimensioned;

(d) The relationship of materials such as concrete, steel, and brick; and wood and stone to each other;

(e) Room arrangements in plans with sections and elevations to illustrate details;

(f) Finishes such as plaster or tile walls; and

(g) Ceiling, floor surfaces, and fixtures.

15.1.3 *Structural drawings*—Structural drawings depict all plans and details necessary to construct the building frame with complete working dimensions and elevations.

15.1.4 *Mechanical drawings*—Mechanical drawings depict piping, heating and air-conditioning ducts, and mechanical equipment. Installation of this work often requires the construction of sumps, pits, and openings in walls and floors. Structural drawings need to depict these concrete details, so the mechanical drawings should be used for reference purposes only. In some instances, the manufacturer of special equipment supplies structural details to accompany mechanical features; in this case, notes should be placed on the structural drawings referring to where the specialty drawings can be found.

15.1.5 *Electrical drawings*—Electrical drawings depict electric wiring, conduits, fixture locations, control panels, and electric pumps. Sometimes concrete encasement of conduits requires reinforcing bars. This encasement should be detailed on the structural drawings, with a reference made to the electrical drawings for specific details.

15.2—Structural drawings

Structural drawings should be divided into general structural drawings and detailing or placing drawings and related schedules. All structural drawings should be signed by the licensed design professional and should include:

(a) Project name;

(b) Date the design was performed;

(c) Name of the licensed design professional;

(d) Name and edition of code to which design conforms;

(e) Live load and other loads assumed in design;

(f) Specified compressive strength of concrete at stated ages of construction for each part of the structure;

(g) Specified strength or grade of reinforcement; and

(h) Statement of the design limitations with respect to: occupancy; maximum number of stories; maximum area per floor; maximum span length; minimum number of spans; maximum cantilever span; maximum slope for slabs, girders, beams, and joists; and maximum slope of the terrain.

15.2.1 *General structural drawings*—Items (1) through (12) should be included in a set of general structural drawings.

(1) Complete and clear dimensions so the building structure can be constructed without reference to other drawings;

(2) Size and shape of all individual structural members, such as footings, columns, walls, beams, joists and slabs, in plan, section, elevation, or schedules, or in combination;

(3) Elevations at:

- The bottom of footings and walls;
- Floor and roof levels;
- Brick ledges on walls; and
- Steps in wall footings;

(4) Flow lines for drainage structures;

(5) Location and details of construction joints;

(6) Reinforcing bar quantity or spacing, position, shape, and size that are often listed in separate schedules for columns, beams, joists, and slabs;

(7) Location and length of all lap splices;

(8) Camber information for horizontal members such as long-span beam and girders, and cantilevers;

(9) Sections of special framing and bar detailing as necessary to clarify such framing;

(10) General notes, such as:

(a) Grade of steel when, if more than one grade is shown, each is located in the structure;

(b) Specified concrete strengths for the various structural members;

(c) Reference to the detailing and placing drawings;

(d) Reference to the code for overall design compliance;

(e) Class of concrete slab finish and bar support types; and

(f) Mill certifications for reinforcing steel, if needed.

(11) Notes where deviation from recognized standards and tolerances exist, or where specific instructions are needed for unusual job conditions; and

(12) Typical diagrams showing bar arrangement for all concrete members and bar support arrangement and spacing.

For further detailing recommendations, refer to ACI SP-66(04).

15.2.2 *Detailing and placing drawings or schedules*—Detailing and placing drawings or schedules should satisfy two purposes that provide:

(1) Details from which the builder or bar fabricator extracts information needed to cut and bend the steel; and

(2) Details and placing instructions for workers to place bars on the job.

When made by the builder or fabricator, they indicate to the design engineer how the builder or fabricator interprets general structural drawings. Approval of the placing drawings by the licensed design professional indicates acceptance of that interpretation.

The recommended procedure for making the detailing and placing drawings or schedules is listed in (1) through (5).

(1) Project specifications and general structural drawings should be used to prepare placing drawings, which show detailed quantities, lengths, bending diagrams, and positioning of reinforcing steel and bar supports.

(2) When electronic drawings are not employed, highly transparent tracing paper should be selected. This paper should be superimposed on structural or architectural drawings. Outlines on these drawings should be traced and used as part of the placing drawings. This will save the detailer time in mechanically reproducing the plans, elevations, and sections needed for reinforcement detailing.

(3) When creating building outlines, only the portion needed for bar detailing should be traced. Brick or tile work, non-reinforced partition walls, and mechanical details should not be shown. Unnecessary sections and elevations should be omitted. Generally, dimensions may be omitted on placing drawings. For construction purposes, structural and architectural drawings should be used.

(4) Several methods can produce transparencies. Reproductions on transparent paper will save the detailer time and ensure accurate work tracings for the details. In reproducing these drawings, it is possible to block out titles, notes, and unneeded sections and details. Reinforcing bar details should then be added to the transparencies to supplement original drawings and produce placing drawings.

(5) Placing drawings or schedules should be clear, complete, and workable, and include all information necessary to place the steel. The operation of placing reinforcement is a key construction phase. Unclear drawings cause field delays, slow the production of the placing crew and, if prolonged, will disrupt construction operations following placing.

15.3—Project specifications

Project specifications should include all requirements, written by the licensed design professional, to supplement and amplify project drawings. Specification details control construction requirements over drawings. They should be divided into sections, beginning with one covering general conditions, and followed in logical construction order by sections pertaining to materials and execution.

General conditions include a form of contract governing the relationship and responsibilities of the registered design professional, the general contractor, subcontractor, suppliers of materials, and workers. This section should also include general instructions governing the distribution of reinforcement placing drawings for approval, and use and procedures for submitting any required material samples for testing or approval.

Following the general conditions are the detailed parts of the project by sections. Each section defines the requirements for one phase of construction or materials.

Typical section titles are:

- Division 01 — General requirements
- Division 02 — Site construction
- Division 03 — Concrete
- Division 04 — Masonry
- Division 05 — Metals
- Division 06 — Wood and plastics
- Division 07 — Thermal and moisture protection
- Division 08 — Doors and windows
- Division 09 — Finishes
- Division 10 — Specialties
- Division 11 — Equipment
- Division 12 — Furnishings
- Division 13 — Special construction
- Division 14 — Conveying systems
- Division 15 — Mechanical
- Division 16 — Electrical

Division 03 on concrete work is the primary division that addresses construction covered by this guide, but each division should be reviewed for items that could affect the structural design.

For example, the masonry section might include steel reinforcement for brick or concrete block construction. The division on structural steel might include reinforcement for concrete fireproofing of structural steel roof trusses, beams, and columns.

Under the concrete division are a number of sections. ACI 301-10, "Specifications for Structural Concrete" is a reference specification for concrete building construction. Some typical sections are listed and explained below:

(a) Proportioning, mixing, handling, placing, quality, and testing of concrete;

(b) Concrete strengths and, if several are specified, where each is to be used in the project;

(c) Formwork, type of materials, erection, bracing, shoring, and removal;

(d) Grade of reinforcing bars and, if more than one, where each is to be used, fabrication of reinforcing bars, and tolerances;

(e) Concrete cover over the bars;

(f) Quality control of reinforcing bars, whether by testing, acceptance of mill test reports, or both;

(g) Class or type of bar supports, bar positioning, and general arrangement of bar supports; and

(h) Concrete finishes.

A carefully prepared project specification should minimize the need for lengthy, detailed notes on the drawings. Certain notes always go on the structural drawings because they are fundamental, such as the specified strength of concrete, even if the information is included in the specification.

CHAPTER 16—CONSTRUCTION

16.1—Introduction

16.1.1 *General*—Concrete is a carefully proportioned mixture of cement, sand and gravel or other aggregate, and water, hardened in forms of the shape and dimensions of the desired structure.

Most concrete material consists of fine and coarse aggregate. Cement and water interact chemically to bind aggregate particles into a solid mass. Additional water, over and above that needed for this chemical reaction, is necessary to give the mixture the workability that enables it to fill forms and surround the embedded reinforcing steel before hardening. Concrete properties depend on, to a substantial degree, mixture proportions, the thoroughness with which the various constituents are intermixed, and humidity and temperature conditions in which the mixture is maintained from the moment it is placed in the forms until it fully hardens. The process of controlling humidity and temperature conditions is known as curing. To protect against unintentional production of substandard concrete, a high degree of skillful control and supervision is necessary throughout the process, from the proportioning by individual component weight, through mixing and placing, until the completion of curing.

16.1.2 *Cement*—A cementing material has the adhesive and cohesive properties necessary to bond inert aggregates into a solid mass of adequate strength and durability.

For making structural concrete, hydraulic cements are used exclusively. Water is needed for the chemical process (hydration) in which the cement powder sets and hardens into one solid mass. Of the various hydraulic cements that have been developed, portland cement is the most common. Portland cement is a finely powdered, grayish material that consists mainly of calcium and aluminum silicates. These are ground, blended, fused to clinkers in a kiln, cooled, and ground to the required fineness. The material is shipped in bulk or in bags containing 94 lb (42.6 kg) of cement in the United States and 110 lb (50 kg) of cement in most other countries. When cement is mixed with water to form a soft paste, it gradually stiffens until it becomes a solid. This process is known as setting and hardening; the cement is said to have set when it has gained sufficient rigidity to support an arbitrarily defined pressure, after which it continues for a long time to harden and gain further strength. Water in the paste dissolves material at the surfaces of the cement grains and forms a gel that gradually increases in volume and stiffness. This leads to a rapid stiffening of the paste 2 to 4 hours after water has been added to the cement. Hydration continues to proceed deeper into the cement grains, at decreasing speed, with continued stiffening and hardening of the mass. In ordinary concrete, the cement is probably never completely hydrated. The gel structure of the hardened paste seems to be the chief reason for the volume changes caused in concrete by variations in moisture, such as the shrinkage of concrete as it dries. For complete hydration of a given amount of cement, an amount of water equal to about 25 percent of that of cement, by weight, is needed chemically. An additional 10 to 15 percent should be present, however, to provide mobility for the water in the cement paste during the hydration process so that it can reach the cement particles. This makes for a total minimum water-cement ratio (w/c) of 0.35 to 0.40 by weight. This corresponds to 3.5 to 4 gal. (13 to 15 L) of water per 94 lb (42.6 kg) sack of cement, the more customary way of expressing w/c. Water-cement ratios in concrete are generally considerably larger than this minimum to provide the necessary workability of the concrete mixture. Any amount of water above the 25 percent consumed in the chemical reaction produces pores in the cement paste. The hardened paste strength decreases in inverse proportion to the fraction of the total volume occupied by pores. This is why the cement paste strength depends primarily on, and decreases directly with, increasing w/c. The chemical process involved in setting and hardening liberates heat, known as heat of hydration. In large concrete masses, this heat is dissipated very slowly and results in a temperature rise and volume expansion of the concrete during hydration, with subsequent cooling and contraction. To avoid serious cracking and weakening that may result from this process, special measures should be taken for its control.

16.1.3 *Aggregates*—Maximum aggregate size should easily fit into the forms and between the reinforcing bars (5.7). The use of lightweight aggregates is beyond the scope of this guide. In ordinary structural concrete, aggregates occupy about 70 to 75 percent of the volume of the hardened mass. The remainder consists of hardened cement paste, uncombined water (water not involved in cement hydration), and air voids. The last two do not contribute to concrete strength. In general, the more densely the aggregate can be packed, the better the strength, weather resistance, and economy of the concrete. For this reason, the gradation of the particle sizes in the aggregate to produce close packing is important. It is also important that the aggregate have good strength, durability, and weather resistance; that its surface be free from impurities such as loam, silt, and organic matter, which may weaken the bond with the cement paste; and that no unfavorable chemical reaction takes place between the cement paste and the cement. Natural aggregates are generally classified as fine and coarse. Fine aggregate or sand is any material that passes a sieve with 1/4 in. (6 mm) openings. Material coarser than this is classified as coarse aggregate or gravel. When favorable gradation is desired, aggregates are separated by sieving into two or three size groups of sand and several size groups of coarse aggregate. These can then be combined according to grading charts to result in a densely packed aggregate.

16.1.4 *Admixtures*—In addition to the main components of concrete, admixtures are often used for specific purposes. There are admixtures to improve workability, accelerate or retard setting and hardening, aid in curing, improve durability, add color, and impart other properties. Whereas the beneficial effects of some admixtures are well established, the claims of others should be viewed with caution.

16.1.5 *Reinforcement*—The reinforcement types covered in this guide are deformed reinforcing bars, plain bars, and welded-wire reinforcement (5.2.5).

For most effective reinforcement action, it is necessary that sufficient bond exists between the two materials so that relative movement between the steel bar and the surrounding concrete does not occur. This bond is provided by the relatively large chemical adhesion that develops at the steel-concrete interface by the natural roughness of the mill scale of hot-rolled reinforcing bars, and by the closely spaced surface bar deformations, which provide a high degree of interlocking of the two materials.

16.1.6 *Formwork*—Forms should result in a final structure that conforms to the shapes, lines, and dimensions of the members indicated by the design drawings and specifications.

Forms should be substantial and sufficiently tight to prevent mortar leakage. They should be properly braced or tied together to maintain position and shape. Forms and their support should be designed so as not to damage a previously placed structure. Formwork design should consider the rate and method of placing concrete and construction loads, including vertical, horizontal, and impact loads.

16.2—Concrete mixture proportioning

16.2.1 *General*—This guide does not intentionally address concrete containing supplementary cementing materials (SCM) including fly ash. It is recognized that in many countries, like the United States, that fly ash and other SCM are used in over 50 percent of all concrete mixtures and are routinely delivered by concrete suppliers as constituents of typical mixtures. This guide only makes reference to concrete design strength as it is primarily directed to engineers with limited exposure to materials technology and training. For a complete definition of proportions, the weight of cement, water, sand, and coarse aggregate should be provided.

Various mixture components are proportioned so that the resulting concrete has adequate strength, proper workability for placing, and low cost. The latter calls for using the minimum amount of cement—the most costly component—to achieve adequate properties. Concrete attributes are generally influenced as indicated in Table 16.2.1. The better the aggregate gradation, the smaller the volume of voids, and the less cement paste is needed to fill these voids. Water is required for hydration and needed for wetting the aggregate surface. As water is added, plasticity and fluidity of the mixture increases and workability improves, but strength decreases because free water creates a larger volume of voids. To reduce free water while retaining workability, cement should be added. Therefore, as for cement paste, the *w/c* is the main factor that controls the concrete strength. For a given *w/c*, the minimum amount of cement by weight that will secure the desired workability should be selected.

Concrete should be proportioned to provide an average compressive strength f_{cr}' that minimizes the frequency of strength tests below f_c'. Determination of f_{cr}' should be based on 28-day age tests on pairs of cylinders made and tested as described in 16.5.3.2. In addition, concrete material proportions should be established to provide:

(a) Workability and consistency to permit concrete to be worked readily into forms and around reinforcement under the placement conditions, without segregation or excessive bleeding;

(b) Resistance to anticipated exposures; and

(c) Conformance with strength test requirements.

Concrete proportions, including *w/c*, should be established based on field experience, trial mixtures, or both, with the materials to be used.

16.2.2 *Durability*

16.2.2.1 *General*—To obtain appropriate concrete durability, a minimum amount of cement should be provided by using *w/c* below specified values and by specifying a minimum compressive strength for the concrete. This guide, however, does not provide mixture recommendations for buildings with special durability concerns.

16.2.2.2 *Calculation of* w/c—Water-cement ratios specified in 16.2.2 should be calculated using water weight in concrete, lb/yd^3 (kg per m^3), divided by the cement in the concrete mixture, in lb/yd^3 (kg per m^3). The use of fly ash, pozzolans, slag, and silica fume is beyond the scope of this guide and if used should be in accordance with referenced standards of Chapter 17.

16.2.2.3 *Freezing-and-thawing exposures*—Concrete exposed to freezing and thawing or deicing chemicals should be air entrained with a 6 percent total air content for severe exposure and 5 percent for moderate exposure. Tolerance on air content in fresh concrete should be ±1.5 percent. Air entrainment should not be used for concrete surfaces with specified hard trowel finish.

16.2.2.4 *Special exposure conditions*—For exposures given in Table 16.2.2.4, concrete should conform to the corresponding maximum *w/c* and minimum specified concrete compressive strength.

16.2.2.5 *Sulfate exposures*—When water-soluble sulfate (SO_4) is present in soil and has a concentration greater than 0.10 percent by weight or is present in water with more than 150 parts per million (150 ppm), concrete exposed to these sulfate-containing solutions or soils should have a *w/c* less than or equal to 0.45 by weight and a minimum compressive strength f_c' of 4500 psi (31 MPa). The use of sulfate-resistant cement is recommended, if available. Calcium chloride as an admixture should not be used in concrete exposed to sulfates.

16.2.2.6 *Chloride-ion exposure*—For corrosion protection of reinforcement in concrete, maximum water-soluble chloride ion concentrations in hardened concrete at ages from 28 to 42 days contributed from ingredients including water, aggregates, cement, and admixtures should not exceed the limits of Table 16.2.2.6.

16.2.3 *Average compressive strength*—Average compressive strength f_{cr}' for concrete should be 1500 psi (10 MPa) greater than specified concrete compressive strength f_c'. This exceeds the ACI 318-11 requirement of 1200 psi (8.3 MPa) because of an assumed lower quality control in remote areas.

16.2.4 *Proportioning the concrete mixture*—The concrete mixture proportions should be established from field experience producer or trial mixtures using combinations of materials for the proposed work, using at least three different *w/c* that comply with 16.2.2, durability, slump limits from Table 16.2.4, and encompass the required average strength f_{cr}'. Trial mixtures should be designed to produce slumps within ±3/4 in. (20 mm) of the maximum permitted. For each *w/c*, at least three cylinders should be made and cured in accordance with ASTM C192/C192M-07 and tested at 28 days in accordance with ASTM C39/C39M-11. From the results of the cylinder tests, a curve should be plotted showing the relationship between the *w/c* and the compressive strength.

A field-experienced concrete producer should have strength test records not more than 12 months old, and a sample standard deviation s_s shall be established. Test records from which s_s are calculated shall:

(a) Represent materials, quality control procedures, and conditions similar to those expected, and changes in materials and proportions within the test records shall not have been more restricted than those for proposed work;

(b) Represent concrete produced to meet a specified compressive strength or strengths within 1000 psi (7 MPa) of f_c';

(c) Consist of at least 30 consecutive tests or two groups of consecutive tests totaling at least 30 tests.

The *w/c* to be used in construction should be the value determined from the graph for the required average strength from 16.2.3. Slump should be measured in accordance with ASTM C143/C143M-10.

Table 16.2.1—Effects on concrete attributes due to an increase of any single mixture component

Attribute	Cement	Fine aggregate	Coarse aggregate	Water	Entrained air	Mixing	Age
Slump	–	–	+	+	+	+	
Cohesion	+	+	–	–	+	+	
Workability	+	+	–	+	+	+	
Segregation	–	–	+	+	–	–	
Sedimentation	–	+	–	+	–	–	
Wet consolidation	–	+	–	+	–	–	
Bleeding	–	–	+	+	–	–	
Entrained air	–	+	–	+		+	
Durability	+	–	+	–	+	+	–
Strength	+	–	+	–	–	+	+
Elastic modulus	+	–	+	–	–	+	+
Freeze resistance	+	–	+	–	+	+	–
Wear resistance	+	–	+	–	+	+	+
Chemical resistance	+	–	+	–	+	+	+
Permeability	–	+	–	+	–	–	–
Wet expansion	+	–	–	–	+	–	–
Dry shrinkage	+	–	–	+	+	–	+
Density	+	–	+	–	–	+	
Form finish	+	+	–	– +	+	+	

Table 16.2.2.4—Concrete properties for exposure conditions

Exposure condition	Maximum *w/c* by weight	Minimum f_c', psi	Minimum f_c', MPa
Concrete intended to have low permeability when exposed to water	0.45	4000	28
Concrete exposed to freezing and thawing in a moist condition or to deicing chemicals	0.45	4500	31
For corrosion protection of reinforcement in concrete exposed to chlorides from deicing chemicals, salt, salt water, brackish water, seawater, or spray from these sources	0.4	5000	35

Table 16.2.2.6—Maximum chloride ion content for corrosion protection of reinforcement

Member type	Maximum water-soluble chloride ion (Cl^-) in concrete, percent by cement weight
Reinforced concrete exposed to chloride in service	0.15
Reinforced concrete that will be dry or protected from moisture in service	1.00
Other reinforced concrete construction	0.30

Table 16.2.4—Slump limits for various construction types

Construction type	Slump, in. (mm)	
	Maximum	Minimum
Reinforced foundation walls footings	3 (75)	1 (25)
Plain footings and caissons substructure walls	3 (75)	1 (25)
Beams and reinforced walls	4 (100)	1 (25)
Columns	4 (100)	1 (25)
Pavements and slabs	3 (75)	1 (25)
Mass concrete	2 (50)	1 (25)

A higher *w/c* can be used with concrete additives, such as super plasticizers or high-range water-reducing admixtures (HRWRA).

16.3—Placing reinforcement

16.3.1 *Bending*—All reinforcement should be bent cold. Reinforcement partially embedded in concrete should not be field bent.

16.3.2 *Surface condition of reinforcement*—At the time concrete is placed, reinforcement should be free from mud, oil, or other nonmetallic coatings that decrease bond. Reinforcement with rust, mill scale, or a combination of both should be considered satisfactory, provided the minimum diameter (including height of deformations) and mass of a hand-wire-brushed test specimen is not less than applicable ASTM standard specification.

16.3.3 *Placing and securing*—Reinforcement should be accurately placed and adequately supported before concrete is placed, and secured against displacement within tolerances permitted in 16.3.4. Welding of crossing bars should not be permitted for assembly of reinforcement.

16.3.4 *Reinforcement placing tolerances*—Reinforcement should be placed within the tolerances given (16.3.4.1 and 16.3.4.2).

Table 16.3.4.1—Tolerance in depth and minimum cover

	Tolerance on depth *d*	Tolerance on specified concrete cover
$d \leq 8$ in. (200 mm)	±3/8 in. (10 mm)	–3/8 in. (–10 mm)
$d > 8$ in. (200 mm)	±1/2 in. (13 mm)	–1/2 in. (–13 mm)

16.3.4.1 *Tolerance for depth and cover*—Depth *d* and specified concrete cover in girders, beams, joists, walls, and columns should be as set in Table 16.3.4.1, except that tolerance for the clear distance to formed soffits should be –1/4 in. (6 mm), and tolerance for cover should not exceed –1/3 the minimum concrete cover indicated in 5.4.

16.3.4.2 *Tolerance for location of bends*—Tolerance for longitudinal location of bends and reinforcement ends should be ±2 in. (50 mm), except at discontinuous ends of members where tolerance should be ±1/2 in. (13 mm).

16.4—Concrete mixing and transportation

16.4.1 *General*—All concrete should be mixed and transported to its final destination following 16.4.

On all but the smallest jobs, batching is carried out in batching plants. Separate hoppers contain cement and various fractions of aggregate. Proportions are controlled, by weight, by manually operated or automatic dial scales connected to the hoppers. Mixing water is hatched either by measuring tanks or by water meters. The principal purpose of mixing is to produce an intimate mixture of cement, water, fine and coarse aggregate, and admixtures of uniform consistency throughout each batch. This is achieved by machine mixers of the revolving-drum type. On large projects, particularly where ample space is available, movable mixing plants are installed and operated at the site. In construction under congested conditions, on smaller jobs, and frequently in highway construction, ready mixed concrete is used. Such concrete is batched in a stationary plant and then hauled to the site in trucks in one of three ways: (1) mixed completely at the stationary plant and hauled in a truck agitator; (2) transit-mixed, that is, hatched at the plant but mixed in a truck mixer; or (3) partially mixed at the plant with mixing completed in a truck mixer.

Concrete should be discharged from the mixer or agitator within a maximum of 1-1/2 hours after water is added to the batch. In hot weather conditions, the maximum time before discharge should be reviewed.

Most building concrete is conveyed from the mixer or truck to the form in wheelbarrows or buggies on horizontal runways or by pumping through steel pipelines. The main concern during conveying is segregation. Individual components of concrete tend to segregate because of their dissimilarity. In overly wet concrete standing in containers or forms, heavier gravel components tend to settle and lighter materials (particularly water) rise. Lateral movement, such as flow within the forms, tends to separate coarse gravel from the finer components of the mixture. The danger of segregation has caused the discarding of chutes and conveyor belts in favor of methods that minimize this tendency.

Placing is the process of transferring fresh concrete from the conveying device to its final place in the forms. Before placing, loose rust should be removed from reinforcement, forms cleaned, and hardened surfaces of previous concrete lifts cleaned and treated appropriately. Placing and compacting are critical in their effect on the final quality of the concrete. Proper placement should avoid segregation, form displacement of forms or of reinforcement in the forms, and poor bond between successive layers of concrete. Immediately upon placing, concrete should be compacted by means of hand tools or vibrators. Compacting prevents honeycombing, ensures close contact with forms and reinforcement, and serves as a partial remedy to possible prior segregation. Compacting is achieved by hand tamping with a variety of tools, and now commonly and successfully with high-frequency, power-driven vibrators. These are of the internal type, immersed in the concrete, or of the external type, attached to the forms. The former are preferable but should be supplemented by the latter where narrow forms or other obstacles make immersion impossible.

16.4.2 *Preparation of equipment and place of deposit*—Preparation before concrete placement should include:

(a) Equipment for mixing and transporting concrete is clean;

(b) Debris and ice is removed from spaces that will be occupied by concrete;

(c) Forms are properly coated;

(d) Masonry filler units that will be in contact with concrete are well saturated;

(e) Reinforcement is thoroughly clean of ice or other deleterious coating;

(f) Water is removed from place of deposit before concrete is placed unless a tremie is to be used; and

(g) All laitance and other unsound material are removed before additional concrete is placed against hardened concrete.

16.4.3 *Mixing*

16.4.3.1 *General*—All concrete should be mixed until there is uniform materials distribution that is discharged completely before the mixer is recharged.

16.4.3.2 *Ready mixed concrete*—Ready mixed concrete should be mixed and delivered in accordance with ASTM C94/C94M-11.

16.4.3.3 *Job-mixed concrete*—When concrete is mixed on site, it should be mixed in accordance with the following:

(a) Mixing is done in a batch mixer;

(b) Mixer is rotated at manufacturer's recommended speed;

(c) Minimum mixing time is 1 minute for mixers of not more than 1 yd^3 (0.75 m^3) capacity, with an additional 30 seconds for each additional 1 yd^3 (0.75 m^3). Mixing can be continued for a considerable time without adverse effect;

(d) Mixing is continued for at least 1-1/2 minutes after all materials are in the drum; and

(e) A detailed record should be kept to identify:

(1) Number of batches produced;

(2) Proportions of materials used;

(3) Approximate location of final deposit in structure; and

(4) Time and date of mixing and placing.

16.4.4 *Conveying*—Concrete should be conveyed from mixer to place of final deposit by methods that prevent separation or loss of material. Conveying equipment should be capable of providing a supply of concrete at the site of placement without separation of ingredients and interruptions that cause loss of plasticity between successive increments.

16.4.5 *Depositing*

16.4.5.1 *Place of deposit*—Concrete should be deposited as nearly as practical in its final position to avoid segregation from rehandling or flowing.

16.4.5.2 *Deposit rate*—Concreting should be carried on at a rate that concrete is plastic during discharge and flows readily into spaces between reinforcement.

16.4.5.3 *Partially hardened concrete*—Concrete that has partially hardened or been contaminated by foreign materials should not be deposited in the structure.

16.4.5.4 *Retempering concrete*—Retempering concrete is not permitted and concrete that has been remixed after initial set should not be used.

16.4.5.5 *Deposit boundaries*—When concreting begins, it should be a continuous operation until placing of a panel or section, as defined by its boundaries or predetermined joints, is completed except as permitted or prohibited by 16.4.6.

16.4.5.6 *Top surfaces*—Top surfaces of vertically formed lifts should be level.

16.4.5.7 *Consolidation*—Concrete should be thoroughly consolidated by suitable means during placement and thoroughly worked around reinforcement and embedded fixtures, as well as into form corners.

16.4.6 *Construction joints*

16.4.6.1 *General*—Construction joint surfaces should be cleaned and laitance removed. Immediately before placing new concrete, all construction joints should be wetted and standing water removed. Construction joints should be made and located so as not to impair the strength of the structure. Provisions should be made for transfer of shear and other loads through construction joints.

16.4.6.2 *In slabs*—Construction joints should be located within the middle third of spans of slabs, beams, and girders (Fig. 16.4.6.2). Joints in girders should be offset a minimum distance of twice the width of intersecting beams.

16.4.6.3 *In beams, girders, and slabs supported by columns or walls*—Concrete should not be cast or erected until concrete in the vertical support members is no longer plastic.

16.4.6.4 *In beams, girders, haunches, drop panels, and capitals*—Concrete in beams, girders, haunches, drop panels, and capitals should be placed monolithically with the slab system.

16.5—Concrete strength evaluation

16.5.1 *General*—Concrete should be tested in accordance with 16.5.

Concrete properties are affected by many factors; thus, systematic quality control should be instituted at the construction site. Compressive strength is the main measure of the structural quality of concrete. Tests for this property are made on cylindrical specimens of height equal to twice the diameter, 6 x 12 in. (150 by 300 mm) or 4 x 8 in. (100 x 200 mm). Impervious molds of this shape are filled with concrete during the placement operation as specified by ASTM C172/C172M-10. Cylinders are moist-cured at 73 ± 3°F (23 ± 2°C) for 28 days according to ASTM C31/C31M-10 and then tested in the laboratory at a specified loading rate as required by ASTM C39/C39M-11. Compressive strength determined from such tests, as the average of two cylinders cast the same day from the same concrete, is the cylinder strength f_c'. The concrete as furnished should be in satisfactory agreement with the specified strength value.

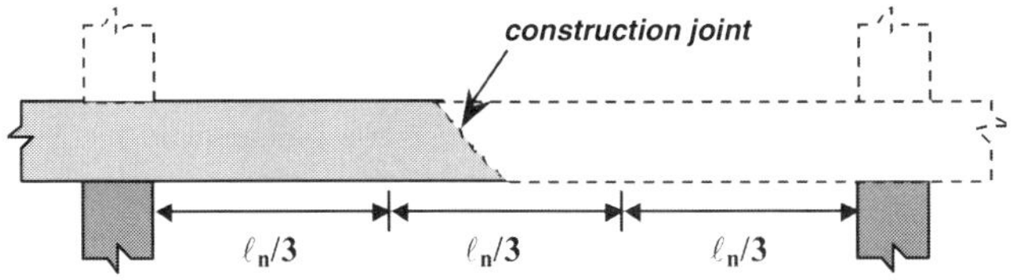

Fig. 16.4.6.2—Construction joints in slabs, girders, and beams.

16.5.2 *Frequency of testing*

16.5.2.1 *Number of samples*—Samples for strength tests of each class of concrete placed each day should be at least two cylinders for 6 x 12 in. (150 x 300 mm) or three cylinders for 4 x 8 in. (100 x 200 mm) not less than once a day, or less than once for each 50 yd^3 (40 m^3) of concrete, or less than once for each 2000 ft^2 (200 m^2) of surface area of slabs or walls. A sample of two cylinders for 6 x 12 in. (150 x 300 mm) or three cylinders for 4 x 8 in. (100 x 200 mm) should, at least, be taken of the concrete of the columns of each floor. A sample of two cylinders for 6 x 12 in. (150 x 300 mm) or three cylinders for 4 x 8 in. (100 x 200 mm) should, at least, be taken every 25 batches of concrete.

16.5.2.2 *Small jobs*—On a given project, if the total concrete volume is such that testing frequency indicated by 16.5.2.1 would provide less than five strength tests (average of two cylinders) of a given class of concrete, tests should be made from at least five randomly selected batches or from each batch if fewer than five batches are used.

16.5.2.3 *Definition of strength test*—A strength test should be the average of the strengths of two cylinders for 6 x 12 in. (150 x 300 mm) or three cylinders for 4 x 8 in. (100 x 200 mm) made from the same sample and tested at 28 days.

16.5.3 *Acceptance of concrete*

16.5.3.1 *Making test specimens*—Cylindrical specimens 6 x 12 in. (150 x 300 mm) or 4 x 8 in. (100 x 200 mm) should be cast in impervious molds of this shape and filled with concrete during the placement operation as specified by ASTM C172/172M-10.

16.5.3.2 *Curing and testing specimens*—Cylinders should be moist-cured at 73 ± 3°F (23 ± 2°C) for 28 days and then tested in the laboratory at a loading rate as required ASTM C39/C39M-11.

Table 16.6.1—Curing time

Item	Minimum duration concrete kept wet after setting (in days)	
	Cold weather	Warm weather
Self-supporting floors and beams	7	10
Walls 8 in. (200 mm) thick or less	7	7
Thick walls, columns, and piers	7	7
Floors and pavements on soil	4	7

Table 16.7a—Minimum form removal time

Item	Forms held in place after concrete has set, in days	
	Cold weather	Warm weather
Self-supporting floors and beams	14	10
Walls 8 in. (200 mm) thick or less	4	3
Thick walls, columns and piers	3	2
Floors and pavements on soil	—	—

16.5.3.3 *Acceptance of concrete*—Strength level of an individual class of concrete should be considered satisfactory when both of the following are met:

(a) The average of any three consecutive pairs for 6 x 12 in. (150 x 300 mm) or three cylinders for 4 x 8 in. (100 x 200 mm) should be at least equal to the specified design strength f_c'; and

(b) No individual strength test (average of two cylinders for 6 x 12 in. [150 x 300 mm] or three cylinders for 4 x 8 in. [100 x 200 mm]) falls below f_c' by more than 500 psi (3.5 MPa).

16.5.4 *Low strength results*—When the likelihood of low-strength concrete exists and calculations indicate that load-carrying capacity is significantly reduced, the affected portion of the structure should be demolished or repaired. A statistical procedure for evaluating the results of concrete strength tests is given in ACI 214R-11. If cores are taken to verify concrete strengths, ACI 214.4R-10 summarizes practices for obtaining cores and interpreting core compressive strength test results.

16.6—Concrete curing

16.6.1 *General*—Concrete should be cured by maintaining it above the temperatures specified and in a moist condition for the period indicated in Table 16.6.1.

Fresh concrete gains strength rapidly during the first few days and weeks. Structural design is generally based on 28-day strength, about 70 percent of which is reached at the end of the first week after placing. Final concrete strength depends greatly on the conditions of moisture and temperature during this initial period. The maintenance of proper conditions during this time is known as curing. Thirty percent or more of the potential strength can be lost by premature drying of the concrete; similar amounts may be lost when concrete temperature drops to 39°F (4°C) or lower during the first few days, unless the concrete is maintained continuously moist for a long time thereafter. Freezing of fresh concrete may reduce strength by as much as 50 percent. To prevent such damage, concrete should be protected from moisture loss for at least 7 days and up to 14 days. Curing can be achieved by keeping exposed surfaces continually wet through sprinkling, ponding, or covering with wet burlap. Water used for curing should be no more than 20°F (12°C) different in temperature than the concrete surface being cured to prevent thermal stresses and cracking. Other methods include using curing compounds that form vapor barriers and waterproof papers. In addition to improved strength, proper moist curing provides better shrinkage control. To protect concrete against low temperature during cold weather, mixing water and, occasionally, aggregates are heated; temperature insulation is used where possible; and special admixtures, particularly calcium chloride, are used. When air temperatures are very low, external heat may have to be supplied in addition to insulation.

16.6.2 *Curing time*—Concrete should be maintained above 50°F (10°C) and in a moist condition not less than the time given in Table 16.6.1.

16.6.3 *Cold weather*—Adequate equipment should be provided for heating concrete materials and protecting concrete during freezing or near-freezing weather. All concrete materials and reinforcement, forms, fillers, and ground that concrete comes in contact with should be frost-free. Frozen materials or materials containing ice should not be used.

16.6.4 *Hot weather*—During hot weather, attention should be given to ingredients, production methods, handling, placing, protection, and curing to prevent excessive concrete temperatures or water evaporation that could impair a member's or structure's design strength or serviceability.

16.7—Form removal

Forms should be removed so as not to impair safety and serviceability of the structure. When removing form supports, the operation should permit concrete to take its share of the load gradually and uniformly. Concrete to be exposed by form removal should have sufficient enough strength to avoid damage during the removal operation. Concrete within the scope of this guide usually attains 70 percent of f_c' after 7 days. Therefore, a 4000 psi (28 MPa) concrete should be at least 2800 psi (20 MPa) at 7 days. Based on this estimation, the minimum removal time given by Table 16.7a should be used. Allow earlier removal of the forms when the concrete strength tested, using field-cured cylinders, has achieved strengths given in Table 16.7b.

16.8—Inspection

Systematic observation and inspection ensures close correspondence between plans and specifications and the finished structure. A competent architect/engineer, preferably the design engineer or one responsible to the designer, should observe the structural system for conformance to the approved construction document at significant construction

Table 16.7b—Concrete strength for safe form removal

Structural classification	Minimum compressive strength, psi (MPa)
Concrete not subjected to appreciable bending or direct stress, nor reliant on forms for vertical support, nor liable to injury from form removal operations or other construction operations, such as vertical or approximately vertical surfaces of thick sections, or top of sloping surfaces.	500 (3.5)
Concrete subject to appreciable bending, direct stress, or both, and partially reliant on forms for vertical support:	
(1) Subject to vertical load only, such as vertical or approximately vertical surfaces of thin sections, or undersides of sloping surfaces steeper than 1:1;	700 (5)
(2) Subject to dead and live loads, such as columns.	1500 (10)
Concrete subjected to high bending stresses and wholly or almost wholly reliant on forms for vertical support, such as roof or floor slabs and beams, or undersides of slopping surfaces flatter than 1:1.	2000 (14)

stages, completion of the structural system, and completion of construction.

Additional inspection should be provided by an independent qualified inspector. The inspector's main functions for materials quality control are:

(a) Sampling;

(b) Examination;

(c) Field testing of materials;

(d) Control of concrete proportioning;

(e) Inspection of batching, mixing, conveying, placing, compacting, and curing; and

(f) Supervision of specimen preparation for laboratory tests.

Additionally, inspector's duties include inspection of foundations, formwork, reinforcing steel placement, other pertinent features of the general progress of work, keeping inspection records, and preparing periodic reports. The importance of thorough inspection to the correctness and quality of the finished structure cannot be overemphasized.

CHAPTER 17—REFERENCES

17.1—Referenced standards and reports

American Concrete Institute

214.4R-10 Guide for Obtaining Cores and Interpreting Compressive Strength Results

214R-11 Guide to Evaluation of Strength Test Results of Concrete

301-10 Specifications for Structural Concrete

318-11 Building Code Requirements for Structural Concrete and Commentary

318-63 Building Code Requirements for Reinforced Concrete

350-06 Code Requirements for Environmental Engineering Concrete Structures and Commentary

IPS-1 Essential Requirements For Reinforced Concrete Buildings

SP-66(04) ACI Detailing Manual

American Society of Civil Engineers

ASCE 7-10 Minimum Design Loads for Buildings and Other Structures

ASTM International

A615/A615M-09 Specification for Deformed and Plain Carbon Steel Bars for Concrete Reinforcement

A706/A706M-09 Specification for Low Alloy Steel Deformed and Plain Bars for Concrete Reinforcement

A1064/A1064M-10 Specification for Steel Wire and Welded Wire Reinforcement, Plain and Deformed, for Concrete

C29/C29M-09 Test Method for Bulk Density (Unit Weight) and Voids in Aggregate

C31/C31M-10 Practice for Making and Curing Concrete Test Specimens in the Field

C33/C33M-10 Specification for Concrete Aggregates

C39/C39M-11 Test Method for Compressive Strength of Cylindrical Concrete Specimens

C94/C94M-11 Specification for Ready Mixed Concrete

C143/C143M-10 Test Method for Slump of Hydraulic Cement Concrete
C150/C150M-11 Specification for Portland Cement
C172/C172M-10 Practice for Sampling Freshly Mixed Concrete
C192/C192M-07 Practice for Making and Curing Concrete Test Specimens in the Laboratory
C260/C260M-10 Specification for Air-Entraining Admixtures for Concrete
C330/C330M-09 Specification for Lightweight Aggregates for Structural Concrete
C494/C494M-11 Specification for Chemical Admixtures for Concrete
C567-05 Standard Test Method for Determining Density of Structural Lightweight Concrete
C595/C595M-11 Specification for Blended Hydraulic Cements
C845-04 Specification for Expansive Hydraulic Cement
C1017/1017M-07 Specification for Chemical Admixtures for Use in Producing Flowing Concrete
C1602/C1602M-06 Specification for Mixing Water Used in the Production of Hydraulic Cement Concrete

International Code Council
2009 International Building Code (IBC)

APPENDIX A—COMPARISON OF ACI 314R-11 TO ACI 318-11, IBC 2009, AND ASCE 7-10

Table A.1 serves as a guide to locating the corresponding topics in the supporting codes and standards (ACI 318-11, IBC 2009, and ASCE 7-10) listed in Section 1.4. Some recommendations are more conservative in this guide; therefore, differences between them and those presented in the supporting codes and standards should be expected. Several recommendations do not have direct correspondence. Appendix A is presented for information only.

Table A.1—Guide by section to corresponding topics in supporting codes and standards

Section in ACI 314R-11 by section	ACI 318-11	IBC 2009	ASCE 7-10
Chapter 1—General			
1.3.1		Chapters 3 and 4	1.5
1.6.2	1.2	1603, 1803	
1.6.3	Chapter 16		
1.7	9.1	1604	1.2
1.8	9.1	1605	1.2, 1.3.1.1
1.8.3	9.3		1.3.1.1
1.9	9.5		1.3.2
Chapter 2—Notation and definitions			
2.1	2.1	1602	1.2, 3.1.1, 4.1 7.1, 8.1, 11.2, 26.3
2.2	2.2	1602	2.2, 11.3, 26.2
Chapter 4—Loads			
4.2	9.2	1605.2	2.3
4.4		1606.1	3.1.2
4.5		1606	3.1
4.6		1607	4.3
4.7		1607.11	4.8
4.8		1611	8.0
4.9		1608	7.0
4.10		1609	27.0, 28.0 and 30.0
4.11		1613	11.0, 12.0,
4.12		1610	3.0
4.13			1.4.3
4.14	21.11		11.0, 27.0,
4.15	21.11		
Chapter 5—General reinforced concrete information			
5.2.2	3.2		
5.2.3	3.3		
5.2.4	3.4		
5.2.5	3.5.1, Appendix E		
5.2.5.1	3.5.3		
5.2.5.2	3.5.3.5 to 3.5.3.7		
5.2.5.3	3.5.4		
5.2.6	3.6		
5.2.7	3.7		
5.4.1	7.7.1		
5.4.2	7.7.8	Chapter 7	
5.4.3	7.7.6		
5.5	7.2		
5.6	7.1		
5.7	3.3.2		
5.8.1.1	12.2		
5.8.1.2	12.7, 12.8		
5.8.2.1	12.14.2, 12.15		
5.8.2.2	12.18, 12.19		
5.8.3	12.5		
5.11	10.2, 10.3, Appendix B		
5.12	10.2, 10.3, Appendix B		
5.13.1	11.1		
5.13.4	11.1, 11.2, 11.4		
5.13.5	11.11		
5.13.6	11.5		
5.14	10.14		
Chapter 6—Floor systems			
6.1.2	8.12		

Section in ACI 314R-11 by section	ACI 318-11	IBC 2009	ASCE 7-10
6.1.3	8.13		
6.1.4	Chapter 13		
6.3	7.13		1.4
6.5.2.1	8.13		
6.5.2.2	9.5.2.1		
6.5.2.3	9.5.2.2 to 9.5.2.6		
6.5.3.1	9.5.2.1		
6.5.3.2	9.5.2.2 to 9.5.2.6		
6.5.4	9.5.3.3		
6.5.5	9.5.3.2		
6.7	8.14		
6.8.1	13.4		
6.8.2	6.3		
Chapter 7—Solid slabs supported on girders, beams, joists, or reinforced concrete walls			
7.3.2	7.6.1		
7.3.3	7.12		
7.3.4.1	7.6.5		
7.3.4.2	10.5.4		
7.3.4.3	10.3.5, B.10.3.3		
7.3.5	12.10, 12.11, 13.3		
7.3.6	12.10, 12.12, 13.3		
7.3.8	13.3.6		
7.3.9	7.5.3		
7.4.2	11.2.1.1		
7.8.2	8.3.3		
7.8.4	8.3.3		
7.9	Method III of ACI 318-63		
Chapter 8—Girders, beams, and joists			
8.1.2	8.3.1, 8.5, 8.7 to 8.11		
8.4.2.1	7.6.1		
8.4.2.2	7.6.2		
8.4.4	10.6.3, 10.6.4		
8.4.5	10.5		
8.4.6	10.3.5, B.10.3.3		
8.4.9.5	7.11.1		
8.4.10.1	8.12.2		
8.4.10.2	8.12.3		
8.4.10.3	8.12.4		
8.4.11.1	10.6.6		
8.4.11.2	8.12.5		
8.4.12	10.6.7		
8.4.14	12.10, 12.11		
8.4.15	12.10, 12.12		
8.5.2	12.13, 11.4.1		
8.5.4.2	11.1.1		
8.5.4.3	11.2		
8.5.4.4	11.4.7		
8.5.4.5	11.4.5 to 11.4.7		
8.6.3.3	8.3.3		
8.6.4.3	8.3.3		
8.7.2.4	10.4.1		
8.7.2.5	8.3.3		
8.7.3.2	8.3.3		
8.7.4.1	8.3.3		
Chapter 9—Slab-column systems			
9.3.2	13.2.1		

Section in ACI 314R-11 by section	ACI 318-11	IBC 2009	ASCE 7-10
9.3.3	13.2.2		
9.3.4	13.2.3		
9.3.8	13.6.1		
9.3.9	13.4		
9.3.10	13.3.7		
9.4	13.3		
9.5, 9.6, 9.7	11.11		
9.8	13.6		
Chapter 10—Columns			
10.3.3	10.10		
10.3.4	10.8.2		
10.4.2.2	10.9.1		
10.4.2.4	10.9.2		
10.4.2.6	7.6.3		
10.4.2.7	7.6.4		
10.4.2.8	12.17		
10.4.2.10	7.8.1		
10.4.3.2	7.10.5		
10.4.3.3	7.10.4, 10.9.3		
10.4.3.4	7.9, 11.10.2		
10.6.2.2	11.1.1		
10.6.2.3	11.2		
10.6.2.4	11.4		
Chapter 11—Seismic resistance			
11.1.2.1	21.5.1.3		14.2
11.1.2.2	21.5.2		
11.1.2.3	21.5.3		
11.1.2.4	21.5.4		
11.1.3.1	21.6.1.1		
11.1.3.2	21.6.3		
11.1.3.3	21.6.2		
11.1.3.4	21.6.4		
11.1.3.5	21.6.4		
11.1.3.6	21.6.5		
11.1.4.2	21.7.2.3		
11.1.4.3	21.7.3		
11.1.4.4	21.7.4		
11.1.4.5	21.7.2.2		
11.1.5.2	21.9.6		
11.1.6	21.3.6		
11.2	21.1.2.1	1614	9.6
Chapter 12—Reinforced concrete walls			
12.3.2.1	14.5.3		
12.3.2.2	10.8.2		
12.4.2	14.3.5		
12.4.3	14.3.4		
12.4.4.2	14.3, 21.9.2.1		
12.4.4.3	10.9.1, 14.3.6		
12.4.4.4	21.9.6.5		
12.4.5	14.3, 21.9.2.1		
12.6.2.1	11.1.1		
12.6.2.2	11.9.5		
12.6.2.3	11.9.9, 21.9.4		
Chapter 13—Other structural members			
13.1		Chapters 10 and 11	
Chapter 14—Foundations			

Section in ACI 314R-11 by section	ACI 318-11	IBC 2009	ASCE 7-10
14.1		1803	
14.2		1803, 1806	
14.5.2	15.2		
14.5.3		1804	
14.5.3.5	15.7		
14.5.3.6	15.9		
14.5.3.7	15.3		
14.5.4.4	15.4.4		
14.5.4.7	15.6		
14.5.4.11	15.8		
14.5.5	11.11		
14.5.6	15.5		
14.7	15.10		
14.10	15.10		
14.11.10.5	14.5.3.2		
14.12	21.12.3		
14.13	21.12.3		
Chapter 15—Drawings and specifications			
15.2	1.2	1603	
15.3	1.2	1603	
Chapter 16—Construction			
16.1.2	3.2		
16.1.3	3.3		
16.1.4	3.6		
16.1.5	3.5		
16.1.6	6.1		
16.2.2.2	4.1		
16.2.2.3	4.3, 4.4		
16.2.2.4	4.3		
16.2.2.5	4.3, 4.5		
16.2.2.6	4.2		
16.2.3	5.4		
16.2.4	5.2.1		
16.3.1	7.3		
16.3.2	7.4		
16.3.3	7.5.1		
16.3.4.1	7.5.2.1		
16.3.4.2	7.5.2.2		
16.4.1	5.7		
16.4.2	5.7		
16.4.3	5.8		
16.4.4	5.9		
16.4.5	5.10		
16.4.6	6.4		
16.5.2	5.6.2		
16.5.3	5.6.3		
16.5.4	5.6.5		
16.6.2	5.11.1		
16.6.3	5.12		
16.6.4	5.13		
16.7	6.2		
16.8	1.3, 3.1.3	Chapter 17	